Statistical Disclosure Control

Statistical Disclosure Control

Anco Hundepool
Statistics Netherlands, The Netherlands

Josep Domingo-Ferrer
Universitat Rovira i Virgili, Catalonia, Spain

Luisa Franconi
Italian National Institute of Statistics, Italy

Sarah Giessing
Federal Statistical Office of Germany, Germany

Eric Schulte Nordholt
Statistics Netherlands, The Netherlands

Keith Spicer
Office for National Statistics, UK

Peter-Paul de Wolf
Statistics Netherlands, The Netherlands

A John Wiley & Sons, Ltd., Publication

Library of Congress Cataloging-in-Publication Data

Hundepool, Anco.
    Statistical disclosure control / Anco Hundepool [and six others].
        pages cm. – (Wiley series in survey methodology)
    Includes bibliographical references and index.
    ISBN 978-1-119-97815-2
    1. Confidential communications–Statistical services.   I. Title.
    HA34.H86 2012
    352.7′52384–dc23

                                            2012015785

A catalogue record for this book is available from the British Library.

ISBN 978-1-119-97815-2

Typeset in 10/12pt Times by Aptara Inc., New Delhi, India
Printed and bound in Malaysia by Vivar Printing Sdn Bhd

1   2012

Contents

Preface   **xi**

Acknowledgements   **xv**

1 Introduction   **1**
   1.1   Concepts and definitions   2
         1.1.1   Disclosure   2
         1.1.2   Statistical disclosure control   3
         1.1.3   Tabular data   3
         1.1.4   Microdata   3
         1.1.5   Risk and utility   4
   1.2   An approach to Statistical Disclosure Control   7
         1.2.1   Why is confidentiality protection needed?   7
         1.2.2   What are the key characteristics and uses of the data?   8
         1.2.3   What disclosure risks need to be protected against?   8
         1.2.4   Disclosure control methods   8
         1.2.5   Implementation   9
   1.3   The chapters of the handbook   9

2 Ethics, principles, guidelines and regulations – a general background   **10**
   2.1   Introduction   10
   2.2   Ethical codes and the new ISI code   11
         2.2.1   ISI Declaration on Professional Ethics   11
         2.2.2   New ISI Declaration on Professional Ethics   12
         2.2.3   European Statistics Code of Practice   15
   2.3   UNECE principles and guidelines   16
         2.3.1   UNECE Principles and Guidelines on Confidentiality
               Aspects of Data Integration   18
         2.3.2   Future activities on the UNECE principles and guidelines   19
   2.4   Laws   19
         2.4.1   Committee on Statistical Confidentiality   20
         2.4.2   European Statistical System Committee   20

3 Microdata   **23**
   3.1   Introduction   23
   3.2   Microdata concepts   24
       3.2.1   Stage 1: Assess need for confidentiality protection   24
       3.2.2   Stage 2: Key characteristics and use of microdata   27
       3.2.3   Stage 3: Disclosure risk   30
       3.2.4   Stage 4: Disclosure control methods   32
       3.2.5   Stage 5: Implementation   34
   3.3   Definitions of disclosure   36
       3.3.1   Definitions of disclosure scenarios   37
   3.4   Definitions of disclosure risk   38
       3.4.1   Disclosure risk for categorical quasi-identifiers   39
       3.4.2   Notation and assumptions   40
       3.4.3   Disclosure risk for continuous quasi-identifiers   41
   3.5   Estimating re-identification risk   43
       3.5.1   Individual risk based on the sample: Threshold rule   44
       3.5.2   Estimating individual risk using sampling weights   44
       3.5.3   Estimating individual risk by Poisson model   47
       3.5.4   Further models that borrow information from
               other sources   48
       3.5.5   Estimating per record risk via heuristics   49
       3.5.6   Assessing risk via record linkage   50
   3.6   Non-perturbative microdata masking   51
       3.6.1   Sampling   51
       3.6.2   Global recoding   52
       3.6.3   Top and bottom coding   53
       3.6.4   Local suppression   53
   3.7   Perturbative microdata masking   53
       3.7.1   Additive noise masking   54
       3.7.2   Multiplicative noise masking   57
       3.7.3   Microaggregation   60
       3.7.4   Data swapping and rank swapping   72
       3.7.5   Data shuffling   73
       3.7.6   Rounding   73
       3.7.7   Re-sampling   74
       3.7.8   PRAM   74
       3.7.9   MASSC   78
   3.8   Synthetic and hybrid data   78
       3.8.1   Fully synthetic data   79
       3.8.2   Partially synthetic data   84
       3.8.3   Hybrid data   86
       3.8.4   Pros and cons of synthetic and hybrid data   98
   3.9   Information loss in microdata   100
       3.9.1   Information loss measures for continuous data   101
       3.9.2   Information loss measures for categorical data   108

3.10 Release of multiple files from the same microdata set 110
3.11 Software 111
    3.11.1 μ-ARGUS 111
    3.11.2 sdcMicro 113
    3.11.3 IVEware 115
3.12 Case studies 116
    3.12.1 Microdata files at Statistics Netherlands 116
    3.12.2 The European Labour Force Survey microdata for research purposes 118
    3.12.3 The European Structure of Earnings Survey microdata for research purposes 121
    3.12.4 NHIS-linked mortality data public use file, USA 128
    3.12.5 Other real case instances 130

4 Magnitude tabular data **131**
4.1 Introduction 131
    4.1.1 Magnitude tabular data: Basic terminology 131
    4.1.2 Complex tabular data structures: Hierarchical and linked tables 132
    4.1.3 Risk concepts 134
    4.1.4 Protection concepts 137
    4.1.5 Information loss concepts 137
    4.1.6 Implementation: Software, guidelines and case study 138
4.2 Disclosure risk assessment I: Primary sensitive cells 138
    4.2.1 Intruder scenarios 138
    4.2.2 Sensitivity rules 140
4.3 Disclosure risk assessment II: Secondary risk assessment 152
    4.3.1 Feasibility interval 152
    4.3.2 Protection level 154
    4.3.3 Singleton and multi cell disclosure 155
    4.3.4 Risk models for hierarchical and linked tables 155
4.4 Non-perturbative protection methods 157
    4.4.1 Global recoding 157
    4.4.2 The concept of cell suppression 157
    4.4.3 Algorithms for secondary cell suppression 158
    4.4.4 Secondary cell suppression in hierarchical and linked tables 161
4.5 Perturbative protection methods 163
    4.5.1 A pre-tabular method: Multiplicative noise 165
    4.5.2 A post-tabular method: Controlled tabular adjustment 165
4.6 Information loss measures for tabular data 166
    4.6.1 Cell costs for cell suppression 166
    4.6.2 Cell costs for CTA 167
    4.6.3 Information loss measures to evaluate the outcome of table protection 167

4.7	Software for tabular data protection	168
	4.7.1 Empirical comparison of cell suppression algorithms	169
4.8	Guidelines: Setting up an efficient table model systematically	173
	4.8.1 Defining spanning variables	174
	4.8.2 Response variables and mapping rules	175
4.9	Case studies	178
	4.9.1 Response variables and mapping rules of the case study	178
	4.9.2 Spanning variables of the case study	179
	4.9.3 Analysing the tables of the case study	179
	4.9.4 Software issues of the case study	181
5	**Frequency tables**	**183**
5.1	Introduction	183
5.2	Disclosure risks	184
	5.2.1 Individual attribute disclosure	185
	5.2.2 Group attribute disclosure	186
	5.2.3 Disclosure by differencing	187
	5.2.4 Perception of disclosure risk	190
5.3	Methods	191
	5.3.1 Pre-tabular	191
	5.3.2 Table re-design	192
	5.3.3 Post-tabular	193
5.4	Post-tabular methods	193
	5.4.1 Cell suppression	193
	5.4.2 ABS cell perturbation	193
	5.4.3 Rounding	194
5.5	Information loss	199
5.6	Software	201
	5.6.1 Introduction	201
	5.6.2 Optimal, first feasible and RAPID solutions	202
	5.6.3 Protection provided by controlled rounding	203
5.7	Case studies	204
	5.7.1 UK Census	204
	5.7.2 Australian and New Zealand Censuses	205
6	**Data access issues**	**208**
6.1	Introduction	208
6.2	Research data centres	209
6.3	Remote execution	209
6.4	Remote access	210
6.5	Licensing	211
6.6	Guidelines on output checking	211
	6.6.1 Introduction	211
	6.6.2 General approach	212
	6.6.3 Rules for output checking	215

6.6.4   Organisational/procedural aspects of output checking   224

6.6.5   Researcher training   233

6.7   Additional issues concerning data access   236

6.7.1   Examples of disclaimers   236

6.7.2   Output description   236

6.8   Case studies   237

6.8.1   The US Census Bureau Microdata Analysis System   237

6.8.2   Remote access at Statistics Netherlands   239

Glossary   **243**

References   **261**

Author index   **279**

Subject index   **282**

Preface

In the last 20 years, work in official statistics has changed drastically due to new possibilities of information technology. While in the old days, the output of statistical offices mainly consisted of a set of tables, limited by the mere size of paper publications, nowadays new information technology makes it possible to publish much more detailed tabular data. Moreover, on the input side, the use of computers has increased the amount and detail of the data collected. The use of external registers, now available for a number of statistical offices, has led to an enormous amount of very detailed microdata. Of course, this is a very positive development as the mission of the statistical offices is to describe the society with as much detail as possible. For example, this makes it possible for policy makers to make well-informed decisions.

On the other hand, publishing a vast amount of detailed information has the risk of disclosing sensitive information both on individual persons as well as on economic entities. Privacy issues have become more and more of a concern for people. The growth of the internet has made people aware of consequences of the concept 'big brother is watching you'. Statistical Disclosure Control (SDC) is thus a rapidly evolving field: disclosure control thinking has to keep pace with increase in computing power, developments in matching software and the proliferation of public and private databases. Statistical offices need to find the right balance between the need to inform society as well as possible, on the one hand, and the need to safeguard the privacy of the respondents on the other.

Though the main audience we target consists of employees of statistical offices that need to apply the methods discussed in this book, it may be of interest to other readers as well. For example, data-archiving institutes and health data-collecting institutes have to deal with similar problems concerning confidentiality. Moreover, users of statistical output should be aware of the reasoning and methodology behind statistical disclosure control. As the issue of confidentiality grows in society, the new generation of graduates at universities should be exposed to the concepts and the methods described in this book. Last but not the least, computer scientists, database experts, data miners, practitioners dealing with medical data, etc., may find the contents of this book useful: indeed, the methods used for data anonymisation and privacy-preserving data mining are in essence the same as used in SDC.

There are several reasons to take privacy protection seriously. Firstly, there are legal frameworks that regulate what is allowed and what is not allowed with regard to publication of private information. However, there are other reasons for statistical offices to take confidentiality protection seriously as well. For example, offices need to maintain good relationships with respondents. After all, they are an essential source of information on which statistical offices build their statistics. Without respondents, there are no statistics. The respondents must be able to trust that their private and often sensitive information is safe in the hands of statistical offices. In Chapter 2, we give an overview of the major international guidelines, regulations and principles. Eric Schulte Nordholt is the author mainly responsible for this chapter.

Although disclosure control of tabular data is the oldest part of SDC, we will first focus on confidentiality issues with microdata. Josep Domingo-Ferrer and Luisa Franconi are the main contributors here. Since powerful computers and powerful statistical software are available to standard researchers, the analysis of statistical information has changed from analysing tabular data towards analysing individual data (microdata). As statistical institutes nowadays have very large databases available to produce their statistical output, these databases could ideally have a second life as the basis for various statistical research projects, e.g. at universities. Indeed, it would be a waste of time and money if researchers had to collect the information themselves again; collaboration is much more efficient. However, before sensitive statistical databases can be made available to universities for research, confidentiality must be guaranteed. Several methods exist here and tools have been developed to implement this. Various aspects of the protection of microdata are described in Chapter 3.

There is a much longer tradition of publishing tabular data. In cases where data are aggregated into tables, it is a misunderstanding to believe that there would be no privacy issues. Indeed, tabular data can disclose individual information as well. For example, if a cell in a table has only one contributor, there is an obvious risk of disclosure. It was as late as the 1970s that the first rules were proposed to assess whether tabular data could be disclosive. Since then, considerable developments have taken place. Sarah Giessing and Peter-Paul de Wolf are the main contributors to Chapter 4 on magnitude tables. This chapter deals mainly with two issues. The first step is to decide which cells in a table are disclosive (the easier part) and the second, how to protect disclosive tables adequately (the more difficult part).

Magnitude tables have attracted the most attention when solving disclosure control issues, but frequency tables can also disclose individual sensitive data. There is still a lot of work to be done to make people aware of the problems here. The approaches in the case of frequency tables are very different from those for magnitude tables. Keith Spicer has contributed to Chapter 5 that deals with frequency tables.

The protection of microdata is the subject of Chapter 3. But when data cannot be adequately protected without adversely affecting the ability to answer key research questions, alternative ways to provide access to the data must be sought. A discussion on this can be found in Chapter 6. The first step is to make moderately protected microdata. This data can be made available to serious, trustworthy researchers. Such moderate protection together with a strict contract may be enough to meet both the needs of the researchers and the need for privacy protection. When this is not an option, offices have opened special secure environments on their premises, where researchers can analyse the microdata, while the data remains under the control of the institute. In case researchers want to take results home, the results have to be checked on disclosure before they are released. For that, a series of guidelines are proposed.

When different research groups cooperate but work in different institutes, a common understanding of the terminology is needed. To facilitate this, we have added a glossary of statistical terms used in SDC. The glossary is based on a glossary proposed in 2005 at the UNECE Work Session on Statistical Disclosure Control.

Anco Hundepool
Josep Domingo-Ferrer
Luisa Franconi
Sarah Giessing
Eric Schulte Nordholt
Keith Spicer
Peter-Paul de Wolf

Acknowledgements

This book is the outcome of work on Statistical Disclosure Control (SDC) that has been carried out in Europe over the past years. Smaller teams at Statistical Offices and universities have cooperated intensively. This very fruitful cooperation has been supported financially by several European projects. It started with the SDC project from 1996 to 1998 in the Fourth Framework Programme of the EU, followed by the CASC project (2000–2003) in the Fifth Framework Programme of the EU. Eurostat, the European statistical office, has since then supported our work via various projects. The authors are grateful for the highly appreciated cooperation with their colleagues from both the statistical offices and the universities.

1

Introduction

National Statistical Institutes (NSIs) publish a wide range of trusted, high-quality statistical outputs. To achieve their objective of supplying society with rich statistical information, these outputs are as detailed as possible. However, this objective conflicts with the obligation NSIs have to protect the confidentiality of the information provided by the respondents. Statistical Disclosure Control (SDC) or Statistical Disclosure Limitation (SDL) seeks to protect statistical data in such a way that they can be released without giving away confidential information that can be linked to specific individuals or entities.

In addition to official statistics, there are several other areas of application of SDC techniques, including:

- *Health information*. This is one of the most sensitive areas regarding privacy.

- *E-commerce*. Electronic commerce results in the automated collection of large amounts of consumer data. This wealth of information is very useful to companies, which are often interested in sharing it with their subsidiaries or partners. Such consumer information transfer is subject to strict regulations.

This handbook aims to provide technical guidance on SDC for NSIs on how to approach this problem of balancing the need to provide users with statistical outputs and the need to protect the confidentiality of respondents. SDC should be combined with other tools (administrative, legal and IT) in order to define a proper data-dissemination strategy based on a risk-management approach.

A data-dissemination strategy offers many different statistical outputs covering a range of different topics for many types of users. Different outputs require different approaches to SDC and different mixtures of tools.

Statistical Disclosure Control, First Edition. Anco Hundepool, Josep Domingo-Ferrer,
Luisa Franconi, Sarah Giessing, Eric Schulte Nordholt, Keith Spicer and Peter-Paul de Wolf.
© 2012 John Wiley & Sons, Ltd. Published 2012 by John Wiley & Sons, Ltd.

- *Tabular data protection.* Tabular data protection is the oldest and best established part of SDC, because tabular data have been the traditional output of NSIs. The goal here is to publish static aggregate information, i.e. tables, in such a way that no confidential information on specific individuals among those to which the table refers can be inferred. In the majority of cases, confidentiality protection is achieved only by statistical tools due to the absence of legal and IT restrictions.

- *Dynamic databases.* The scenario here is a database to which the user can submit statistical queries (sums, averages, etc.). The aggregate information obtained as a result of successive queries should not allow him to infer information on specific individuals. The mixture of tools here may vary according to the setting and the data provided.

- *Microdata protection.* In recent years, with the widespread use of personal computers and the public demand for data, microdata (that is data sets containing for each respondent the scores on a number of variables) are being disseminated to users in universities, research institutes and interest groups (Trewin 2007). Microdata protection is the youngest sub-discipline and has experienced continuous evolution in the last years. If microdata are freely disseminated then SDL methods will be very severe to protect confidentiality of respondents; if, on the other hand, legal restrictions are in place (such as Commission Regulation 831/2002; see Section 2.4.2) a different amount of information may be released.

- *Protection of output of statistical analyses.* The need to allow access to microdata has encouraged the creation of Microdata Laboratories (Safe Centres) in many NSIs. Due to an IT-protected environment, legal and administrative restrictions users may analyse detailed microdata. Checking the output of these analyses to avoid confidentiality breaches is another field which is developing in SDC research. This handbook provides guidance on how to protect confidentiality for all of these types of output using statistical methods.

  This first chapter provides a brief introduction to some of the key concepts and definitions involved with this field of work as well as a high-level overview of how to approach problems associated with confidentiality.

1.1   Concepts and definitions

1.1.1   Disclosure

A disclosure occurs when a person or an organisation recognises or learns something that they did not know already about another person or organisation, via released data. There are two types of disclosure risk: (1) identity disclosure and (2) attribute disclosure. Identity disclosure occurs with the association of a respondents' identity with a disseminated data record containing confidential information (see Duncan *et al.* 2001).

Attribute disclosure occurs with the association of either an attribute value in the disseminated data or an estimated attribute value based on the disseminated data with the respondent (see Duncan *et al.* 2001).

Some NSIs may also be concerned with the perception of disclosure risk. For example, if small values appear in tabular output users may perceive that no (or insufficient) protection has been applied. More emphasis has been placed on this type of disclosure risk in recent years because of declining response rates and decreasing data quality.

1.1.2   Statistical disclosure control

SDC techniques can be defined as the set of methods to reduce the risk of disclosing information on individuals, businesses or other organisations. SDC methods minimise the risk of disclosure to an acceptable level while releasing as much information as possible. There are two types of SDC methods; perturbative and non-perturbative methods. Perturbative methods falsify the data before publication by introducing an element of error purposely for confidentiality reasons. Non-perturbative methods reduce the amount of information released by suppression or aggregation of data. A wide range of different SDC methods are available for different types of outputs.

1.1.3   Tabular data

There are two types of tabular output:

1. *Magnitude tables.* In a magnitude table, each cell value represents the sum of a particular response, across all respondents that belong to that cell. Magnitude tables are commonly used for business or economic data providing, for example turnover of all businesses of a particular industry within a region.

2. *Frequency tables.* In a frequency table, each cell value represents the number of respondents that fall into that cell. Frequency tables are commonly used for Census or social data providing, for example the number of individuals within a region who are unemployed.

1.1.4   Microdata

A microdata set **V** can be viewed as a file with n records, where each record contains m *variables* (also called *attributes*) on an individual respondent, who can be a person or an organisation (e.g. a company). Microdata are the form from which all other data outputs are derived and they are the primary form that data are stored in. While in the past, NSIs simply derived outputs of other forms, more and more, microdata are becoming a key output by themselves.

Depending on their sensitivity, the variables in an original unprotected microdata set can be classified into four categories which are not necessarily disjoint:

1. *Identifiers.* These are variables that *unambiguously* identify the respondent. Examples are passport number, social security number, full name, etc. Since

the objective of SDC is to prevent confidential information from being linked to specific respondents, SDC normally assumes that identifiers in **V** have been removed/encrypted in a pre-processing step.

2. *Quasi-identifiers or key variables*. Borrowing the definition from Dalenius (1986), Samarati (2001), a quasi-identifier is a set of variables in **V** that, in combination, can be linked with external information to re-identify (some of) the respondents to whom (some of) the records in **V** refer. Unlike identifiers, quasi-identifiers cannot be removed from **V**. The reason is that any variable in **V** potentially belongs to a quasi-identifier (depending on the external data sources available to the user of **V**). Thus, one would need to remove all variables (!) to make sure that the data set no longer contains quasi-identifiers.

3. *Confidential outcome variables*. These are variables which contain sensitive information on the respondent. Examples are salary, religion, political affiliation, health condition, etc.

4. *Non-confidential outcome variables*. Those variables which contain non-sensitive information on the respondent. Examples are town and country of residence, etc. Note that variables of this kind cannot be neglected when protecting a data set, because they can be part of a quasi-identifier. For instance, if 'Job' and 'Town of residence' can be considered non-confidential outcome variables, but their combination can be a quasi-identifier, because everyone knows who is the doctor in a small village.

Depending on their data type, the variables in a microdata set can be classified as:

- *Continuous*. A variable is considered continuous if it is numerical and arithmetical operations can be performed on it. Examples are income and age.

- *Categorical*. A variable is considered categorical when it takes values over a finite set and standard arithmetical operations do not make sense. Two main types of categorical variables can be distinguished:

  - *Ordinal*. An ordinal variable takes values in an ordered range of categories. Thus, the $\leq$, max and min operators are meaningful with ordinal data. The instruction level and the political preferences (left-right) are examples of ordinal variables.

  - *Nominal*. A nominal variable takes values in an unordered range of categories. The only possible operator is comparison for equality. The eye color and the address of an individual are examples of nominal variables.

1.1.5   Risk and utility

SDC seeks to optimise the trade-off between the disclosure risk and the utility of the protected released data. Yet, SDC is a discipline which was born from daily statistical practice and any theory trying to bestow a unified scientific standing to it should be flexible enough to deal with the risk-utility trade-off in a rather vast number of situations (different data structures, different contexts of previously released information

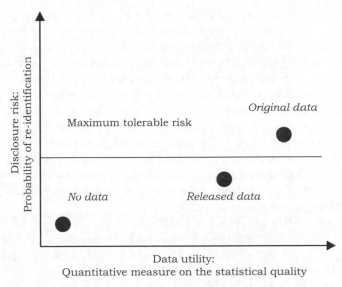

Figure 1.1   *R-U confidentiality map.*

and intruders' side knowledge, etc.). How to accurately measure risk and utility, how to distinguish legitimate users from intruders and how far to implement transparency are some of the open challenges SDC faces today, most of which are insightfully highlighted in Cox *et al.* (2011) and discussed in Domingo-Ferrer (2011).

We next review briefly some proposed models to deal with the risk-utility tension. One of them is merely descriptive (R-U maps depicting risk and utility), while the other two (*k*-anonymity and differential privacy) adopt a minimax approach: subject to a minimum guaranteed privacy, utility is to be maximised.

R-U maps

NSIs should aim to determine optimal SDC methods and solutions that minimise disclosure risk while maximising the utility of the data. Figure 1.1 contains an R-U confidentiality map developed by Duncan *et al.* (2001), where R is a quantitative measure of disclosure risk and U is a quantitative measure of data utility.

In the lower left hand quadrant of the graph, low disclosure risk is achieved but also low utility, where no data is released at all. In the upper right hand quadrant of the graph, high disclosure risk is realised but also high utility, represented by the point where the original data is released. The NSI must set the maximum tolerable disclosure risk based on standards, policies and guidelines. The goal in this disclosure risk, data utility decision problem is then to find the balance in maintaining the utility of the data but reducing the risk below the maximum tolerable risk threshold.

k-anonymity

k-anonymity is a concept that was proposed by Samarati (2001); Samarati and Sweeney (1998); Sweeney (2002a, 2002b) as a different approach to face the conflict between information loss and disclosure risk.

A data set is said to satisfy k-anonymity for $k > 1$ if, for each combination of values of quasi-identifiers (e.g. name, address, age, gender, etc.), at least k records exist in the data set sharing that combination.

Note that, if a protected data set $\mathbf{V}'$ satisfies k-anonymity, an intruder trying to link $\mathbf{V}'$ with an external non-anonymous data source will find at least k records in $\mathbf{V}'$ that match any value of the quasi-identifier the intruder uses for record linkage. Thus re-identification, i.e. mapping a record in $\mathbf{V}'$ to a non-anonymous record in the external data source, is not possible; the best the intruder can hope for is to map groups of k records in $\mathbf{V}'$ to each non-anonymous external record.

If, for a given k, k-anonymity is assumed to be enough protection, one can concentrate on minimising information loss with the only constraint that k-anonymity should be satisfied. This is a clean way of solving the tension between data protection and data utility.

In Samarati (2001) and Sweeney (2002b) the approach suggested to reach k-anonymity is to combine generalisation and local suppression; in Domingo-Ferrer and Torra (2005), the use of microaggregation was proposed as an alternative; see Section 3.6 for a description of those methods.

k-anonymity has been criticised as being necessary but sometimes not sufficient to guarantee privacy. For example, if a group of k records sharing a quasi-identifier combination also shares the same value for a confidential variable (e.g. AIDS='YES'), it suffices for an intruder to know that John Smith's record is one of k records in the group to learn that John Smith suffers from AIDS. To remedy this, other privacy properties evolving from k-anonymity have been proposed; see Domingo-Ferrer and Torra (2008) for a survey of critiques and evolved models.

Differential privacy

In general, SDC methods take the data to be published as input and modify them with the aim of reducing the risk of disclosure by removing/changing combinations of variables likely to be re-identifiable. Differential privacy (Dwork 2006, 2011) tackles privacy preservation from a different perspective. On one side, differential privacy is not based on the precise understanding that certain combinations of variables might be re-identifiable. Instead, it seeks to guarantee, in a probabilistic sense, that after the addition of a new record to a database any information extracted from the database will remain close to what it was before. On the other side, differential privacy does not focus on a specific database. It seeks to protect all the possible databases that may arise from record addition.

The following formal definition of differential privacy can be found in Dwork (2006). A randomised function κ gives ε-differential privacy if, for all data sets D and D' differing in at most one row, and all $S \subset Range(\kappa)$

$$P[\kappa(D) \in S] \leq e^{\varepsilon} P[\kappa(D') \in S]. \tag{1.1}$$

Similarly to what happened with k-anonymity, the idea is that, among the randomisations offering differential privacy for a certain parameter value ε, one should choose the one causing minimal information loss.

Computationally, differential privacy is achieved by output perturbation; the responses are computed on the real data and the result is perturbed before release. In most of the literature on differential privacy, the noise added for perturbation is taken to follow a Laplace distribution. However, it has been shown in Soria-Comas and Domingo-Ferrer (2011) that better noise distributions exist, in the sense that, given a parameter ε, they guarantee ε-differential privacy and have a smaller variance than the Laplace distribution (which implies less distortion for the data and hence less information loss).

However elegant, differential privacy has been criticised as not caring about data utility. In particular, if the original variables have bounded ranges, ε-differentially private data are likely to go off-range if ε is small (high protection); see Sarathy and Muralidhar (2011) and Soria-Comas and Domingo-Ferrer (2011) for details.

1.2   An approach to Statistical Disclosure Control

This section describes the approach that a data provider within an NSI should take in order to meet data users needs while managing confidentiality risks. A general framework for addressing the question of confidentiality protection for different statistical outputs is proposed based on the following five key stages, and we outline how the handbook provides guidance on the different aspects of this process:

1. Why is confidentiality protection needed?

2. What are the key characteristics and uses of the data?

3. What disclosure risks need to be protected against?

4. Disclosure control methods.

5. Implementation.

1.2.1   Why is confidentiality protection needed?

There are three main reasons why confidentiality protection is needed for statistical outputs:

1. It is a fundamental principle for Official Statistics that the statistical records of individual persons, businesses or events used to produce Official Statistics are strictly confidential, and are to be used only for statistical purposes. Principle 6 of the UN Economic Commission report 'Fundamental Principles for Official Statistics', April 1992 states: 'Individual data collected by statistical agencies for statistical compilation, whether they refer to natural or legal persons, are to be strictly confidential and used exclusively for statistical purposes'. The disclosure control methods applied for the outputs from an NSI should meet the requirements of this principle.

2. There may be legislation that places a legal obligation on an NSI to protect individual business and personal data. In addition, where public statements are made about the protection of confidentiality or pledges are made to respondents of business or social surveys these place a duty of confidence on the NSI that the NSI must legally comply with.

3. One of the reasons why the data collected by NSIs is of such high quality is that data suppliers or respondents have confidence and trust in the NSI to preserve the confidentiality of individual information. It is essential that this confidence and trust is maintained and that identifiable information is held securely, only used for statistical purposes and not revealed in published outputs.

More information on regulations and legislation is provided in Chapter 2.

1.2.2   What are the key characteristics and uses of the data?

When considering confidentiality protection of a statistical output it is important to understand the key characteristics of the data since all of these factors influence both disclosure risks and appropriate disclosure control methods. This includes knowing the type of data, e.g. full population or sample survey; sample design, an assessment of quality, e.g. the level of non-response and coverage of the data; variables and whether they are categorical or continuous; and type of outputs, e.g. microdata, magnitude or frequency tables. Producers of statistics should design publications according to the needs of users, as a first priority. It is, therefore, vital to identify the main users of the statistics, and understand why they need the figures and how they will use them. This is necessary to ensure that the design of the output is relevant and the amount of disclosure protection used has the least possible adverse impact on the usefulness of the statistics. Section 3.2 addresses some examples on how to carry out this initial analysis.

1.2.3   What disclosure risks need to be protected against?

Disclosure risk assessment combines the understanding gained above with a method to identify situations where there is a likelihood of disclosure. Risk is a function of likelihood (related to the design of the output), and impact of disclosure (related to the nature of the underlying data). In order to be explicit about the disclosure risks to be managed, one should consider a range of potentially disclosive situations or scenarios and take action to prevent them. A disclosure scenario describes (i) which information is potentially available to an intruder and (ii) how the intruder would use the information to identify an individual. A range of intruder scenarios should be determined for different outputs to provide an explicit statement of what the disclosure risks are, and what elements of the output pose an unacceptable risk of disclosure. Issues in developing disclosure scenarios are provided in Section 3.3.1. Risk-assessment methods for microdata are covered in Section 3.4 and different rules applied to assess the risk of magnitude and frequency tables are described in Chapters 4 and 5, respectively.

1.2.4   Disclosure control methods

Once an assessment of risk has been undertaken, an NSI must then take steps to manage any identified risks. The risk within the data is not entirely eliminated but is reduced to an acceptable level, this can be achieved either through the application of SDC methods or through the controlled use of outputs, or through a combination of both. Several factors must be balanced through the choice of approach. Some measure of information loss and impact on main uses of the data can be used to compare alternatives. Any method must be implemented within a given production system so available software and efficiency within demanding production timetables must be considered. SDC methods used to reduce the risk of microdata, magnitude tables and frequency tables are covered in Chapters 3–5, respectively. Chapter 6 provides information on how disclosure risk can be managed by restricting access.

1.2.5   Implementation

The final stage in this approach to a disclosure control problem is implementation of the methods and dissemination of the statistics. This will include identification of the software to be used along with any options and parameters. The proposed guidance will allow data providers to set disclosure control rules and select appropriate disclosure control methods to protect different types of outputs. The most important consideration is maintaining confidentiality but these decisions will also accommodate the need for clear, consistent and practical solutions that can be implemented within a reasonable time and using available resources. The methods used will balance the loss of information against the likelihood of individuals information being disclosed. Data providers should be open and transparent in this process and document their decisions and the whole risk-assessment process so that these can be reviewed. Users should be aware that a data set has been assessed for disclosure risk, and whether methods of protection have been applied. For quality purposes, users of a data set should be provided with an indication of the nature and extent of any modification due to the application of disclosure control methods. Any technique(s) used may be specified, but the level of detail made available should not be sufficient to allow the user to recover disclosive data. Each chapter of the handbook provides details of software that can be used to assess and manage disclosure risk of the different statistical outputs.

1.3   The chapters of the handbook

This book starts with an overview of regulations describing the legal underpinning of SDC in Chapter 2. Microdata are covered in Chapter 3, magnitude tables are addressed in Chapter 4 and Chapter 5 provides guidance for frequency tables. Chapter 6 describes the confidentiality problems associated with microdata access issues. Within each chapter, different approaches to assessing and managing disclosure risks are described and the advantages and disadvantages of different SDC methods are discussed. Where appropriate recommendations are made for best practice. In Chapter 7, a glossary of statistical terms used in SDC has been included.

2

Ethics, principles, guidelines and regulations – a general background

2.1   Introduction

Information from statistics becomes available for the public in tabular and microdata
form. Historically, only tabular data were available and National Statistical Institutes
(NSIs) had a monopoly on the microdata. Since the 1980s, the PC revolution led to
the end of this monopoly. Now also other users of statistics have the possibility of
using microdata. These microdata can be conveyed with CD-ROMs, USB sticks and
other means. Recently, also other possibilities of getting statistical information have
become more popular: remote access and remote execution. With these techniques,
researchers can get access to data that remain in a statistical office or can execute
set-ups without having the data on their own PC. For very sensitive information,
some NSIs have the possibility to let bona fide researchers work on-site within the
premises of the NSI.

The task of statistical offices is to produce and publish statistical information
about society. The data collected are ultimately released in a suitable form to policy
makers, researchers and the general public for statistical purposes. The release of such
information may have the undesirable effect that information on individual entities
instead of on sufficiently large groups of individuals is disclosed. The question then
arises how the information available can be modified in such a way that the data
released can be considered statistically useful and do not jeopardise the privacy of
the entities concerned. The Statistical Disclosure Control theory is used to solve
the problem of how to publish and release as much detail in these data as possible

Statistical Disclosure Control, First Edition. Anco Hundepool, Josep Domingo-Ferrer,
Luisa Franconi, Sarah Giessing, Eric Schulte Nordholt, Keith Spicer and Peter-Paul de Wolf.
© 2012 John Wiley & Sons, Ltd. Published 2012 by John Wiley & Sons, Ltd.

without disclosing individual information (see, e.g. Willenborg and de Waal 1996, Willenborg and de Waal 2001, Lenz 2010 or Duncan *et al*. 2011).

In the current chapter, the ethical codes (Section 2.2), UNECE principles and guidelines (Section 2.3) and laws (Section 2.4) will be described.

2.2 Ethical codes and the new ISI code

Many countries have an ethical code that forms the basis of the production of official statistics. An internationally recognised ethical code is the declaration on professional ethics by the International Statistical Institute (ISI).

2.2.1 ISI Declaration on Professional Ethics

After an intense preparation process taking place from 1979 to 1985, ISI adopted the ISI Declaration on Professional Ethics in 1985 (see ISI 1985). For statistical disclosure control, clauses 4.5 and 4.6 of the declaration are of importance, and therefore, they are cited below.

4.5 Maintaining confidentiality of records
Statistical data are unconcerned with individual identities. They are collected to answer questions such as 'how many?' or 'what proportion?', not 'who?'. The identities and records of cooperating (or non-cooperating) subjects should, therefore, be kept confidential, whether or not confidentiality has been explicitly pledged.
4.6 Inhibiting disclosure of identities
Statisticians should take appropriate measures to prevent their data from being published or otherwise released in a form that would allow any subject's identity to be disclosed or inferred.

There can be no absolute safeguards against breaches of confidentiality, that is the disclosure of identified or identifiable data in contravention of an implicit or explicit obligation to the source. Many methods exist for lessening the likelihood of such breaches, the most common and potentially secure of which is anonymity. Its virtue as a security system is that it helps to prevent unwitting breaches of confidentiality. As long as data travel incognito, they are more difficult to attach to individuals or organisations.

There is a powerful case for identifiable statistical data to be granted 'privileged' status in law so that access to them by third parties is legally blocked in the absence of the permission of the responsible statistician (or his or her subjects). However, even without such legal protection it is the statistician's responsibility to ensure that the identities of subjects are protected.

Anonymity alone is by no means a guarantee of confidentiality. A particular configuration of attributes can, like a fingerprint, frequently identify its owner beyond reasonable doubt. So statisticians need to counteract the opportunities for others to infer identities from their data. They may decide to group data in such a way as to disguise identities or to employ a variety of available measures that seek to impede the detection of identities without inflicting very serious damage to the aggregate data set. Some damage to analysis possibilities is unavoidable in these circumstances, but it needs to be weighted against the potential damage to the sources of data in the absence of such action.

The widespread use of computers is often regarded as a threat to individuals and organisations because it provides new methods of disclosing and linking identified records. On the other hand, the statistician should attempt to exploit the impressive capacity of computers to disguise identities and to enhance data security.

2.2.2   New ISI Declaration on Professional Ethics

A declaration on professional ethics has to be renewed from time to time. After a number of years of preparation, the new declaration on professional ethics was adopted by the ISI Council on 22 and 23 July 2010 in Reykjavik (Iceland) (see ISI 2010).

The New ISI Declaration on Professional Ethics consists of a statement of shared professional values and a set of ethical principles that derive from these values.

For the purposes of the new declaration, the definition of who is a statistician goes well beyond those with formal degrees in the field, to include a wide array of creators and users of statistical data and tools. Statisticians work within a variety of economic, cultural, legal and political settings, each of which influences the emphasis and focus of statistical inquiry. They also work within one of several different branches of their discipline, each involving its own techniques and procedures and, possibly, its own ethical approach.

The aim of the new declaration is to enable the statistician's individual ethical judgements and decisions to be informed by shared values and experience, rather than by rigid rules imposed by the profession.

The declaration recognises that the operation of one principle may impede the operation of another. Statisticians thus have competing obligations not all of which can be fulfilled simultaneously. Thus, statisticians will sometimes have to make choices between principles. The declaration does not attempt to resolve these choices or to establish priorities among the principles. Instead, it offers a framework within which the conscientious statistician should be able to work comfortably. It is urged that departures from the framework of principles be the result of deliberation rather than of ignorance.

The declaration's first intention is to be informative and descriptive rather than authoritarian or prescriptive. Second, it is designed to be applicable as far as possible to the wide and changing areas of statistical methodology and application. For this reason, its provisions are drawn quite broadly. Third, although the principles are framed so as to have wider application to decisions than to the issues it specifically mentions, the declaration is by no means exhaustive. It is designed in the knowledge that it will require periodic updating and amendment. Fourth, the values, principles and the commentaries acknowledge with the general written or unwritten rules or norms, such as compliance with the law or the need for probity. However, the declaration restricts itself insofar as possible to matters of specific concern to statistical inquiry.

The principles of the new declaration inherently reflect the obligations and responsibilities of – as well as the resulting conflicts faced by – statisticians to forces and pressures outside of their own performance, namely to and from:

1. society;

2. employers, clients and funders;

3. colleagues;

4. subjects.

In carrying out his or her responsibilities, each statistician must be sensitive to the need to ensure that his or her actions are, first, consistent with the best interests of each group and, second, do not favour any group at the expense of any other, or conflict with any of the principles.

The shared professional values are described in the new declaration as respect, professionalism, and truthfulness and integrity:

1. *Respect.* We respect the privacy of others and the promises of confidentiality given to them. We respect the communities where data are collected and guard against harm coming to them by misuse of the results. We should not suppress or improperly detract from the work of others.

2. *Professionalism.* The value professionalism implies responsibility, competence and expert knowledge and informed judgement. We work to understand our users' needs. We use our statistical knowledge, data and analyses for the Common Good to serve the society. We strive to collect and analyse data of the highest quality possible. We are responsible for the fitness of data and methods for the purpose at hand. We discuss issues objectively and strive to contribute to the resolution of problems. We obey the law and work to change laws we believe impede good statistical practice. We are continuously learning both about our own field as well as those to which we apply our methods. We develop new methods as appropriate. We do not take assignments in which we have a clear conflict of interest. We act responsibly with our employers.

3. *Truthfulness and integrity.* By truthfulness and integrity, we mean independence, objectivity and transparency. We produce statistical results using our science and are not influenced by pressure from politicians or funders. We are transparent about the statistical methodologies used and make these methodologies public. We strive to produce results that reflect the observed phenomena in an impartial manner. We present data and analyses honestly and openly. We are accountable for our actions. We have respect for intellectual property. As

scientists, we pursue promising new ideas and discard those demonstrated to be invalid. We work towards the logical coherence and empirical adequacy of our data and conclusions. We value well-established objective criteria of assessment.

According to the ISI Declaration, the ethical principles that derive from the stated professional values are as follows:

1. *Pursuing objectivity.* Statisticians should pursue objectivity without fear or favour, only selecting and using methods designed to produce the most accurate results. They should present all findings openly, completely, and in a transparent manner regardless of the outcomes. Statisticians should be particularly sensitive to the need to present findings when they challenge a preferred outcome. The statistician should guard against predictable misinterpretation or misuse. If such misinterpretation or misuse occurs, steps should be taken to inform potential users. Findings should be communicated for the benefit of the widest possible community, yet attempt to ensure no harm to any population group.

2. *Clarifying obligations and roles.* The respective obligations of employer, client or funder and statistician in regard to their roles and responsibility that might raise ethical issues should be spelled out and fully understood. In providing advice or guidance, statisticians should take care to stay within their area of competence, and seek advice, as appropriate, from others with the relevant expertise.

3. *Assessing alternatives impartially.* Available methods and procedures should be considered and an impartial assessment provided to the employer, client, or funder of the respective merits and limitations of alternatives, along with the proposed method.

4. *Conflicting interests.* Statisticians avoid assignments where they have a financial or personal conflict of interest in the outcome of the work. The likely consequences of collecting and disseminating various types of data and the results of their analysis should be considered and explored.

5. *Avoiding pre-empted outcomes.* Any attempt to establish a pre-determined outcome from a proposed statistical inquiry should be rejected, as should contractual conditions contingent upon such a requirement.

6. *Guarding privileged information.* Privileged information is to be kept confidential. This prohibition is not to be extended to statistical methods and procedures utilised to conduct the inquiry or produce published data.

7. *Exhibiting professional competence.* Statisticians shall seek to upgrade their professional knowledge and skills, and shall maintain awareness of technological developments, procedures and standards which are relevant to their field, and shall encourage others to do the same.

8. *Maintaining confidence in statistics.* In order to promote and preserve the confidence of the public, statisticians should ensure that they accurately and correctly describe their results, including the explanatory power of their data. It is incumbent upon statisticians to alert potential users of the results to the limits of their reliability and applicability.

9. *Exposing and reviewing methods and findings.* Adequate information should be provided to the public to permit the methods, procedures, techniques, and findings to be assessed independently.

10. *Communicating ethical principles.* In collaborating with colleagues and others in the same or other disciplines, it is necessary and important to ensure that the ethical principles of all participants are clear, understood, respected, and reflected in the undertaking.

11. *Bearing responsibility for the integrity of the discipline.* Statisticians are subject to the general moral rules of scientific and scholarly conduct: they should not deceive or knowingly misrepresent or attempt to prevent reporting of misconduct or obstruct the scientific/scholarly research of others.

12. *Protecting the interests of subjects.* Statisticians are obligated to protect subjects, individually and collectively, insofar as possible, against potentially harmful effects of participating. This responsibility is not absolved by consent or by the legal requirement to participate. The intrusive potential of some forms of statistical inquiry requires that they be undertaken only with great care, full justification of need and notification of those involved. These inquiries should be based, as far as practicable, on the subjects' freely given, informed consent. The identities and records of all subjects or respondents should be kept confidential. Appropriate measures should be utilised to prevent data from being released in a form that would allow a subject's or respondent's identity to be disclosed or inferred.

2.2.3   European Statistics Code of Practice

On 24 February 2005, the Statistical Programme Committee adopted the European Statistics Code of Practice. The Code was revised by the European Statistical System Committee in September 2011 (see Eurostat 2011a). This Code of Practice has the dual purpose of:

- improving trust and confidence in the independence, integrity and accountability of both National Statistical Authorities and Eurostat, and in the credibility and quality of the statistics they produce and disseminate (i.e. an external focus);

- promoting the application of best international statistical principles, methods and practices by all producers of European Statistics to enhance their quality (i.e. an internal focus).

The Code of Practice is based on 15 principles. Governance authorities and statistical authorities in the European Union commit themselves to adhering to the principles fixed in this code and reviewing its implementation periodically by the use of Indicators of Good Practice for each of the 15 principles, which are to be used as references. Principle 5 concerns statistical confidentiality and is cited below.

Principle 5: Statistical Confidentiality

The privacy of data providers (households, enterprises, administrations and other respondents), the confidentiality of the information they provide and its use only for statistical purposes must be absolutely guaranteed.

The indicators the authorities could use to review this principle are as follows:

- Statistical confidentiality is guaranteed in law.

- Statistical authority staff signs legal confidentiality commitments on appointment.

- Substantial penalties are prescribed for any wilful breaches of statistical confidentiality.

- Instructions and guidelines are provided on the protection of statistical confidentiality in the production and dissemination processes. These guidelines are spelled out in writing and made known to the public.

- Physical and technological provisions are in place to protect the security and integrity of statistical databases.

- Strict protocols apply to external users accessing statistical microdata for research purposes.

2.3   UNECE principles and guidelines

The 2003 Conference of European Statisticians (CES) of the United Nations Statistical Commission and Economic Commission for Europe (UNECE) installed a Task Force, chaired by Dennis Trewin (at that time the Australian Statistician), to draft Principles and Guidelines of Good Practice for Managing Statistical Confidentiality and Microdata Access. The UNECE region does not only include all European countries but also North America and all former Soviet Republics. In their final report of 2007 (see Trewin 2007), the following two key objectives of these guidelines are mentioned:

1. To foster greater uniformity of approach by countries while facilitating better access to microdata by the research community for worthwhile papers.

2. Through these guidelines and supporting case studies, to enable countries to improve their arrangements for providing access to microdata.

The Sixth United Nations Fundamental Principle of Official Statistics, which was mentioned in Section 1.2 of this handbook, is very clear on statistical confidentiality: 'Individual data collected by statistical agencies for statistical compilation, whether they refer to natural or legal persons, are to be strictly confidential and used exclusively for statistical purposes'. Any principles for microdata access must be consistent with this Fundamental Principle. However, making microdata available for research is not in contradiction with the Sixth UN Fundamental Principle as long as it is not possible to identify data referring to an individual.

According to Trewin (2007) four principles should be used for managing the confidentiality of microdata:

- *Principle 1*. It is appropriate for microdata collected for official statistical purposes to be used for statistical analysis to support research as long as confidentiality is protected.

- *Principle 2*. Microdata should only be made available for statistical purposes.

- *Principle 3*. Provision of microdata should be consistent with legal and other necessary arrangements that ensure that confidentiality of the released microdata is protected.

- *Principle 4*. The procedures for researcher access to microdata, as well as the uses and users of microdata, should be transparent and publicly available.

Principle 1 does not constitute an obligation to provide microdata. The NSI should be the one to decide whether to provide microdata or not. There may be other concerns (for example, quality) that make it inappropriate to provide access to microdata. Or there may be specific persons or institutions to which it would be inappropriate to provide microdata.

For Principle 2, a distinction has to be made between statistical or analytical uses and administrative uses. In the case of statistical or analytical use, the aim is to derive statistics that refer to a group (be it of persons or legal entities). In the case of administrative use, the aim is to derive information about a particular person or legal entity to make a decision that may bring benefit or harm to the individual. For example, some requests for data may be legal (a court order) but inconsistent with this principle. It is in the interest of public confidence in the official statistical system that these requests are refused. If the use of the microdata is incompatible with statistical or analytical purposes, then microdata access should not be provided. Ethics committees or a similar arrangement may assist in situations where there is uncertainty whether to provide access or not.

Researchers are accessing microdata for research purposes, but to support this research they may need to compile statistical aggregations of various forms, compile statistical distributions, fit statistical models or analyse statistical differences between sub-populations. These uses would be consistent with statistical purposes. To the extent that this is how the microdata are being used, it could also be said to support research purposes.

With respect to Principle 3, legal arrangements to protect confidentiality should be in place before any microdata are released. However, the legal arrangements have to be complemented with administrative and technical measures to regulate the access to microdata and to ensure that individual data cannot be disclosed. The existence and visibility of such arrangements (whether in law or supplementary regulations, ordinances, etc.) are necessary to increase public confidence that microdata will be used appropriately. Legal arrangements are clearly preferable, but in some countries this may not be possible and some other form of administrative arrangement should be put in place. The legal (or other) arrangements should also be cleared with the privacy authorities of countries where they exist before they are established by law. If such authorities do not exist, there may be National Governmental Organisations (NGOs) who have a 'watchdog' role on privacy matters. It would be sensible to get their support for any legal or other arrangements, or at least to address any serious concerns they might have.

In some countries, authorising legislation does not exist. At a minimum, release of microdata should be supported by some form of authority. However, an authorising legislation is a preferable approach.

Principle 4 is important to increase public confidence that microdata are being used appropriately and to show that decisions about microdata release are taken on an objective basis. It is up to the NSI to decide whether, how and to whom microdata can be released. But their decisions should be transparent. The website of an NSI is an effective way of ensuring compliance and also for providing information on how to access research reports based on released microdata.

The guidelines of the report were endorsed by the CES plenary session in 2006. They addressed the need to unify the approaches internationally and to agree on core principles for dissemination of microdata. They also suggested moving towards a risk management rather than a risk avoidance approach in the provision of microdata.

The report originally contained an annex with 22 case studies describing good practices in different countries. It is a dynamic document that is updated from time to time.

2.3.1   UNECE Principles and Guidelines on Confidentiality Aspects of Data Integration

In 2007 and 2008, a CES Task Force chaired by Brian Pink, the current Australian Statistician, drafted Principles and Guidelines on Confidentiality Aspects of Data Integration (see Pink 2009).

Data integration is concerned with integrating unit record data from different administrative and/or survey sources to compile new official statistics which can then be released in their own right. In addition, these integrated data sets may be used to support a range of economic and social research not possible using traditional sources.

The drafted principles and associated guidelines expand on the Sixth UN Fundamental Principle by providing a common framework for assessing and mitigating legislative and other confidentiality aspects of the creation and use of integrated data sets for statistical and associated research purposes. In particular, they recognise that the fundamental principles of official statistics apply equally to integrated data sets as to any other source of official statistics.

In developing these principles, it is recognised that integration of statistical data sets has become a normal part of the operations of a number of statistical offices and is generally most advanced in those countries where a heavy reliance is placed on obtaining statistical information from administrative registers. Countries that regularly undertake statistical integration usually already have a strong legislative basis and clear rules about protection of the confidentiality of personal and individual business data irrespective of whether the data has been integrated from different sources or not.

However, for many other countries, the notion of integrating data from different sources for statistical and related research purposes is relatively new. The drafted principles and associated guidelines are designed to provide some clarity and consistency of application.

These Principles and Guidelines were endorsed by the CES at their June 2009 meeting.

2.3.2   Future activities on the UNECE principles and guidelines

At its plenary session in June 2011, CES considered two proposals from the UNECE related to statistical confidentiality (see Vale 2011).

The first decision concerned the 'Principles and Guidelines of Good Practice for Managing Statistical Confidentiality and Microdata Access', adopted by the Conference in 2006. The UNECE expressed concern that while the principles and guidelines themselves remain relevant, some of the case studies in this document were becoming out of date, and proposed that the UNECE secretariat should ask countries to update them, or provide new ones.

The second decision concerned the 'Principles and Guidelines on Confidentiality Aspects of Data Integration Undertaken for Statistical or Related Research Purposes' endorsed by the Conference in June 2009. These principles and guidelines were developed by a Task Force chaired by the Australian Bureau of Statistics. The Conference decided that the Guidelines should be tested over a period of 2 years, and should be reviewed in 2011. In a written consultation exercise, countries were invited to inform the secretariat about any problems encountered with the practical implementation of the principles and guidelines. The secretariat was requested by the CES in 2011 to retain the comments for use in a future review of the principles and guidelines.

2.4   Laws

This section gives some background by discussing a few European laws that are relevant for Statistical Disclosure Control. It does not contain any national specialties.

Up to the late 1980s, microdata were rarely sent to Eurostat, the European statistical office. There was a general reliance on submission by NSIs of agreed tabular data. National confidentiality rules in some of the European countries made it impossible to harmonise European statistics. This was an unwanted situation for all NSIs, and especially for Eurostat. Therefore, a regulation on the transmission of confidential data to Eurostat has been prepared and was finally adopted by the Council in June 1990 as Regulation 1588/90 (see Eurostat 1990).

2.4.1   Committee on Statistical Confidentiality

In January 1994, these measures have been defined and formally adopted by the Member States through the Committee on Statistical Confidentiality (CSC). This committee met at least once a year at the Eurostat office in Luxembourg. This Committee discussed the implementation and evaluation of European Regulations on the dissemination of microdata and tabular data. Also revisions to the basic statistical legal framework were considered. The last meeting of the CSC was held in 2008.

Another relevant Council Regulation is No. 322/97 of February 1997 (see Eurostat 1997). This regulation defined the general principles governing community statistics, the processes for the production of these statistics and established detailed rules on confidentiality. This regulation could be considered as the general statistical law of the European Union.

2.4.2   European Statistical System Committee

A new statistical legal framework was introduced in 2009. One of the new aspects concerns statistical confidentiality: the need to enhance the role of the NSIs and Eurostat for organisational, coordination and representation purposes was noted. In this context, the former Statistical Programme Committee was replaced by a new Committee, the European Statistical System Committee (ESSC). This new Committee is also entrusted with the functions of the CSC, which thus ceased to exist.

The ESSC appointed the Working Group on Statistical Confidentiality (WGSC) to advise them on issues concerning confidentiality. The first meeting of the WGSC took place in October 2009.

The European Statistical System (ESS) is defined by Regulation 223/2009 on European statistics of 1 April 2009 (see Eurostat 2009), as the partnership between the community statistical authority (the Commission (Eurostat)) and all national authorities responsible for the development, production and dissemination of European statistics (ES).

The availability of confidential data for the needs of the ESS is of particular importance in order to maximise the benefits of the data with the aim of increasing the quality of European statistics and to ensure a flexible response to the newly emerging community statistical needs.

The transmission of confidential data between ESS partners is allowed if necessary for the production, development and dissemination of ES and also for increasing the quality of these statistics. The conditions for their further transmission, in particular for scientific purposes, are also strictly defined.

The ESSC is consulted on all draft comitology measures submitted by the Commission in the domain of statistical confidentiality.

For statistical disclosure control in the European Union the following two laws are nowadays of importance:

1. **Commission Regulation (EC) No. 831/2002 (Eurostat 2002) of 17 May 2002 implementing Council Regulation (EC) No. 322/97 (Eurostat 1997) on Community Statistics, concerning access to confidential data for scientific purposes.**
   This regulation aims to serve scientific users (mainly researchers of universities and legally established research institutes) of European microdata. Admission of new users requires explicit approval of the Member States. Discussions about the level of detail for the microdata (to which the researchers get restricted access) take place in the relevant Working Groups. Originally, this regulation covered four surveys: (1) Labour Force Survey (LFS), (2) Continuing Vocational Training Survey (CVTS), (3) European Community Household Panel (ECHP) and (4) Community Innovation Survey (CIS). From 2003, the European Union Statistics on Income and Living Conditions (EU-SILC) replaced the ECHP. The Structure of Earnings Survey (SES) was added to this regulation by Commission Regulation (EC) No. 1104/2006 of 18 July 2006. Later also the Adult Education Survey (AES) and the Farm Structure Survey (FSS) became part of this regulation. Article 3 of this Commission Regulation foresees a fairly straightforward and simple request process for researchers from the following categories or organisations:

   - Universities and other higher education organisations established under community law or by the law of a Member State.

   - Organisations or institutions for scientific research established under community law or by the law of a Member State.

   - NSIs of the Member States.

   - The European Central Bank and the national central banks of the Member States (since 2007).

For other bodies, Article 3 of the regulation lays down the condition that they must first be approved by the CSC[1] if they wish to make requests to access confidential data for scientific purposes. The pre-requisite to achieve admissibility is that the institution has demonstrated that it fulfils a set of criteria.

The existing Commission Regulation 831/2002 continues to apply but its revision is necessary in order to better correspond to the new regulatory framework and to take account of other developments since 2002.

ESSC suggested at its first meeting in May 2009 to leave the decision on the way the revision of the regulation will be carried out to the WGSC. The

---

[1] Since 2009, the approval formally has to be taken by the ESSC.

first meeting of the WGSC took place in October 2009 and it was decided there to appoint a Task Force to deal with that subject. In December 2009, the first meeting of this Task Force on the revision of Commission Regulation 831/2002 took place. After five meetings, the work of this Task Force has been finished in 2011 and now the new regulation has to be endorsed.

2. **Commission Regulation (EC) No. 223/2009 of the European Parliament and Council of 11 March 2009 on European statistics.**
This regulation (see Eurostat 2009) has entered into force on 1 April 2009 and replaced the Regulation 322/97 on Community Statistics (see Eurostat 1997) and Regulation 1101/2008 (previously 1588/90, see Eurostat 1990) on the transmission of data subject to statistical confidentiality.

Regulation 223/2009 establishes the legal framework for the development, production and dissemination of European statistics, including the rules on confidentiality.

Article 2, clause 1(e) defines statistical confidentiality as:

'[T]he protection of confidential data related to single statistical units which are obtained directly for statistical purposes or indirectly from administrative or other sources and implying the prohibition of use for non-statistical purposes of the data obtained and of their unlawful disclosure'.

Confidential data are defined as:

'[D]ata which allow statistical units to be identified, either directly or indirectly, thereby disclosing individual information. To determine whether a statistical unit is identifiable, account shall be taken of all relevant means that might reasonably be used by a third party to identify the statistical unit'.

Chapter V of the regulation, 'Statistical Confidentiality', describes in detail rules and measures that shall apply to ensure that confidential data are exclusively used for statistical purposes and how their unlawful disclosure shall be prevented (Articles 20–26). Article 23, in particular, makes provision for the access to confidential data for scientific purposes.

3

Microdata

3.1 Introduction

The purpose of Statistical Disclosure Control (SDC) for microdata is to prevent confidential information from being linked to specific respondents when releasing a microdata file. Therefore, we will assume in what follows that original microdata sets to be released have been pre-processed so as to remove identifiers and quasi-identifiers with low ambiguity.

More formally, we can say that, given an original microdata set V, the goal of SDC is to release a protected microdata set V' in such a way that:

- Disclosure risk (i.e. the risk that a user or an intruder can use V' to determine confidential variables on a specific individual among those in V) is low.

- User analyses (regressions, means, etc.) on V' and V yield the same or at least similar results.

We start the chapter by viewing the whole process of production of microdata for the users according to the five stages approach to SDC provided in Section 1.2. A description of each phase, translated into the microdata release framework, will guide the reader through the process starting with the original microdata and ending with the release of files. Section 3.2 contains such an overview.

The subsequent three sections deal with the risk-assessment phase. In Section 3.3, various definitions of disclosure are provided; the relationships of these with protection methods are also outlined. Disclosure scenarios are presented in Section 3.3.1. In Section 3.4 disclosure risk is defined and in the subsequent Section 3.5 approaches to the estimation of disclosure risk are presented.

Then, the following three sections are devoted to statistical disclosure control methods. In Section 3.6, methods that limit the risk of disclosure without perturbing

Statistical Disclosure Control, First Edition. Anco Hundepool, Josep Domingo-Ferrer,
Luisa Franconi, Sarah Giessing, Eric Schulte Nordholt, Keith Spicer and Peter-Paul de Wolf.
© 2012 John Wiley & Sons, Ltd. Published 2012 by John Wiley & Sons, Ltd.

the original data are described. Perturbative methods are outlined in Section 3.7. The approach to SDC based on the production of synthetic data is discussed in Section 3.8.

In Section 3.9, ways to analyse information loss due to the application of SDC procedures are described. Issues related to the release of different files for different uses but stemming from the same data set are briefly discussed in Section 3.10. A short presentation of the software dedicated to SDC for microdata is given in Section 3.11.

To conclude, Section 3.12 describes a few case studies and indicates supplementary materials available.

3.2   Microdata concepts

This section aims at introducing the reader to the process that, starting from the original microdata file as it is produced by a survey or by an administrative process, ends with the creation of a microdata file for external users. In Section 1.2, we introduced the five stages characterising the statistical disclosure control process, i.e.:

1. Why is confidentiality protection needed?

2. What are the key characteristics and use of the data?

3. Disclosure risk definition and assessment.

4. Disclosure control methods.

5. Implementation.

In this section, we put them into the context of the release of a microdata file. Each stage is divided into simple actions that guide the reader through the process. Table 3.1 provides an overview of the whole process by presenting, in each stage, the necessary input and the corresponding output of each action.

The idea is to identify for each stage of the process the choices to be made, the analyses to be done, the problems to be addressed and the definitions to be selected. This section is organised as follows. A subsection is dedicated to each stage of the process and for each stage the corresponding actions are described. References to the relevant sections where the technical topics are discussed in depth will guide the reader through the process.

3.2.1   Stage 1: Assess need for confidentiality protection

The interpretation of the law is the main actor of Stage 1. Given the relevant legislative framework, the aim here is to analyse the statistical units involved in the surveyed characteristics in order to establish whether confidentiality protection is needed.

Action 1.1

> *Input*: Analysis of the types of statistical units and variables present in the microdata file to be released.
>
> *Expected result*: Establish whether confidentiality protection is needed for releasing the file.

Table 3.1 Overview of SDC process for microdata. For each stage necessary actions are identified with corresponding input, I, and expected results, O.

Stage of disclosure control process	Actions	Input of action and expected results (output)
1. Why is confidentiality protection needed	1.1	I: Analysis of the types of units and variables in microdata O: Decision on the need for confidentiality protection
2. What are the key characteristics and use of the data	2.1	I: Analysis of survey methodology and questionnaire O: Identification of data structure
	2.2	I: Review of information on users' needs O: Priorities from the users' point of view
	2.3	I: Statement on dissemination policy of the agency and users' needs O: Decision on types of release
	2.4	I: Analysis of dissemination plan of the survey and types of releases O: Constraints from the data producer's side
3. Disclosure risk: definition and assessment	3.1	I: Decision on type of microdata to be released O: Definition of disclosure scenarios (i.e. how a breach of confidentiality could be carried out) and disclosure risk
	3.2	I: Definition of risk, disclosure scenarios, data analysis O: Identification of methods to estimate disclosure risk
4. Disclosure control methods	4.1	I: Users' needs, dissemination policy, types of variables involved O: Identification of SDC methods to be used, choice of parameters involved
	4.2	I: Analysis of variables involved, disclosure control methods, users' needs O: Identification of methods to measure information loss
5. Implementation	5.1	I: Analysis of available software/routines O: Choice of instruments to be used to produce protected microdata
	5.2	I: Method to estimate risk (as defined in Action 3.2), original microdata, software/routines O: Risk assessment of microdata

(*Continued*)

Table 3.1   (*Continued*)

Stage of disclosure control process	Actions	Input of action and expected results (output)
	5.3	I: Original microdata, protection methods (as defined in Action 4.1), software/routines O: Protected microdata
	5.4	I: Original microdata, protected microdata, methods to measure information loss (as defined in Action 4.2) and software/routines O: Analysis of information loss and audit of the SDC process
	5.5	I: Description of survey methods used in the statistical production process O: Documentation of the survey for the users
	5.6	I: Description of methods used in the SDC process O: Documentation of SDC methods used and presentation of information loss statistics

The starting point of the process deals with the need to establish whether there are provisions to protect the confidentiality of the units involved in the microdata set to be released. The first step is the analysis of such units. In general, confidentiality needs to be protected when the statistical units involved are individual persons or legal entities. Typical examples are social or business surveys where respondents are either individuals/households or enterprises/farms/local units, etc. An example of microdata that do not need protection is the amount of rainfall in a region.

Some microdata might contain several types of statistical units. More importantly, such units might present a particular structure like a hierarchical structure: individuals inside a household, graduates inside universities, students inside schools, local units inside enterprises inside groups, etc. For example, the European Structure of Earnings Survey (SES), being a linked employer-employee data set, has a hierarchical structure employee inside enterprise/local unit: enterprises provide general information on both the structure of the business and detailed information (age, education, earnings, etc.) on a sample of their employees. In this case, both statistical units involved (enterprise and employee) need to be processed as far as confidentiality protection is concerned and the hierarchical structure issue needs to be addressed in the statistical disclosure limitation process. The complete analysis if the SES data is provided in Section 3.12.3.

The analysis of the statistical units involved might not be sufficient by itself to establish whether confidentiality protection is needed. For example, in the survey carried out in official statistics on road accidents, the statistical unit is neither an individual nor a legal entity but the accident: location, type of road, time of the day, number of vehicles involved, etc. However, such a survey contains also variables related to the drivers of the vehicles (gender, age, information on health status, etc.) which refer to individuals and therefore, if included in the file to be released, need to be dealt with through the statistical disclosure control process. The analysis of the variables involved needs also to take into account the national legislation. For example, according to some national legislation, the causes of death are to be considered extremely sensitive, whereas in others the need to know is stronger than the right to privacy and such data receive a completely different treatment. Therefore, care is needed to evaluate special cases, taking into account country-specific legislation.

The analysis of the types of units involved and the variables present in the data set to be released will lead to the conclusion on whether a confidentiality treatment is needed. In the road accident example, a choice can be made to release a microdata set where only the information on the accident is present and no information on individuals involved is provided.

3.2.2   Stage 2: Key characteristics and use of microdata

In this stage of the SDC process, crucial elements are gathered in order to develop the most useful product for final users. Both types of actors, data users and data producers, are involved in this stage. The file to be released needs both to adhere to the data users' needs and satisfy the data producer's policy, while keeping in mind the key characteristics and structure of the microdata under study.

Action 2.1

Input: Analysis of survey methodology and questionnaire.

Expected results: Identification of data structure, list of quasi-identifiers to be removed.

Features of the microdata that are essential in the statistical disclosure limitation process are to be gathered from the analysis of the survey methodology and questionnaire. The starting point is the information on the data sources for the survey. If the source of the survey are administrative data or any type of registers, then the level of public availability of those source data is important to understand to what extent SDC is needed before release of the survey data.

The second crucial information is whether the microdata represent a census of the reference population or just a sample. In the former case, it might be necessary, before starting the process of statistical disclosure limitation, to draw a sample from the census and analyse and release just that sample. The reason for this may be a legal constraint or the mere recognition that the risk of identification of a statistical unit might be too high when releasing a complete census. The size of the sample to be drawn depends on the type of file and the type of users. Other access channels are more suitable to analyse the whole census microdata, for example a research data centre or a virtual data laboratory (for more information on this access mode the reader is referred to Chapter 6).

The analysis of the sampling design used in the survey reveals information useful for all the phases of the statistical disclosure limitation process. When estimating the risk of disclosure, the sampling design and information on the estimation variables and estimation domains might be helpful. For example, when dealing with business microdata, it is very common that some strata, related to large enterprises, have been completely enumerated. This information is extremely important as it implies different and higher risks for those strata than for the rest of the sample. Information on the sampling design might also be used to adapt the disclosure limitation methodology. Also, any accuracy requirements on particular estimates should be taken into account to define, whenever possible, a disclosure control method that preserves such accuracy. Survey weights represent a particular type of information in the released file, as they may provide information on the specific stratum where the unit stands.

An analysis of the questionnaire is useful to determine the type of information present in the file with respect to confidentiality: identifiers, quasi-identifiers, confidential variables and sensitive variables. The direct identifiers such as names, addresses, fiscal code, etc., if at all present, should be removed. Also, working variables should be removed: variables used as internal checks, flags for imputation, variables that were not validated, variables deemed not useful because containing too many missing values, information on the design stratum from which the unit comes from, etc. Categories of quasi-identifiers with too significant identifying power are commonly aggregated into broader categories. This is particularly true when releasing public use files (PUFs), because certain variables, when too detailed, could retain a level of 'sensitivity'. For example, in a household expenditure survey, we might avoid releasing for the PUF very detailed information on the housing expenditure (mortgage and rent) or detailed information on the age of the house or its number of rooms when these are very high, as they might be considered as giving too much information for particular outlying cases.

The analysis of the structure of the questionnaire reveals if there are links or relationships among variables; in this case, care should be exerted when applying SDC methods in order to avoid inconsistencies. For example, there are variables related to other variables; if a person is beyond a certain age then he/she might work, or he/she might be retired. In these cases, if a perturbation of the variable Age is applied to the microdata, then an audit should be made in order to preserve coherence. There might be cases where it is sufficient to statistically preserve relationships.

Some variables are mathematical functions of other variables. For example, expenditure is the sum of the items, average monthly earning is defined by means of the annual earning and so on. If perturbation methods are applied to such variables then the mathematical function should be preserved. In general, adjustments are necessary to the perturbation in order to preserve relationships. In the latter case of earnings, it is sufficient to perturb the annual value and the recover from this the monthly one while maintaining original percentages.

Action 2.2

Input: Analysis of users' needs.

Expected result: List of priorities on the importance of variables to be included, levels of detail in classifications, types of statistical analyses on data and types of release for microdata from the point of view of the users.

In this action, the objectives from the users' viewpoint are defined. First, the identification of the type of users is necessary. Some microdata maybe useful mostly in a scientific research environment. Other types of data could be interesting also for marketing purposes, teaching, policy making, etc. For each type of user, the needs should be investigated. For scientific researchers, information on their needs can be gained from several sources. In some countries, for each theme or survey a user group has been created in order to share information. Such groups could be contacted in order to gain knowledge on importance of variables, minimum level of the classifications needed, etc. Another source of useful information are the papers published in scientific journals or presented at conferences; knowledge extraction from electronic libraries of scientific papers could reveal the most usual types of statistical analyses made using the survey under investigation, the most common methods, the relevant models and the statistics of interest. All these should be listed and considered when defining the protection methods to be used in the data set. Another source of information can come from the descriptions of the projects that researchers have to produce in order to have access to research data centres. These could be studied as well to identify types of analyses required and methodologies used. Finally, survey experts could be involved as they know the phenomenon under study and they might have contacts with users. As results from this analysis, knowledge needs to be gained on the priority from the point of view of the users of structural variables of the survey, identification of minimum level of details for such variables and types of analyses to be performed (weighted totals, regressions, ratios, etc.).

Once the analysis on users' needs has been conducted, it can be decided which type of file is more relevant for users, which general characteristics need to be present and which minimum requirements in the classification used are necessary in order for the file to be useful for further analyses.

Action 2.3

Input: Dissemination policy of the agency producing the data and users' needs.

Expected result: Decision on types of release.

The dissemination policy of the agency coupled with the needs from the users lead to a decision on the types of release to be carried out. Different types of release could be possible: microdata under contract (MUC) (a.k.a. microdata file for research purposes, MFR), file directly available on the website (PUF), file available for teaching purposes, etc. The agency decides which distribution channels are possible for the type of data according to its legislation and national practice. Some agencies may easily allow the dissemination of some kind of microdata (e.g. social), while strictly prohibiting the dissemination of other data types (e.g. enterprises), or indeed the other way around. Therefore, the dissemination policy plays a crucial role in the decision on types of release.

In order to satisfy the needs of different users, many agencies adopt the policy of producing different types of releases from the same survey microdata. In Section 3.10, we outline the problems that arise with multiple releases from the same microdata file and present possible approaches to address them.

Action 2.4

Input: Analysis of dissemination plan and types of release.

Expected result: Constraints from the data producer's side.

Generally, any microdata release is anticipated by the publication of a set of tables also in the form of a data warehouse containing aggregated information on the survey variables. When releasing microdata, coherence with already published statistics ought to be considered; in fact, the release of microdata that allows for recalculation of published totals as much as possible is considered a good practice.

The analysis of the dissemination plan should be focussed on the type of classification/aggregation used for the most important variables. This is to avoid different classifications in different releases. The geographical breakdown in the released microdata, as well as the classifications adopted for other important variables (e.g. age, type of work, etc.), should be coherent with those adopted for published aggregates. For example, if a certain classification of the variable age is published in some tables then the microdata file should use a classification which has compatible break points so that to avoid incoherences and the possibility of gaining information by differencing.

The analysis of the dissemination plan will provide as an output the list of weighted totals that should be maintained as much as possible as well as other statistics that should be preserved (in the example of Section 3.12.3 the variations of the variable earnings with respect to the original data ought to be lower than a predefined threshold). This list may represent the set of constraints that the protection methods, to be applied in subsequence stages of the SDC process, should preserve.

In some cases, it is not possible to exactly preserve what has already been published without a significant information loss from the point of view of other statistics. In such cases, at least an assessment of the differences one might obtain between the published totals and the ones computed using the released microdata file is highly recommended, in order to alert users on this issue and provide some information on this.

3.2.3   Stage 3: Disclosure risk

Once the characteristics and use of the survey data have been clarified, it is time to start the proper analysis of the disclosure risk. This implies first the definition of disclosure and then a description of possible situations that might bring to disclosure (disclosure scenario definition) coupled with the detection of the quasi-identifiers. Finally, with these elements, we can define the disclosure risk and establish/choose methods to estimate such quantity.

Action 3.1

Input: Decision on type of microdata file to be released (PUF and MFR).

Expected results: Definition of disclosure scenarios (i.e. how a breach of confidentiality could be carried out) and disclosure risk.

The idea of disclosure in microdata is related to the possibility of identification of a respondent in the released microdata and, consequently, the possibility of gaining confidential information about him or her. Section 3.3 presents various definitions of disclosure.

A disclosure scenario is the definition of realistic assumptions about what an intruder might know about respondents and what information would be available to him to match against the microdata to be released and potentially make an identification and disclosure. The definition of a disclosure scenario involves two steps. On the one hand, the identification of the variables which might be available to users and that are present in the file to be released: these are called quasi-identifiers. On the other hand, the ways in which such variables might be used. Different types of microdata release (PUF or microdata for research purposes) may require different disclosure scenarios. An example of disclosure scenario for MFR purposes is the so called 'Spontaneous recognition' where the researchers might unintentionally recognise some units while studying the data. This may happen in enterprise microdata where it is publicly known that the largest enterprises are generally included in the sample because of their significant impact on the studied phenomenon. Moreover, the largest enterprises are also the most identifiable ones as recognisable by all (the largest car producer factory, the national mail delivery enterprise, etc.). Consequently, a spontaneous identification or recognition might occur just by knowing the economic activity of the enterprise and its size. These are in this case the quasi-identifiers, i.e., the variables allowing indirect identification of the statistical units. In this scenario, only recognisable enterprises would be considered at risk. Therefore, a definition of risk in this case would be based on identifying outlying units. A description of most used scenarios is presented in Section 3.3.1.

Intuitively, a unit is at risk of disclosure if we are able to single it out from the rest. The idea at the base of the definition of disclosure risk is a way to measure rareness of a unit either in the sample or in the population. Such measure of rareness is a function of the values taken by quasi-identifiers. For example, if we decide that age, gender and marital status are quasi-identifiers for our data to be released, then a unit that assumes rare values for such quasi-identifiers (e.g. 101 years old) or rare combinations of them (e.g. an 18-years-old widow) would be considered at risk.

The risk can be either individual, when it is referred to a single statistical unit, or global, i.e., for the entire file. Usually, global risk is a function of individual risks; see Section 3.4.2 for some definitions. Different approaches are used in the definition of individual risk if the quasi-identifiers are categorical or continuous. In the former case, the concept of a key, i.e., the combination of categories of the identifying variables, is at the basis of the definition; in the latter the concept of density around units is of interest. The problem is more complex when a mixture of categorical and numerical key variables is present; see Section 3.4 for a classification of different definitions.

Action 3.2

Input: Definition of risk, disclosure scenarios, data analysis.

Expected results: Identification of methods to estimate or measure disclosure risk.

Once a formal definition of risk of disclosure has been adopted, the next step is the identification of methods to estimate it.

For microdata containing whole censuses or registers the disclosure risk can be calculated if we make the assumption that such census or register represents the whole population. In fact, as the disclosure risk is a function of the values of the quasi-identifiers, the knowledge of such variables for the whole population, as in a census, allows for a direct calculation. However, most surveys contain only a sample of the population and the values of the quasi-identifiers are unknown or only partially known through marginal distributions for the whole population. In such cases, probabilistic modelling or heuristics based on the information available in the sample need to be developed to estimate disclosure risk measures.

Besides the method to estimate the risk, the definition of thresholds to define when a unit or a file presents an acceptable risk or when, on the contrary, the unit or the file has to be considered at risk, depends of course on the type of measure adopted. Choice of scenarios and level of acceptable risk are extremely dependent on different situations in different countries, different policies applied by different agencies, different approaches to statistical analysis, different perceived risk and different legislations. To this end, it must be stressed that different countries may have extremely different situation/phenomenon therefore different scenarios and risk methods are indeed necessary.

Each risk estimation methodology has its pros and cons; in many cases, the hypotheses at the basis of the theory make one more suitable than others according to the situation encountered. Pros and cons of each method are described in the relevant subsections and can be used as a guideline for the most appropriate choice of the risk estimation method in different situations. A proper analysis of methods to estimate the risk of disclosure is presented in Section 3.5.

In some agencies, risk definition and proper risk estimation via statistical methods are replaced by thresholds and checklists based on experience and rule of thumb (see, for example, U.S. Bureau of Census 2007).

3.2.4   Stage 4: Disclosure control methods

If the risk-assessment phase indicates that disclosure control methods are needed then again we use findings from Stage 1 above to identify appropriate methods to be applied to the data. To choose them, we need to recall the users' needs (maintain utility in the data), the constraints inherent to data structure and those derived from policy and the types of disclosure we want to prevent (data producer policy). Once the protection methods are chosen, ways to measure the information loss induced by SDC procedures as well as the detection of audit procedures are needed.

Action 4.1

Input: Users' needs, dissemination policy and analysis of the type of variables involved.

Expected results: Identification of disclosure limitation (protection) methods to be applied, choice of parameters, thresholds, etc.

Choice of the most suitable disclosure limitation method is based on previous analyses regarding the type of users and the statistical methods applied to analyse the microdata, the objectives to be investigated, the constraints imposed by the production process and the policy of the agency.

Microdata protection methods can generate the protected microdata set V':

- either by *masking original data*, i.e. generating a modified version V' of the original microdata set V;

- or by *generating synthetic data* V' that preserve some statistical properties of the original data V.

Regarding masking methods, described in Sections 3.6 and 3.7, these can in turn be divided in two categories depending on their effect on the original data (Willenborg and de Waal 2001):

1. *Non-perturbative masking.* Non-perturbative methods do not distort data; rather, they produce partial suppressions or reductions of detail in the original data set. Global recoding, local suppression and sampling are examples of non-perturbative masking.

2. *Perturbative masking.* The microdata set is distorted before publication. In this way, unique combinations of scores in the original data set may disappear and new unique combinations may appear in the perturbed data set; such confusion is beneficial for preserving statistical confidentiality. The perturbation method used should be such that statistics computed on the perturbed data set do not differ significantly from the statistics that would be obtained on the original data set.

Regarding synthetic data, at a first glance, they seem to have the philosophical advantage of circumventing the re-identification problem: since published records are invented and do not derive from any original record, some authors claim that no individual having supplied original data can complain from having been re-identified. At a closer look, other authors (e.g. Winkler 2004b and Reiter 2005a) note that even synthetic data might contain some records that allow for re-identification of confidential information. In short, synthetic data overfitted to original data might lead to disclosure just as original data would. On the other hand, a clear problem of synthetic data is data utility: only the statistical properties explicitly selected by the data protector are preserved, which leads to the question of whether the data protector should not directly publish the statistics he wants preserved rather than a synthetic microdata set. We will see in Section 3.8 that synthetic data have some advantages over the mere publication of preserved statistics.

To complete this introduction to microdata concepts, we must signal some differences between protection of continuous and categorical data (see Section 1.1.4 for a discussion on those types of data):

- When designing methods to protect continuous data, one has the advantage that arithmetic operations are possible, and the drawback that every combination of numerical values in the original data set is likely to be unique, which leads to disclosure if no action is taken.

- In the case of categorical data, the situation is different depending on whether the data are ordinal or nominal. With ordinal data, there is an order relationship and the maximum, minimum and median operators make sense. With nominal data, only pairwise comparison is possible. When designing methods to protect categorical data, the inability to perform arithmetic operations is certainly inconvenient, but the finiteness of the value range is one property that can be successfully exploited.

Action 4.2

Input: Analysis of the types of variables involved, disclosure control methods used, users' needs.

Expected results: Identification of methods to measure the information loss.

Every SDC method implies some information loss. It is important to quantify the loss incurred following the SDC method applied as to help users in understanding possible limitations to their analysis. Section 3.9 presents approaches to measure information loss.

3.2.5   Stage 5: Implementation

When all decisions have been taken on the methods to be applied for the whole SDC process in order to create the file to be released we proceed to implementation.

Action 5.1

Input: Analysis of available software or estimation of the effort needed to write new routines or modify available ones.

Expected results: Choice of instruments to be used to produce protected microdata.

A study of available software and routines that implement chosen methods should be investigated in order to efficiently plan the SDC process. Only when such routines are not available or do not satisfy constraints and requirements, in-house implementation should be carried out. Section 3.11 presents available software in SDC for microdata.

Action 5.2

Input: Original microdata, risk estimation method (as defined in Action 3.2), software/routines;

Expected results: Risk assessment of microdata.

The application of risk estimation methods to the microdata under study allows to perform risk assessment to evaluate whether statistical disclosure control methods are needed.

Action 5.3

Input: Original microdata, disclosure limitation methods (as defined in Action 4.1)
and software/routines.

Expected results: Protected data.

The application of the protection methods on the data results in the protected micro-data.

Action 5.4

Input: Original microdata, protected microdata, methods for measuring informa-tion loss (as defined in Action 4.2), software/routines.

Expected results: Analysis of information loss and audit of the SDC process.

Especially when perturbative methods are applied to the microdata, it is important to verify that no distortion or aberrant effects have been imposed on the original microdata. For this reason, it is important to analyse information loss.

It is advisable to apply also audit routines to verify whether the perturbation has achieved its aim. If perturbative methods have been applied to the data, it is a good practice to verify that records that have been recognised as at risk during the risk-assessment process are indeed not at risk anymore after the SDC process. An audit phase analysing whereas unusual values have originated from the SDC procedure might be crucial especially when quasi-identifiers are continuous variables and when the applied SDC methods are data driven (such as individual ranking, see Section 3.7.3).

Action 5.5

Input: Description of the survey methods (sampling design, data editing, imputa-tion, etc.).

Expected results: Documentation on survey methodology for the users.

Documentation is an essential part of any dissemination strategy. For a deep un-derstanding of the data, information on methodologies used at various stages of the survey process (sampling, imputation, validation, etc.) together with information on magnitude of sampling errors, estimation domains, etc. is necessary. The released microdata need to be accompanied by all necessary metadata. The availability of reading programs for the released data in the most common statistical software are very appreciated by users. In times of global access via the web, to increase the number of possible users an effort should be considered in translating in English at least the most essential information. If the microdata are not immediately available on the web because of accreditation procedures metadata should be. Consider also the possibility of making available to users on the web a microdata set with exactly the same structure as the one of the file to be released but with no statistical meaning in order for the user to start thinking of possible programs to be run on the real data.

In recent time, metadata standards like Statistical Data and Metadata eXchange (SDMX) or the Data Documentation Initiative (DDI) are becoming more widespread. The former is used to document and exchange aggregated data whereas the latter is more devoted to microdata and the description of its life cycle. In the future, rather than requiring two different sets of metadata, one to describe the statistical production process and the other to support secure access, the development of a harmonised profile of the DDI metadata to support both functions could be developed and implemented; see Thomas *et al.* (2011) for further details.

Action 5.6

> *Input*: Description of methods used in the SDC process: microdata protection and information loss.
>
> *Expected results*: Documentation on SDC methods used and presentation of information loss statistics for the users.

Besides the documentation on the survey methods, the documentation on the SDC process is essential both for auditing from external authorities and for transparency towards users. The former may include description of legal and administrative steps for a risk management policy together with the technical solution applied. The latter is essential for the user to understand what has been changed or limited in the data because of confidentiality constraints and the consequences of such actions.

It could be useful for the user to have available the list of variables modified because of the SDC process together with a brief description of changes introduced by such process (suppression, modification, perturbation, insertion of missing values, etc.) or by other reasons (e.g. lack of quality in original data).

If a data-reduction method has been applied with some local suppressions then the distribution of such suppressions should be given for a series of different dimensions of interest (distribution by variables, by household size, household type, etc.) and any other statistics that are deemed relevant for the user. If a data-perturbation method has been applied then, for transparency reasons, this should be clearly stated. Information on which statistics have been preserved by the SDC method and which have been modified and some order of magnitude of possible changes should be provided as far as possible.

3.3   Definitions of disclosure

Different definitions of disclosure are available to describe different types of breaches.

We talk about identity disclosure when a specific individual record can be recognised in a released microdata file (see Paas 1988). As usually large size enterprises are always included in business surveys, it is possible to recognise the largest in a particular industrial sector. In general an identity disclosure takes place when a correct record re-identification operation is achieved by an intruder by comparing a target individual in a sample with an available list of units that contains individual identifiers such as name and address (see Willenborg and de Waal 2001).

We have attribute disclosure when sensitive information contained in the micro-data about a specific individual unit is revealed through the release of the file. In the previous example, if no perturbation is applied to the original value, then it is possible to recover exactly the turnover or export of the enterprise in the released file. This means that, if confidential/sensitive variables are present in the released file, identity disclosure leads to attribute disclosure.

Inferential disclosure, proposed by Dalenius (1977), occurs when from the re-leased microdata one can determine the value of some characteristics of an individual unit more accurately than otherwise would have been possible. In other words, infer-ential disclosure occurs when information can be inferred with high confidence from statistical properties of the released data. This definition is not commonly used in the microdata setting.

In a microdata setting, the most used definitions are those of identity and attribute disclosure.

Note that if an SDC method (e.g. a masking method) is applied to quasi-identifiers, we aim at preserving identity disclosure. Whereas, if we apply SDC methods to sensitive/confidential variables, we preserve attribute disclosure. In fact, for example, perturbation or aggregation of categories of quasi-identifiers perform changes in the variables that allow for identification. The use of top coding on the quasi-identifier age makes it difficult to identify a very old person; top coding of the variable earnings has the function of avoiding attribute disclosure for high earnings.

According to the policy of the statistical agency, either one or the other approach (or indeed both) might be used. In the example in Section 3.12.1, the local suppression is applied to quasi-identifiers to avoid identity disclosure. In Section 3.12.3, strategy A for employees, attribute disclosure is taken care of by perturbing the earnings and identity disclosure is reduced by grouping the variable age.

3.3.1   Definitions of disclosure scenarios

A scenario synthetically describes (i) which is the potentially available information to identify a unit, i.e., the quasi-identifiers and (ii) how such information could be used to identify an individual unit i.e. means and strategies for a disclosure. Often, in the microdata setting, defining more than one scenario might be convenient, because different sources of information might be alternatively or simultaneously available. Moreover, disclosure risk can be assessed keeping into account different scenarios at the same time.

We refer to the information available to the intruder as an external archive, where information is provided at individual level, jointly with identifiers, such as name, surname, etc. The disclosure scenario is based on the assumption that, besides the identifiers, such a register contains some other variables. Some of these further variables are available also in the microdata to be released. A strategy would be to use this overlapping information to breach confidentiality by matching direct identifiers to a record in the microdata. This set of matching variables are the quasi-identifiers for that microdata and they vary according to the survey variables. In general, for social surveys such as the labour force survey or the household expenditure survey the quasi-identifiers are mainly demographic variables as those in general freely available electoral lists: gender, age, marital status, municipality of residence, place of birth, etc. In education surveys, the information freely available might be different and related to the type of school or university degree (see, for example, Casciano *et al.* 2011). In business surveys, they would be the main economic activity, the size of the enterprise and the location as these are information freely available from administrative sources (see Section 3.12.3 for an example of this case).

According to the survey, the development of a scenario is devoted to identify possible variables freely available (at individual level) to the type of users targeted by the release.

Elliot and Dale (1999) describe a system for analysing disclosure scenarios. Elliot *et al.* (2010) developed a prototype tool the *Key Variable Mapping System* which is designed to produce lists of quasi-identifiers.

Once a list of possibly available variables for identification purposes has been defined the other important factor in the definition of the scenario is the strategy of 'search' for identifiable units. Definition of such strategy is closely related to the type of users of the microdata and the type of release that has been planned. If we are dealing with a microdata for research purposes where the users sign an agreement and several conditions need to be met in order to receive the file, the reasonable scenario seems the one where only unintentional disclosure might happen. This is the so called *spontaneous recognition scenario* where only personal knowledge about few target units would be reasonable. On the contrary for a PUF, the number of quasi-identifiers might be higher than the one adopted in the MFR purposes. Also re-identification via record linkage could be taken into consideration as a possibility.

Further example of scenarios can be found in Section 3.12.

3.4   Definitions of disclosure risk

Intuitively, a unit is at risk of disclosure when it cannot be confused with several other units, i.e. if it can be singled out from the rest. Record-level definitions of risk, called individual risk, may be useful for two purposes. On the one hand, they can be exploited to protect data selectively, i.e. apply SDC procedures only to those records being at risk. On the other hand, they can be used to built an overall definition of risk for the whole microdata file to be released, i.e. a global risk measure (see Lambert 1993).

As the process to single a unit out of the others depends on the type of quasi-identifiers, also the definition of risk depends heavily on the type of the quasi-identifiers. We distinguish three cases:

1. All the quasi-identifiers are *categorical*.

2. The quasi-identifiers are *continuous*.

3. Some quasi-identifier are categorical and some are continuous.

The first case is the common situation found when microdata stem from social surveys, the other two cases are mainly present when dealing with business surveys.

The last case is the most complex one. One way to tackle the problem is to stratify the data by the categories of the categorical quasi-identifiers and then consider in this sub-population the continuous quasi-identifiers case. In the later sections, we analyse various definitions that have been proposed in the literature according to whether they have been proposed for categorical (Section 3.4.1) or continuous (Section 3.4.2) quasi-identifiers.

3.4.1 Disclosure risk for categorical quasi-identifiers

The classical approach to risk definition considers identity disclosure and in particular is based on the concept of *re-identification* (see, for example, Duncan and Lambert 1986, Fienberg and Makov 1998 and Skinner and Holmes 1998). A correct re-identification is by matching a target individual in the sample to be released with an available list of units, according to the chosen scenario. The risk is then defined as the probability that the match is correct, conditional on the data assumed available to the intruder. There are three variations of this approach:

1. A *re-identification risk* based on the microdata to be released.

2. A *re-identification risk* based on population characteristics.

3. A *re-identification risk* based on real external files.

When the quasi-identifiers are all categorical and no perturbation is imposed on the microdata the first case above defines disclosure risk in terms of the cells of the contingency table built by cross-tabulating the quasi-identifiers. Combinations of categories of such variables that present frequencies below a given threshold are at risk.

Consequently, the records presenting those combinations are all at risk. Note that this approach implies a scenario where the intruder has knowledge on whether or not a unit belongs to the sample. This is the approach used by Statistics Netherlands to release microdata files (see Section 3.5.1 for further details).

The second case foresees an inference step and, when solely categorical quasi-identifiers are present, the risk of disclosure is a function of the cells of the contingency table built by cross tabulating the quasi-identifiers in the population. Records presenting combinations of quasi-identifiers that are unusual or rare in the population have high disclosure risk whereas those presenting even unique combination in the sample are not necessarily at risk. The risk is defined at record level or, more precisely, a cell specific measure as it only depends on the combinations of values of the quasi-identifiers. For the second type of approach, then, we are concerned with the risk of a unit as determined by its probability of re-identification in the population. The idea then is that a unit is at risk if such probability is above a given threshold. Such probability depends heavily on the frequencies of the scores of the quasi-identifiers in the population, a quantity which, apart from special cases (census, registers), is unknown. The aim then is that of estimating such probability. We now provide a formulation of this class of re-identification risk. See the end of Section 3.4.2 for a discussion of variation 3.

3.4.2   Notation and assumptions

The following assumptions are usually made to express the ways in which a disclosure could be carried out. Most of them are conservative and contribute to the definition of the worst case scenario:

- A sample s from a population $\mathcal{P}$ is to be released, and sampling design weights are available.

- The external file available to the intruder covers the whole population $\mathcal{P}$; consequently for each $i \in s$, the matching unit i^* does always exist in $\mathcal{P}$.

- The external file available to the intruder contains the individual direct identifiers and a set of categorical quasi-identifiers that are also present in the sample.

- The intruder tries to match a unit i in the sample with a unit i^* in the population register by comparing the values of the quasi-identifiers in the two files.

- The intruder has no extra information other than that contained in the external file.

- A re-identification occurs when a link between a sample unit i and a population unit i^* is established and i^* is actually the individual of the population from which the sampled unit i was derived; i.e., the match has to be a correct match before an identification takes place.

Moreover we add the following assumptions:

- The intruder tries to match all the records in the sample with a record in the external file.

- The quasi-identifiers agree on correct matches. That is no errors, missing values or time changes occur in recording such variables in the two microdata file.

The following notation is introduced here and used throughout the section when describing different methods for estimating the disclosure risk of microdata with categorical quasi-identifiers.

We focus on the K cells $\mathcal{C}_1, \ldots, \mathcal{C}_K$ of the contingency table spanned by the quasi-identifiers: the keys. We consider the frequencies $\mathbf{F} = (F_1, \ldots, F_K)$ and $\mathbf{f} = (f_1, \ldots, f_K)$ in the population and sample table, respectively. The population counts $\mathbf{F}$ are unknown. The risk estimation is closely related to the estimation of the elements of the vector $\mathbf{F}$.

Following the assumption made in the scenario above, the units belonging to the same cell $\mathcal{C}_k$ are exchangeable for the intruder. Then the probability of re-identification of individual i being in cell $\mathcal{C}_k$, when F_k individuals in the population are known to belong to it, is $1/F_k$, $k = 1, \ldots, K$.

According to the concept of re-identification disclosure given above, we define the individual risk of disclosure of unit i in the sample as its probability of re-identification under the worst case scenario. Therefore, the risk we get is certainly not smaller than the actual risk; the individual risk is a conservative estimate of the actual risk:

$$r_k = P(i \text{ correctly linked to } i^* | s, \mathcal{P}, \text{ worst case scenario}). \qquad (3.1)$$

This re-identification risk definition is at record level. A global file-level disclosure risk measure τ_1, can be defined by aggregating the individual disclosure risk measures over the sample:

$$\tau_1 = \sum_k \frac{1}{F_k}. \qquad (3.2)$$

Since the uniques in the population, $F_k = 1$, are the dominant factor in the disclosure risk, we may focus our attention only on sample uniques $f_k = 1$:

$$\tau_2 = \sum_k I(f_k = 1)\frac{1}{F_k}, \qquad (3.3)$$

where I represents an indicator function. This second measure is the expected value of correct guesses if each sample unique is matched to an individual chosen at random from the same population cell. An alternative definition is the number of sample uniques which are also population uniques following the approach by Skinner and Holmes (1998):

$$\tau = \sum_k I(f_k = 1, F_k = 1).$$

All of these global risk measures can also be presented as rates if we divide them by the corresponding quantities: sample size n or the number of uniques in the sample.

In Section 3.5, we analyse different approaches to estimate such population frequencies. They may be estimated through a modelling process; examples of this reasoning are the individual risk of disclosure based on weights initially developed by Benedetti and Franconi (1998), which is described in Section 3.5.2, and the Poisson distribution and log-linear models developed by Skinner and Holmes (1998) and Elamir and Skinner (2006) which is outlined in Section 3.5.3. Further models are described in Section 3.5.4. The risk might also be estimated through heuristic methods; an example of this reasoning is given by the Special Uniques Detection Algorithm (SUDA) developed by Elliot *et al.* (2002) which is outlined in Section 3.5.5.

Finally, the last approach corresponding to the third variation of Section 3.4.1, is to identify available data sets, possibly on the web, that can be used to re-identify statistical units. Then the definition of risk is expressed on the base of record linkage experiments. Whenever a unit is correctly matched to a record, then such unit is at risk; see Sections 3.12.4 and 3.12.3 (strategy B) for real case examples. Further details on the assessment of risk via record linkage is presented in Section 3.5.6.

3.4.3   Disclosure risk for continuous quasi-identifiers

As already stated, continuous quasi-identifiers are mainly present in business microdata. For such data, in order to reach a clear picture of the phenomenon under study, the large and most identifiable enterprises are always present in the sample.

For such reason, most of the definitions of disclosure risk in this setting are based on the microdata to be released as this may coincide, at least for large size enterprises, with the population.

When identifying variables are continuous, we cannot exploit the concept of rareness of the combinations of categories of the quasi-identifiers and we transform such concept into rareness in the neighbourhood of the record.

Proposed definitions of risk can be classified as follow:

- risk based on outliers detection strategies;

- risk based on clustering techniques;

- *a posteriori re-identification risk* based on record linkage.

A basic definition of risk for continuous quasi-identifiers based on the outlier detection framework identifies at risk all the units for which the univariate continuous quasi-identifier takes a value greater than a pre-defined quantile of the observed values of such variable. This outlier approach generates a fixed percentage of units at risk, depending on the subjective choice of the quantile. Moreover, the units at risk would be, by definition, on the right tail of the distribution of the variable.

A second approach along the line of outlier detection stems from the rank-based intervals as presented in Mateo-Sanz *et al.* (2004b). In Truta *et al.* (2006), a standard-deviation-based interval approach is proposed. Using this strategy, the proportion of original values that fall into the interval centred around their corresponding masked values is a measure of disclosure risk. If rank-based intervals are used, the risk measure is called RID; if the standard-deviation-based intervals are used, the risk measure is called SDID. The latter measure depends also on the percentage of the standard deviation that is used for the computation of the width of the interval. A robust version of these measures is discussed in Templ and Meindl (2008). This approach depends on the masking method applied and defines *a posteriori risk*, i.e., the risk after the application of SDC methods, and does not act as a guideline to check where the problems are.

The issues arising when several continuous quasi-identifiers are present in the microdata have received little attention so far. Foschi (2011) in the framework of multivariate outliers detections proposes robust fitting of a finite multivariate Gaussian mixture.

Approaches based on clustering techniques are sometimes used when continuous quasi-identifiers are present in the data. In Bacher *et al.* (2002), a standard hierarchical clustering algorithm was applied to decide whether a unit is safe or not. When using aggregating clustering algorithms, a cut-off value on the aggregation distance defines the units at risk; the last aggregated units are considered at risk. The clustering algorithms used in the statistical disclosure control setting are distance-based algorithms. A limitation of such approach is that the standard algorithms tend to find clusters with equal variance. Such algorithms would find units at risk, only on the tails of the distribution of the continuous quasi-identifier. For this reason, the clustering algorithms may not be suitable for quasi-identifiers with very skew distributions as they are present in business microdata. With respect to rank-based intervals (or standard-deviation-based intervals), an advantage of the clustering algorithms is that they may be adapted to multivariate settings.

Ichim (2009) proposes a density-based approach to the definition of disclosure risk. The clusters are based on local density, i.e., on the estimates of the density around units by mean of the local outlier factor. The statistical agency may set a threshold and define at risk of identification those units presenting a value of the local outlier factor greater than such threshold. Such factor measures the degree of relative isolation of a unit with respect to its nearest neighbours. In practice, the absolute measurement error on continuous economic variables is proportional to the values taken by the variables; the risk of disclosure by means of local measures results then being a meaningful choice. A further advantage of this local measure is its independence on the location of the units at risk: tails or centre of the distribution.

As for the last class of definition of risk devoted to continuous quasi-identifiers, the idea is to use a perturbative method to protect the microdata, as presented in Section 3.7, and then apply a record linkage procedure between the original unmasked data and the perturbed data to assess how many records have been correctly matched. The risk is defined by the proportion of matches which are correct. In the statistical disclosure control literature, distance-based record linkage procedures are commonly applied (see for example, Winkler 2004b and Domingo-Ferrer and Torra 2003). However, as noted by Ichim (2009), for the way in which these risk measures are conceived, they seem more adequate for evaluating the efficacy of different masking methods rather than for measuring the risk of disclosure. Indeed, the resulting set of units to be considered at risk depends on the applied masking method and not on their proper characteristics. The process outlined in Section 3.2 can be fitted in this framework with many difficulties: the characteristics and structure of the data can hardly be taken into account and the only feasible disclosure scenario is the one where the information available for identification is exclusively contained in the original microdata. These assumptions seem unrealistic, especially for variables related to enterprises, and highly conservative. To avoid such limitation commercially available files could be used. A more conceptual issue with this approach, as noted in Skinner (2008), is that it fails to reflect adequately the type of information available to the intruder. Section 3.5.6 provides some examples of the use of such an approach.

3.5   Estimating re-identification risk

This section provides an overview of methods that are available in order to estimate disclosure risk measures therefore providing risk assessment for the microdata file to be released. As we have already defined in Section 3.4, a re-identification occurs when the unit in the released file and a unit in the external file belong to the same individual in the population. It is likely that the intruder will be interested in identifying those sample units that are unique or very rare with respect to combinations of quasi-identifiers.

All of the methods based on keys in the population described in this section aim at estimating the individual per-record disclosure risk (Equation (3.1)) that, under the assumptions in Section 3.4.2, can be formulated as $1/F_k$. The population frequencies F_k are unknown parameters, and therefore need to be estimated from the sample.

3.5.1   Individual risk based on the sample: Threshold rule

The threshold rule is based on information on the sole sample. In a disclosure scenario keys, i.e. combinations of quasi-identifiers, are supposed to be used by an intruder to re-identify a respondent. Re-identification of a respondent can occur when this respondent is rare in the population with respect to a certain key value, i.e. a combination of values of identifying variables. Hence, rarity of respondents with respect to certain key values should be avoided. When a respondent appears to be rare with respect to a key value, then disclosure control measures should be taken to protect this respondent against re-identification. Following the Nosy Neighbour scenario that assumes knowledge of which person belongs to the sample, the aim of the threshold rule is to avoid the occurrence of combinations of scores that are rare in the sample and not only to avoid population uniques. To define what is meant by rare the data protector has to choose a threshold value for each key. If a key occurs more often than this threshold, the key is considered safe, otherwise the key must be protected because of the risk of re-identification. The level of the threshold and the number and size of the keys to be inspected depend on the level of protection to be achieved. PUFs require much more protection than microdata files under contract that are only available to researchers under a contract. How this rule is used in practice is given in the example of Section 3.12.1.

3.5.2   Estimating individual risk using sampling weights

In order to infer the population frequencies F_k of a given combination from the corresponding sample frequency f_k, a Bayesian approach may be pursued, in that the posterior distribution of F_k given $\mathbf{f}$ is exploited. In this manner, the risk is defined as the expected value of $1/F_k$ under this distribution. The original model proposed in Benedetti and Franconi (1998) is equivalent (see, for example, Polettini 2004 and Di Consiglio and Polettini 2006 and citation therein) to the following hierarchical model:

$$F_k|\pi_k \sim \text{Poisson}(N\,\pi_k), \quad F_k = 0, 1, \ldots, \tag{3.4}$$
$$f_k|F_k, \pi_k, p_k \sim \text{Binom}(F_k, p_k), \quad f_k = 0, 1, \ldots, F_k$$

independently across cells. Moreover, π_k in Equation (3.4) follows an improper prior distribution, proportional to $1/\pi_k$, $k = 0, 1, \ldots, K$. The parameters p_k, each one representing the probability that a member of the population cell $\mathcal{C}_k$ falls into the sample, are not further modelled. Following a kind of Empirical Bayesian approach Benedetti and Franconi (1998) proposes to plug in the quantity $\hat{p}_k = f_k/\hat{F}_k^D$ in the final modelling step, where $\hat{F}_k^D = \sum_{i \in \mathcal{C}_k} w_i$ indicates the direct design unbiased estimator of the population counts F_k based on the sampling design weights w_i.

Under the model (3.4) the individual risk is given by:

$$\hat{r}_k = \frac{\hat{p}_k^{f_k}}{f_k} \, _2F_1(f_k, f_k; f_k + 1; 1 - \hat{p}_k),$$

where $_2F_1(a, b; c; z)$ is the Hypergeometric function.

The procedure relies on the assumption that the available data are a sample from a larger population. If the sampling weights are not available, or if data represent the whole population, the strategy used to estimate the individual risk is not meaningful. This risk estimation relies heavily on the use of sampling weights; this because statistical agencies usually make use of complex sampling designs that utilise calibration procedures imposing sound characteristics to the sampling weights. If the microdata are a result of a simple random sampling, then it is advisable not to use such estimation procedure.

Threshold setting for individual risk

The individual risk provides a measure of risk at the individual level. A global measure of disclosure risk for the whole file can be expressed in terms of the expected number of re-identifications in the file. If we interpret r_k as an estimate of the probability of re-identification, then the expected number of re-identification in the file is given by $\sum_k r_k f_k$. This measure of disclosure depends on the number of records, n. The re-identification rate instead is independent of n:

$$R = \frac{1}{n} \sum_k r_k f_k. \tag{3.5}$$

One use of this global risk measure is the following: if the re-identification rate is below a level η which is considered safe, i.e. $R < \eta$, then the file can be released.

Consider the re-identification rate R: a key k contributes to R an amount $r_k f_k$ of expected re-identifications. Since units belonging to the same key k have the same individual risk, keys can be arranged in increasing order of risk r_k. Let the subscript (k) denote the kth element in this ordering. A threshold r^* on the individual risk can be set. Consequently, unsafe cells are those for which $r_k \geq r^*$ that can be indexed by $(k) = k^* + 1, \ldots, K$. The key k^* is in a one-to-one correspondence to r^*. This allows setting an upper bound R^* on the re-identification rate of the released file (after data protection) substituting $r_k f_k$ with $r^* f_k$ for each (k).

For the mathematical details, see Franconi and Polettini (2004) and reference therein and the μ-ARGUS manual, Hundepool *et al.* (2008). This method is implemented in both μ-ARGUS software and the *sdcMicro* package (see Section 3.11).

Hierarchical structure and the household risk

A relevant characteristic of social microdata is its inherent hierarchical structure, which allows us to recognise groups of individuals in the file, the most typical case being the household. When defining the re-identification risk, it is important to take into account this dependence among units: indeed re-identification of an individual in the group may affect the probability of disclosure of all its members. So far, implementation of a hierarchical risk has been performed only with reference to households, i.e., a household risk. Allowing for dependence in estimating the risk enables us to attain a higher level of safety than when merely considering the case of independence.

The household risk makes use of the same framework defined for the individual risk in Section 3.4. In particular, the concept of re-identification holds with the additional assumption that the intruder attempts a confidentiality breach by re-identification of individuals in households. The household risk is defined as the probability that at least one individual in the household is re-identified. For a given household g of size $|g|$, whose members are labelled $i_1, \ldots, i_{|g|}$, the household risk is:

$$r_g^h = P(i_1 \cup i_2 \cup \ldots \cup i_{|g|} \text{re-identified}). \tag{3.6}$$

All the individuals in household g share the same value of the household risk r_g^h.

Threshold setting for the household risk

A similar reasoning applies when setting the threshold for the household risk. Since all the individuals in a given household have the same household risk, the expected number of re-identified records in household g equals $|g| r_g^h$. Denote by G the total number of households in the file and define the expected number of re-identified households in the file as $\sum_{g=1}^{G} r_g^h$. The household re-identification rate, i.e.

$$R^h = \frac{1}{G} \sum_{g=1}^{G} r_g^h,$$

can be exploited to select a household risk threshold, along the same lines as above. Once a threshold on the household risk has been selected, the procedure described in the paragraph above can be applied to transform that value into an individual risk threshold. Note that the household risk r_g^h of household g is computed by the individual risks of its household members. For a given household, it might happen that a household is unsafe (i.e. r_g^h exceeds the threshold) because just one of its members, say i, has a high value r_i of the individual risk. Therefore, to protect the households, the followed approach is to protect individuals in households, first protecting those individuals who contribute most to the household risk. For this reason, inside unsafe households, detection of unsafe individuals is needed. In other words, the threshold on the household risk r^h has to be transformed into a threshold on the individual risk r_i. To this aim, it can be noticed that the household risk is bounded by the sum of the individual risks of the members of the household. $r_g^h \leq \sum_{j=1}^{|g|} r_{i_j}$.

Consider to apply a threshold r_{h^*} on the household risk. In order for household g to be classified safe (i.e. $r_g^h < r_{h^*}$), it is sufficient that all of its components have individual risk less than $\delta_g = r_{h^*}/|g|$. This is clearly an approach possibly leading to overprotection, as we check whether a bound on the household risk is below a given threshold. It is important to remark that the threshold δ_g just defined depends on the size of the household to which individual i belongs. This implies that for two individuals that are classified in the same key k (and therefore have the same individual risk r_k), but belong to different households with different sizes, it might happen that one is classified safe, while the other unsafe (unless the household size is included in the set of identifying variables). In practice, denoting by $g(i)$ the household to which record i belongs, the approach pursued so far consists in turning a threshold r_{h^*} on the household risk into a vector of thresholds on the individual risks r_i, $i = 1, \ldots, n$:
$$\delta_g = \delta_{g(i)} = r_{h^*}/|g(i)|.$$

Individuals are finally set to unsafe whenever $r_i \geq \delta_{g(i)}$; local suppression might then be applied to those records, if requested. Suppression of these records ensures that after protection the household risk is below the threshold δ_g.

Choice of quasi-identifiers in household risk

For household data, it is important to include in the list of quasi-identifiers that are used to estimate the household risks also the available information on the household, such as the number of components or the household type. Suppose one computes the risk using the household size as the only identifying variable in a household data file, and that such file contains households whose risk is above a fixed threshold. Since information on the number of components in the household cannot be removed from a file with household structure, these records cannot be safely released, and no suppression can make them safe. This permits to check for presence of very peculiar households (usually, the very large ones) that can be easily recognised in the population just by their size and whose main characteristic, namely their size, can be immediately computed from the file. For a discussion on this issue, see Hundepool *et al.* (2008).

3.5.3 Estimating individual risk by Poisson model

Elamir and Skinner (2006) assume that the F_k are independently Poisson distributed with means λ_k. If there is the further assumption of a Bernoulli sampling scheme with equal selection probably π, then f_k and $F_k - f_k$ are independently Poisson distributed as: $f_k|\lambda_k \sim \text{Poisson}(\pi\lambda_k)$ and $F_k - f_k|\lambda_k \sim \text{Poisson}((1 - \pi)\lambda_k)$.

Under this assumption, the individual risk measure for a sample unique is defined as $r_k = E_\lambda(1/F_k|f_k = 1)$ which is equal to:

$$r_k = \frac{1}{\lambda_k(1 - \pi)}(1 - \exp^{-\lambda_k(1-\pi)}). \qquad (3.7)$$

In this approach, the parameters λ_k are estimated by taking into account the structure and dependencies in the data through log-linear modelling. Assuming that the sample frequencies f_k are independently Poisson distributed with a mean of $u_k = \pi \lambda_k$, a log-linear model for the u_k can be expressed as: $\log(u_k) = x_k' \beta$ where x_k is a design vector denoting the main effects and interactions of the model for the quasi-identifiers. Using standard procedures, such as iterative proportional fitting, we obtain the Poisson maximum-likelihood estimates for the vector β and calculate the fitted values: $\hat{u}_k = \exp(x_k' \hat{\beta})$. The estimate for $\hat{\lambda}_k$ is equal to $\hat{u}_k / \pi$ which is substituted for λ_k in the above formula (3.7). The individual disclosure risk measures can be aggregated to obtain a global (file-level) measure:

$$\hat{\tau}_2 = \sum_{k \in SU} \hat{r}_k = \sum_{k \in SU} \frac{1}{\hat{\lambda}_k (1 - \pi)} (1 - \exp^{-\hat{\lambda}_k (1 - \pi)}),$$

where SU is the set of all sample uniques. More details on this method are available from Skinner and Shlomo (2008) and Shlomo and Young (2006b).

Skinner and Shlomo (2008) have developed goodness-of-fit criteria for selecting the most robust log-linear model that will provide accurate estimates for the global disclosure risk measure detailed above. The method begins with a log-linear model where a high test statistic indicates under-fitting (i.e. the disclosure risk measures will be overestimated). Then a forward search algorithm is employed by gradually adding higher order interaction terms to the model, until the test statistic approaches the level (based on a normal distribution approximation) where the fit of the log-linear model is accepted.

The method is based on theoretical well-defined disclosure risk measures and goodness of fit criteria which ensure the fit of the log-linear model and the accuracy of the disclosure risk measures. It requires a model search algorithm which takes some computer time and requires intervention. This method is implemented in the *sdcMicro* package.

3.5.4   Further models that borrow information from other sources

Other methods for probabilistic risk assessment have been developed based on a generalised Negative Binomial smoothing model for sample disclosure risk estimation which subsumes both the model in Section 3.5.2 and the Poisson log-linear model above. A first method, see Rinott and Shlomo (2006), is useful for key variables that are ordinal where local neighbourhoods can be defined for inference on cell k. The Bayesian assumption of $\lambda_k \sim \text{Gamma}(\alpha_k, \beta_k)$ is added independently to the Poisson model above which then transforms the marginal distribution to the generalised Negative Binomial Distribution: $f_k \sim \text{NB}(\alpha_k, p_k = 1/[1 + N \pi_k \beta_k])$ and

$$F_k | f_k \sim \text{NB}\left(\alpha_k + f_k, \rho_k = \frac{1 + N \pi_k \beta_k}{1 + N \beta_k} \right),$$

where π_k is the sampling fraction. In each local neighbourhood of cell k, a smoothing polynomial regression model is carried out to estimate α_k and β_k, and a global disclosure risk measure is estimated based on the Negative Binomial Distribution,

$$\hat{\tau}_2 = \sum_{k \in SU} \hat{r}_k = \sum_{k \in SU} \frac{\hat{\rho}_k(1 - \hat{\rho}_k)^{\hat{\alpha}_k}}{\hat{\alpha}_k(1 - \hat{\rho}_k)}.$$

For more details on this method, see Rinott and Shlomo (2006). For the estimation of the re-identification risk, a local neighbouring approach to the analysis of contingency tables using log-liner models is presented in Rinott and Shlomo (2007). The idea is that a sample unique for which its neighbouring cells have small values is more likely a population unique than other sample uniques.

Ichim (2008) recognises the importance of smoothing as a way to borrow information from the neighbouring cells in large sparse tables as those arising from the estimation of re-identification risk. Moreover, it has to be stressed that the relationship between the cross classifying variables might determine the number of sample uniques that are also population uniques. In Ichim (2008), a penalised likelihood approach is proposed to deal with smoothness. The penalty function is expressed in terms of independence constraints. An extension presented in Ichim (2008) deals with the proposal of a log-rate model using the sampling weights as an offset variable. The characteristics of such model are unbiasedness and validity of goodness of fit tests. A further extension of this model is presented in Ichim and Foschi (2011) where smoothing strategies are combined with the use of graphical log-liner model decomposition.

Further approaches that try to borrow strength in the risk estimation phase are proposed by Di Consiglio and Polettini (2006) and Di Consiglio and Polettini (2008). In the former, the use of auxiliary information arising from the previous census, as done in the context of small area estimation, is performed via SPREE type estimators that use the association structure observed at the previous census. In the latter, models are considered that use the structure of a population contingency table while allowing for smooth variations.

3.5.5   Estimating per record risk via heuristics

Besides modelling approaches to the estimation of re-identification risk measures also heuristics strategies have been developed. Elliot *et al.* (1998) define a *special unique* as a record that is a sample unique on a set of variables and that is also unique on a subset of those variables. Empirical work shows that special uniques are more likely to be population unique than random uniques. Special uniques have been classified according to the size and number of the smallest subset of quasi-identifiers that defines the record as unique, known as minimal sample uniques (MSU). A software system, SUDA, Elliot *et al.* (2006), a windows application available as freeware under restricted licence, has been developed that provides disclosure risk broken down by record, variable, variable value and by interactions of those. SUDA grades and orders records within a microdata file according to the level of risk. The method assigns a per record matching probability to a sample unique based on the number and size of minimal sample uniques.

Further development along these lines brought to the definition of the DIS Measure (Elliot 2000), the conditional probability of a correct match given a unique match, which is estimated by a simple sample-based measure without modelling assumptions that is approximately unbiased. Elliot *et al.* (2006) describes a heuristic which combines the DIS measure with scores resulting from the algorithm (i.e. SUDA scores). This method known as DIS-SUDA produces estimates of an intruder's confidence in a match against a given record being correct. This is related to the probability that the match is correct and is heuristically linked to the estimate of the global measure of risk τ_2 (see Equation (3.3)).

The advantage of this method is that it relates to a practical model of data intrusion, and it is possible to compare different values directly. The disadvantages are that it is sensitive to the level of the max MSU parameter and is calculated in a heuristic manner. In addition, it is difficult to compare disclosure risk across different files. However, the method has been extensively tested and was used for the detection of high-risk records in the UK Sample of Anonymised Records (SAR) drawn from the 2001 Census. The assessment showed that the DIS-SUDA measure calculated from the algorithm provided a good estimate for the individual disclosure risk measure. The algorithm also identifies the variables and the values of variables that are contributing most to the disclosure risk of the record.

3.5.6   Assessing risk via record linkage

As discussed in Section 3.4.2 and 3.4.3, in many studies a way to measure *a posteriori* the disclosure risk of a microdata file is via record linkage: the measure of risk is the proportion of matches among the total number of records. Although this instrument seems more adequate to evaluate masking methods, it has also been used to asses disclosure risk.

Distance-based record linkage consists in linking each record a in file A to its nearest record b in file B. Therefore, this method requires a definition of a distance function for expressing closeness between records. This record-level distance can be constructed from distance functions defined at the level of variables. Construction of record-level distances requires standardising variables to avoid scaling problems and assigning each variable a weight on the record-level distance.

Winkler (2004b) provides methods for constructing re-identification metrics defined according to procedures adopted to limit disclosure (micro aggregation, rank swapping, etc.). The record linkage algorithm introduced in Bacher *et al.* (2002) is similar in spirit to distance-based record linkage. This is so because it is based on cluster analysis, and therefore, links records that are near to each other.

Also probabilistic record linkage, Fellegi and Sunter (1969) and Jaro (1989), have been used in SDC; see, for example, Domingo-Ferrer and Torra (2003) and a presentation of the framework for using probabilistic record linkage in the SDC setting by Skinner (2008).

3.6   Non-perturbative microdata masking

Non-perturbative masking does not rely on distortion of the original data but on partial suppressions or reductions of detail. Some of the methods in this class are usable on both categorical and continuous data, but others are not suitable for continuous data.

Table 3.2 lists the non-perturbative methods described as follows. For each method, the table indicates whether it is suitable for continuous and/or categorical data.

3.6.1   Sampling

Instead of publishing the original microdata file, what is published is a sample S of the original set of records.

Sampling methods are suitable for categorical microdata, but their adequacy for continuous microdata is less clear in a general disclosure scenario. The reason is that such methods leave a continuous variable V_i unperturbed for all records in S. Thus, if variable V_i is present in an external administrative public file, unique matches with the published sample are very likely: indeed, given a continuous variable V_i and two respondents o_1 and o_2, it is highly unlikely that V_i will take the same value for both o_1 and o_2 unless $o_1 = o_2$ (this is true even if V_i has been truncated to represent it digitally).

If, for a continuous identifying variable, the score of a respondent is only approximately known by an attacker (as assumed in Willenborg and de Waal 1996), it might make sense to use sampling methods to protect that variable. However, assumptions on restricted attacker resources are perilous and may prove definitely too optimistic if good-quality external administrative files are at hand. For the purpose of illustration, the following example gives the technical specifications of a real-world application of sampling.

Example 3.6.1 *Statistics Catalonia released in 1995 a sample of the 1991 population census of Catalonia. The information released corresponds to 36 categorical variables (including the recoded versions of initially continuous variables); some of the variables are related to the individual person and some to the household. The technical specifications of the sample were as follows:*

- *Sampling algorithm.* Simple random sampling.

- *Sampling unit.* Individuals in the population whose residence was in Catalonia as of March 1, 1991.

Table 3.2   Non-perturbative methods vs. data types.

Method	Continuous data	Categorical data
Sampling		X
Global recoding	X	X
Top and bottom coding	X	X
Local suppression		X

- *Population size.* 6 059 494 inhabitants.
- *Sample size.* 245 944 individual records.
- *Sampling fraction.* 0.0406.

With the above sampling fraction, the maximum absolute error for estimating a maximum-variance proportion is 0.2 %.

3.6.2   Global recoding

For a categorical variable V_i, global recoding, a.k.a. generalisation, combines several categories to form new (more general) categories, thus resulting in a new V_i' with $|D(V_i')| < |D(V_i)|$, where $| \cdot |$ is the cardinality operator. For a continuous variable, global recoding means replacing V_i by another variable V_i' which is a discretised version of V_i. In other words, a potentially infinite range $D(V_i)$ is mapped onto a finite range $D(V_i')$. This is the technique used in the μ-ARGUS SDC package (Hundepool *et al.* 2008).

This technique is more appropriate for categorical microdata, where it helps disguise records with strange combinations of categorical variables. Global recoding is used heavily by statistical offices.

Example 3.6.2 *If there is a record with 'Marital status = Widow/er' and 'Age = 17', global recoding could be applied to 'Marital status' to create a broader category 'Widow/er or divorced', so that the probability of the above record being unique would diminish. Global recoding can also be used on a continuous variable, but the inherent discretisation leads very often to an unaffordable loss of information. Also, arithmetical operations that were straightforward on the original V_i are no longer easy or intuitive on the discretised V_i'.*

Example 3.6.3 *We can recode the variable 'Occupation', by combining the categories 'Statistician' and 'Mathematician' into a single category 'Statistician or Mathematician'. When the number of female statisticians in Urk (a small town in The Netherlands) plus the number of female mathematicians in Urk is sufficiently high, then the combination 'Place of residence = Urk', 'Gender = Female' and 'Occupation = Statistician or Mathematician' is considered safe for release. Note that instead of recoding 'Occupation' one could also recode 'Place of residence', for instance.*

It is important to realise that global recoding is applied to the whole data set, not only to the unsafe part of the set. This is done to obtain a uniform categorisation of each variable. Suppose, for instance, that we recode the 'Occupation' in the above way. Suppose furthermore that both the combinations 'Place of residence = Amsterdam', 'Gender = Female' and 'Occupation = Statistician', and 'Place of residence = Amsterdam', 'Gender = Female' and 'Occupation = Mathematician' are considered safe. However, to obtain a uniform categorisation of 'Occupation' we would not publish these combinations, but only the combination 'Place of residence = Amsterdam', 'Gender = Female' and 'Occupation = Statistician or Mathematician'.

3.6.3   Top and bottom coding

Top and bottom coding is a special case of global recoding which can be used on variables that can be ranked, that is, variables that are continuous or categorical ordinal. The idea is that top values (those above a certain threshold) are lumped together to form a new category. The same is done for bottom values (those below a certain threshold) (see Hundepool *et al*. 2008 or Templ 2008)

3.6.4   Local suppression

If a combination of key variable values (quasi-identifier values) is shared by too few records, it is called an *unsafe combination*, because it may lead to re-identification. Certain values of individual variables are suppressed, that is, replaced with missing values, with the aim of eliminating unsafe combinations by increasing the set of records agreeing on each combination of key values.

Example 3.6.4   *In Example 3.6.3, we can protect the unsafe combination 'Place of residence = Urk', 'Gender = Female' and 'Occupation = Statistician' by suppressing the value of 'Occupation', assuming that the number of females in Urk is sufficiently high. The resulting combination is then given by 'Place of residence = Urk', 'Gender = Female' and 'Occupation = missing'. Note that, instead of suppressing the value of 'Occupation', one could also suppress the value of another variable of the unsafe combination. For instance, if the number of female statisticians in the Netherlands is sufficiently high then one could suppress the value of 'Place of residence' instead of the value of 'Occupation' in the above example to protect the unsafe combination. A local suppression is only applied to a particular value. When, for instance, the value of 'Occupation' is suppressed in a particular record, then this does not imply that the value of 'Occupation' has to be suppressed in another record. The freedom that one has in selecting the values that are to be suppressed allows one to minimise the number of local suppressions.*

Ways to combine local suppression and global recoding are discussed in de Waal and Willenborg (1995) and implemented in the μ-ARGUS SDC package (see Hundepool *et al*. 2008).

If a continuous variable V_i is part of a set of key variables, then each combination of key values is probably unique. Since it does not make sense to systematically suppress the values of V_i, we conclude that local suppression is rather oriented to categorical variables.

3.7   Perturbative microdata masking

Perturbative microdata masking methods allow for the release of the entire microdata set, although perturbed values rather than exact values are released. Not all perturbative methods are designed for continuous data; this distinction is addressed further below for each method.

Table 3.3   Perturbative methods vs. data types. 'X' denotes
applicable and '(X)' denotes applicable to ordinal
categories only.

Method	Continuous data	Categorical data
Noise masking	X	
Micro-aggregation	X	(X)
Rank swapping	X	(X)
Data shuffling	X	(X)
Rounding	X	
Re-sampling	X	
PRAM		X
MASSC		X

Most perturbative methods reviewed below (including noise masking (additive
and multiplicative), rank swapping, microaggregation and post-randomisation) are
special cases of matrix masking. If the original microdata set is $\mathbf{X}$, then the masked
microdata set $\mathbf{Z}$ is computed as

$$\mathbf{Z} = \mathbf{AXB} + \mathbf{C},$$

where $\mathbf{A}$ is a record-transforming mask, $\mathbf{B}$ is a variable-transforming mask and $\mathbf{C}$ is
a displacing mask or noise (Duncan and Pearson 1991).

Table 3.3 lists the perturbative methods described as follows. For each method,
the table indicates whether it is suitable for continuous and/or categorical data.

3.7.1   Additive noise masking

The main noise addition algorithms in the literature are:

- Masking by uncorrelated noise addition.

- Masking by correlated noise addition.

- Masking by noise addition and linear transformation.

- Masking by noise addition and non-linear transformation (see Sullivan 1989).

As argued in Brand (2002a), in practice only simple noise addition (two first
variants) or noise addition with linear transformation are used. When using linear
transformations, a decision has to be made whether to reveal to the data user the
parameter c (see page 56) determining the transformations to allow for bias adjustment
in the case of sub-populations.

With the exception of the not very practical method of Sullivan (1989), noise
addition is not suitable to protect categorical data. On the other hand, it is well suited
for continuous data for the following reasons:

- It makes no assumptions on the range of possible values for V_i (which may be
infinite).

- The noise being added is typically continuous and with mean zero, which suits well continuous original data.

- No exact matching is possible with external files. Depending on the amount of noise added, approximate (interval) matching might be possible.

We next review in some detail the above four variants of noise addition; see Brand (2002a) for further details.

Masking by uncorrelated noise addition

Masking by additive noise assumes that the vector of observations x_j for the jth variable of the original data set X_j is replaced by a vector

$$z_j = x_j + \epsilon_j,$$

where ϵ_j is a vector of normally distributed errors drawn from a random variable $\varepsilon_j \sim N(0, \sigma_{\varepsilon_j}^2)$, such that $\text{Cov}(\varepsilon_t, \varepsilon_l) = 0$ for all $t \neq l$ (white noise).

The general assumption in the literature is that the variances of the ε_j are proportional to those of the original variables. Thus, if $\text{Var}(X_j) = \sigma_j^2$ is the variance of X_j, then $\sigma_{\varepsilon_j}^2 := \alpha \sigma_j^2$.

In the case of a p-dimensional data set, simple additive noise masking can be written in matrix notation as

$$Z = X + \varepsilon,$$

where $X \sim (\mu, \Sigma)$, $\varepsilon \sim N(0, \Sigma_\varepsilon)$ and

$$\Sigma_\varepsilon = \alpha \; diag(\sigma_1^2, \sigma_2^2, \ldots, \sigma_p^2), \quad \text{for } \alpha > 0$$

with $diag(\cdots)$ denoting a diagonal matrix.

This method preserves means and covariances, i.e.

$$E(Z) = E(X) + E(\varepsilon) = E(X) = \mu,$$

$$\text{Cov}(Z_j, Z_l) = \text{Cov}(X_j, X_l) \; \forall j \neq l.$$

Unfortunately, neither variances nor correlation coefficients are preserved:

$$\text{Var}(Z_j) = \text{Var}(X_j) + \alpha \text{Var}(X_j) = (1 + \alpha)\text{Var}(X_j),$$

$$\rho_{Z_j, Z_l} = \frac{\text{Cov}(Z_j, Z_l)}{\sqrt{\text{Var}(Z_j)\text{Var}(Z_l)}} = \frac{1}{1 + \alpha}\rho_{X_j, X_l} \; \forall j \neq l.$$

Masking by correlated noise addition

Correlated noise addition also preserves means and additionally allows preservation of correlation coefficients. The difference with the previous method is that the covariance matrix of the errors is now proportional to the covariance matrix of the original data, i.e., $\varepsilon \sim N(0, \Sigma_\varepsilon)$, where $\Sigma_\varepsilon = \alpha \Sigma$.

With this method, we have that the covariance matrix of the masked data is

$$\Sigma_Z = \Sigma + \alpha \Sigma = (1 + \alpha)\Sigma. \tag{3.8}$$

Preservation of correlation coefficients follows, since

$$\rho_{Z_j, Z_l} = \frac{1+\alpha}{1+\alpha} \frac{\text{Cov}(X_j, X_l)}{\sqrt{\text{Var}(X_j)\text{Var}(X_l)}} = \rho_{X_j, X_l}.$$

Regarding variances and covariances, we can see from Equation (3.8) that masked data only provide biased estimates for them. However, it is shown in Kim (1990) that the covariance matrix of the original data can be consistently estimated from the masked data as long as α is known.

As a summary, masking by correlated noise addition outputs masked data with higher analytical validity than masking by uncorrelated noise addition. Consistent estimators for several important statistics can be obtained as long as α is revealed to the data user. However, simple noise addition as discussed in this section and in the previous one is seldom used because of the very low level of protection it provides (see Tendick 1991 and Matloff 1994).

Masking by noise addition and linear transformations

In Kim (1986), a method is proposed that ensures by additional linear transformations that the sample covariance matrix of the masked variables is an unbiased estimator for the covariance matrix of the original variables. The idea is to use simple additive noise on the p original variables to obtain overlayed variables

$$Z_j = X_j + \varepsilon_j, \text{ for } j = 1, \ldots, p.$$

As in the previous section on correlated masking, the covariances of the errors ε_j are taken proportional to those of the original variables. Usually, the distribution of errors is chosen to be normal or the distribution of the original variables, although in Roque (2000) mixtures of multivariate normal noise are proposed.

In a second step, every overlayed variable Z_j is transformed into a masked variable G_j as

$$G_j = cZ_j + d_j.$$

In matrix notation, this yields

$$Z = X + \varepsilon,$$

$$G = cZ + D = c(X + \varepsilon) + D,$$

where $X \sim (\mu, \Sigma)$, $\varepsilon \sim (0, \alpha\Sigma)$, $G \sim (\mu, \Sigma)$ and D is a matrix whose jth column contains the scalar d_j in all rows. Parameters c and d_j are determined under the restrictions that $E(G_j) = E(X_j)$ and $\text{Var}(G_j) = \text{Var}(X_j)$ for $j = 1, \ldots, p$. In fact, the first restriction implies that $d_j = (1-c)E(X_j)$, so that the linear transformations depend on a single parameter c.

Due to the restrictions used to determine c, this method preserves expected values and covariances of the original variables and is quite good in terms of analytical validity. Regarding analysis of regression estimates in sub-populations, it is shown in Kim (1990) that (masked) sample means and covariances are asymptotically biased estimates of the corresponding statistics on the original sub-populations. The magnitude of the bias depends on the parameter c, so that estimates can be adjusted by the data user as long as c is revealed to her. Revealing c to the user has a fundamental disadvantage, though: the user can undo the linear transformation, so that this method becomes equivalent to plain uncorrelated noise addition (Domingo-Ferrer *et al.* 2004).

The most prominent shortcomings of this method are that it does not preserve the univariate distributions of the original data and that it cannot be applied to discrete variables due to the structure of the transformations.

Masking by noise addition and non-linear transformations

An algorithm combining simple additive noise and non-linear transformations is proposed in Sullivan (1989). The advantages of this proposal are that it can be applied to discrete variables and that univariate distributions are preserved.

The method consists of several steps:

- Calculate the empirical distribution function for every original variable.

- Smooth the empirical distribution function.

- Convert the smoothed empirical distribution function into a uniform random variable and this into a standard normal random variable.

- Add noise to the standard normal variable.

- Back-transform to values of the distribution function.

- Back-transform to the original scale.

In the European project CASC (IST-2000-25069), the practicality and usability of this algorithm was assessed. Unfortunately, the internal CASC report by Brand (2002b) concluded that:

> 'All in all, the results indicate that an algorithm as complex as the one proposed by Sullivan can only be applied by experts. Every application is very time consuming and requires expert knowledge on the data and the algorithm.'

3.7.2   Multiplicative noise masking

One main challenge regarding additive noise with constant variance is that on the one hand small values are strongly perturbed and on the other large values are weakly perturbed. For instance, in a business microdata set the large enterprises – which are much easier to re-identify than the smaller ones – remain still high at risk. A possible way out is given by the multiplicative noise approaches explained as follows.

Let X be the matrix of the original numerical data and W the matrix of continuous perturbation variables with expectation 1 and variance $\sigma_W^2 > 0$. The corresponding anonymised data X^a is then obtained by

$$X^a = W \odot X,$$

where $\odot$ is the so-called Hadamard product (element-wise matrix multiplication). That is

$$(X^a)_{ij} := w_{ij} X_{ij}$$

for each pair (i, j).

Multiplicative noise preserving first- and second-order moments

A specific approach is given by the method in Höhne (2004). In a first step, for each record it is randomly decided whether its values are diminished or enlarged, each with probability 1/2. This scaling is done using the main factors $1 - f$ and $1 + f$. In order to avoid that all values of some record are perturbed with the same noise, these main factors are themselves perturbed with some additive noise s (where $s < \frac{f}{2}$). The following transformation is needed to preserve the first- and second-order moments of the distribution:

$$x_i^{a^R} := \frac{\sigma_X}{\sigma_{X^a}} (x_i^a - \mu_{X^a}) + \mu_X,$$

where μ_X and μ_{X^a} define the average of the original and anonymised variables, and σ_X and σ_{X^a} the corresponding standard deviations, respectively.

Particularly, if the original data follow a strongly skewed distribution, the deviations using this method may strongly depend on the configuration of the noise factors for a few, but large values. That is, despite consistency, means and sums might be unsatisfactorily reproduced. For this reason, Höhne (2004) suggests a slight modification of the method. At first, normal distributed random variables W_i with expectation greater than zero and 'small' variance are generated such that the realisation of W_i yields a positive value. Afterwards, the data are sorted in descending order by the considered variable. Then, the record with the largest entry in this variable is reduced to

$$X_1^a = (1 - W_1)X_1.$$

The records $X_2, \ldots, X_{n-1}$ are now perturbed as follows:

$$X_i^a = (1 - W_i)X_i, \text{ if } \sum_{k=1}^{i-1} X_k^a > \sum_{k=1}^{i-1} X_k,$$

$$X_i^a = (1 + W_i)X_i, \text{ if } \sum_{k=1}^{i-1} X_k^a \leq \sum_{k=1}^{i-1} X_k.$$

Hence, means and sums are preserved and the diminishing and enlarging effects of single values cancel out each other. For the remaining record X_n, we set

$$X_n^a = X_n - \left(\sum_{k=1}^{n-1} X_k^a - \sum_{k=1}^{n-1} X_k \right)$$

in order to preserve the overall sum.

Multiplicative noise preserving various constraints

In Oganian (2010, 2011), several multiplicative noise masking schemes are presented. These schemes are designed to preserve positivity constraints, inequality constraints, and the first two moments – the mean vector and the covariance matrix – of the original data.

Multiplicative noise is often implemented by applying log-normal noise to the original data, that is, by taking logarithms of the original data, applying additive, normally distributed noise and exponentiating. If the data set contains not only non-negative variables but variables with negative values as well, log-normal noise addition preserving means and covariance matrix cannot be applied directly; see Oganian (2011) for more details.

One possible solution is to convert all the variables to z-scores and make these z-scores non-negative by adding some value (or vector – for multivariate data), denote it lag, such that $lag \geq |\min(Z)|$. Denote these non-negative z-scores by Z_p. Then we can apply log-normal noise as sketched above to Z_p and after that return the resulting data to the original scale: let k be a parameter chosen by the statistical institute and

$$Z_m = \frac{(\sqrt{1+k} - 1)lag + [Z_p \odot \exp(E^{z_p})]}{\sqrt{1+k}}, \qquad (3.9)$$

$$X_m = (Z_m - lag) \odot \sigma_o + E(X_o), \qquad (3.10)$$

where $\odot$ is the Hadamard product, X_m denotes masked data, X_o denotes original data, σ_o is the main diagonal of Σ_o (the covariance matrix of original data) and E^{z_p} has the following covariance matrix and mean vector, respectively:

$$\Sigma_{E_{z_p}}(i, j) = \log \left(1 + \frac{k \Sigma_{z_p}(i, j)}{E[Z_p(i) Z_p(j)]} \right), \qquad i, j = 1, \ldots, d, \qquad (3.11)$$

$$\mu_{E_{z_p}}(i) = -\sigma_{E_{z_p}}(i)/2, \qquad i = 1, \ldots, d, \qquad (3.12)$$

where $\Sigma_{z_p}(i, j)$ is the (i, j) element of the covariance matrix of positive z-scores.

It is shown in Oganian (2011) that masked data X_m obtained with the above scheme preserves the means and the covariance matrix of original data X_o. It is also shown that positivity of the original data is preserved by the masked data if

$$lag \leq \frac{E(X_o)}{\sigma_o} \sqrt{1+k},$$

where division is done componentwise.

Suppose that original data in addition to positivity constraints also have inequality constraints of the form $X > Y$. For example, masking an income data with the variables 'Gross income' and 'Federal taxes' should produce a masked data such that 'Gross income > Federal taxes'. The protocols described above can be used as building blocks of a new scheme which would guarantee the preservation of inequality constraints. This scheme is the following:

- Apply the multiplicative noise scheme to $(Y_o, [X_o - Y_o])$. Denote the result by $(Y^*, [X_o - Y_o]^*)$.

- The masked data corresponding to (X_o, Y_o) are $(X_m, Y_m) = (Y^* + [X_o - Y_o]^*, Y^*)$.

It is shown in Oganian (2011) that this scheme preserves means and the covariance matrix.

3.7.3 Microaggregation

Microaggregation is a family of SDC techniques for continuous microdata. The rationale behind microaggregation is that confidentiality rules in use allow publication of microdata sets if records correspond to groups of k or more individuals, where no individual dominates (i.e. contributes too much to) the group and k is a threshold value. Strict application of such confidentiality rules leads to replacing individual values with values computed on small aggregates (microaggregates) prior to publication. This is the basic principle of microaggregation.

Univariate microaggregation was proposed in the Eurostat paper Defays and Nanopoulos (1993) and multivariate microaggregation was introduced in Domingo-Ferrer and Mateo-Sanz (1999, 2002). Microaggregation has been used for practical official statistics in Italy (see Pagliuca and Seri 1999), Germany (see Rosemann 2003 and Lenz and Vorgrimler 2005) and several other countries (see UNECE 2001).

To obtain microaggregates in a microdata set with n records, these are combined to form g groups of size at least k. For each variable, the average value over each group is computed and is used to replace each of the original averaged values. Groups are formed using a criterion of maximal similarity. Once the procedure has been completed, the resulting (modified) records can be published.

By construction, microaggregated variables preserve the means of original variables.

Microaggregation exists in several variants:

- Univariate (Defays and Nanopoulos 1993) vs. multivariate (Domingo-Ferrer and Mateo-Sanz 1999, 2002; Domingo-Ferrer and Torra 2005; Domingo-Ferrer *et al.* 2008 and others).

- Fixed (Defays and Nanopoulos 1993; Domingo-Ferrer and Torra 2005) vs. variable-size group (Domingo-Ferrer and Mateo-Sanz 2002; Domingo-Ferrer *et al.* 2008; Laszlo and Mukherjee 2005; Mateo-Sanz and Domingo-Ferrer 1999; Sande 2002; Solanas and Martínez-Ballesté 2006 and others).

- Exact optimal (Hansen and Mukherjee 2003; Oganian and Domingo-Ferrer 2001) vs. heuristic microaggregation (Defays and Nanopoulos 1993; Domingo-Ferrer and Mateo-Sanz 2002; Mateo-Sanz and Domingo-Ferrer 1999; Sande 2002 and others) vs. approximations to optimal microaggregation (Domingo-Ferrer *et al.* 2008, Laszlo and Mukherjee 2009).

- Categorical microaggregation (Torra 2004).

We briefly review those variants in what follows. Consider a microdata set with p continuous variables and n records (i.e. the result of recording p variables on n individuals). The record of each individual can be viewed as an instance of $\mathbf{X}^t = (X_1, \ldots, X_p)$, where the X_i are the variables. A k-partition of the set of records is a partition into g groups with n_i records in the ith group such that $n = \sum_{i=1}^{g} n_i$ and $n_i \geq k$ for $1 \leq i \leq g$. Denote by x_{ij} the jth record in the ith group; denote by $\bar{x}_i$ the average record over the ith group, and by $\bar{x}$ the average record over the whole set of n individuals.

The optimal k-partition (from the information loss point of view) is defined to be the one that maximises within-groups homogeneity; the higher the within-groups homogeneity, the lower the information loss, since microaggregation replaces values in a group by the group centroid. The sum of squares criterion is common to measure homogeneity in clustering. The within-groups sum of squares SSE is defined as

$$SSE = \sum_{i=1}^{g} \sum_{j=1}^{n_i} (x_{ij} - \bar{x}_i)^t (x_{ij} - \bar{x}_i).$$

The lower SSE, the higher the within-groups homogeneity. The total sum of squares is

$$SST = \sum_{i=1}^{g} \sum_{j=1}^{n_i} (x_{ij} - \bar{x})^t (x_{ij} - \bar{x}).$$

In terms of sums of squares, the optimal k-partition is the one that minimises SSE.

For a microdata set consisting of p variables, these can be microaggregated together or partitioned into several groups of variables. Also the way to form groups may vary. We next review the main proposals in the literature.

Example 3.7.1 *This example illustrates the use of microaggregation for SDC and, more specifically, for k-anonymisation (see Samarati and Sweeney 1998, Samarati 2001, Sweeney 2002b and Domingo-Ferrer and Torra 2005). A k-anonymous data set allows no re-identification of a respondent within a group of at least k respondents. We show in Table 3.4 a data set giving, for 11 small or medium enterprises (SMEs) in a certain town, the company name, the surface in square meters of the company's premises, its number of employees, its turnover and its net profit. Clearly, the company name is an identifier. We will consider that turnover and net profit are confidential outcome variables. A first SDC measure is to suppress the identifier 'Company name' when releasing the data set for public use. However, note that the surface of the company's premises and its number of employees can be used by a snooper as key*

Table 3.4   Example: SME data set. 'Company name' is an identifier to be suppressed before publishing the data set.

Company name	Surface (m²)	Number of employees	Turnover (Euros)	Net profit (Euros)
A&A Ltd	790	55	3 212 334	313 250
B&B SpA	710	44	2 283 340	299 876
C&C Inc	730	32	1 989 233	200 213
D&D BV	810	17	984 983	143 211
E&E SL	950	3	194 232	51 233
F&F GmbH	510	25	119 332	20 333
G&G AG	400	45	3 012 444	501 233
H&H SA	330	50	4 233 312	777 882
I&I LLC	510	5	159 999	60 388
J&J Co	760	52	5 333 442	1 001 233
K&K Sarl	50	12	645 223	333 010

variables. Indeed, it is easy for anybody to gauge to a sufficient accuracy the surface and number of employees of a target SME. Therefore, if the only privacy measure taken when releasing the data set in Table 3.4 is to suppress the company name, a snooper knowing that company K&K Sarl has about a dozen employees crammed in a small flat of about 50 m², will still be able to use the released data to link company K&K Sarl with turnover 645 223 Euros and net profit 333 010 Euros.

Table 3.5 is a 3-anonymous version of the data set in Table 3.4. The identifier 'Company name' was suppressed and optimal bivariate microaggregation with k = 3 was used on the key variables 'Surface' and 'No. employees' (in general, if there are p

Table 3.5   Example: 3-anonymous version of the SME data set after optimal microaggregation of key variables.

Surface (m²)	Number of employees	Turnover (Euros)	Net profit (Euros)
747.5	46	3 212 334	313 250
747.5	46	2 283 340	299 876
747.5	46	1 989 233	200 213
756.67	8	984 983	143 211
756.67	8	194 232	51 233
322.5	33	119 332	20 333
322.5	33	3 012 444	501 233
322.5	33	4 233 312	777 882
756.67	8	159 999	60 388
747.5	46	5 333 442	1 001 233
322.5	33	645 223	333 010

key variables, multivariate microaggregation with dimension p should be used to mask all of them). Both variables were standardised to have mean 0 and variance 1 before microaggregation, in order to give them equal weight, regardless of their scale. Due to the small size of the data set, it was feasible to compute optimal microaggregation by exhaustive search. The information or variability loss incurred for those two variables in standardised form can be measured by the within-groups sum of squares $SSE_{opt} = 7.484$. Dividing by the total sum of squares $SST = 22$ (i.e. the sum of squared Euclidean distances from all 11 pairs of standardised (surface, number of employees) to their average) yielded a variability loss measure $SSE_{opt}/SST = 0.34$ bounded between 0 and 1.

It can be seen that the 11 records were microaggregated into three groups: one group with the 1st, 2nd, 3rd and 10th records (companies with large surface and many employees), a second group with the 4th, 5th and 9th records (companies with large surface and few employees) and a third group with the 6th, 7th, 8th and 11th records (companies with a small surface). Upon seeing Table 3.5, a snooper knowing that company K&K Sarl crams a dozen employees in a small flat hesitates between the four records in the third group. Therefore, since turnover and net profit are different for all records in the third group, the snooper cannot be sure about their values for K&K Sarl.

Univariate vs. multivariate microaggregation

The original microaggregation method proposed in the Eurostat paper (Defays and Nanopoulos 1993) was univariate, that is, it formed microaggregates in a single dimension. This can be achieved in two alternative ways:

1. *Individual ranking.* Microaggregation is independently conducted for each continuous variable. In Domingo-Ferrer and Torra (2001b) and Domingo-Ferrer *et al.* (2002), individual ranking microaggregation was shown to incur very high disclosure risk: if each variable is independently microaggregated, it is very easy to link microaggregated records to original records. If unsafe masking can be tolerated (e.g. because individual ranking is used as a complement of other SDC methods), good points of individual ranking are that:

    (a) It causes very little information loss.

    (b) There are theoretical studies of its impact on several inferences: linear model estimation (see Schmid 2006 or Schmid *et al.* 2007), moment estimation (see Schmid and Schneeweiss 2009) and quantile estimation (see Schneeweiss *et al.* 2011).

2. *One-dimensional projection.* Records are projected onto a single dimension prior to microaggregation. This can be done in a number of ways: projecting on the first principal component, on the sum of z-scores of the variables in the data set, on a particular variable, etc. In Domingo-Ferrer and Torra (2001b), univariate microaggregation by one-dimensional projection was shown to be pretty safe (low disclosure risk) but to incur a very high information loss.

To sum up, in general, individual ranking is not recommended due to high disclosure risk and one-dimensional projection is not recommended due to high information loss. Hence, *univariate* microaggregation as a whole is usually regarded as a poor approach, as noted in Rosemann (2003).

Multivariate microaggregation, first proposed in Mateo-Sanz and Domingo-Ferrer (1999) and Domingo-Ferrer and Mateo-Sanz (2002) and consisting of microaggregating all or at least several variables together, *remains an excellent microdata protection method*. With the exception of the one described in Hansen and Mukherjee (2003), all methods mentioned in the rest of this section are multivariate.

Fixed vs. variable group size

The first microaggregation algorithms (see Defays and Nanopoulos 1993) required that all groups except perhaps one be of size k; allowing groups to be of size $\geq k$ depending on the structure of data was termed *data-oriented microaggregation* (see Mateo-Sanz and Domingo-Ferrer 1999 and Domingo-Ferrer and Mateo-Sanz 2002). Figure 3.1 illustrates the advantages of variable-sized groups. If fixed-size microaggregation with $k = 3$ is used, we obtain a partition of the data into three groups, which looks rather unnatural for the data distribution given. On the other hand, if variable-sized groups are allowed then the five data on the left can be kept in a single group and the four data on the right in another group; such a variable-size grouping yields more homogeneous groups, which implies lower information loss. However, except for specific cases such as the one depicted in Figure 3.1, the small gain in within-groups homogeneity obtained with variable-sized groups hardly justifies the higher computational overhead of this option with respect to fixed-sized groups. This is particularly evident for multivariate data, as noted by Sande (2002). In fact, there are well known and commonly used multivariate microaggregation heuristics which are fixed-size: maximum distance to average vector (MDAV) (see Domingo-Ferrer

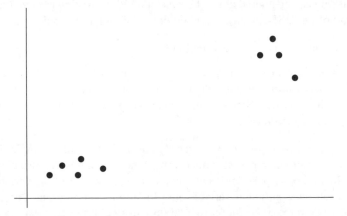

Figure 3.1   Variable-sized groups vs. fixed-sized groups.

and Torra 2005), described later, is a fixed-size heuristic far more used than its variable-size version V-MDAV (see Solanas and Martínez-Ballesté 2006).

Exact optimal vs. heuristic microaggregation vs. approximations to optimal microaggregation

For $p = 1$, i.e., a univariate data set or a multivariate data set where variables are microaggregated one at a time, an exact polynomial-time shortest path algorithm exists to find the k-partition that optimally solves the microaggregation problem (Hansen and Mukherjee 2003). See its description as follows:

For $p > 1$, finding an exact optimal solution to the microaggregation problem, i.e. finding a grouping where groups have maximal homogeneity and size at least k, has been shown to be NP-hard in Oganian and Domingo-Ferrer (2001).

Unfortunately, the univariate optimal algorithm by Hansen and Mukherjee (2003) is not very useful in practice and this for two reasons:

1. Microdata sets are normally multivariate and using univariate microaggregation to microaggregate them one variable at a time is not good in terms of disclosure risk (see discussion above on individual ranking microaggregation).

2. Although polynomial-time, the optimal algorithm is quite slow when the number of records is large.

Thus, practical methods in the literature are multivariate heuristics:

* Multivariate methods directly deal with unprojected data (see Mateo-Sanz and Domingo-Ferrer 1999, Domingo-Ferrer and Mateo-Sanz 2002 and others). When working on unprojected data, one can microaggregate all variables of the data set at a time, or independently microaggregate groups of two variables at a time, three variables at a time, etc. In any case, it is preferable that variables within a group of variables which is microaggregated at a time be correlated (Winkler 2004a), in order to preserve as much as possible the analytic properties of the file.

* Approximation heuristics to optimal microaggregation obtain a k-partition whose SSE is guaranteed to be no more than a fixed multiple of the minimum SSE that would be achieved by optimal microaggregation. The first microaggregation algorithm with approximation bounds was μ-Approx (Domingo-Ferrer et al. 2008), for which a $O(k^3)$ guarantee vs. the minimal SSE was given; later, Mukherjee (2009) gave $O(k^2)$ approximation bounds for their heuristic described in Laszlo and Mukherjee (2005).

The complexity of all above-mentioned multivariate heuristics is quadratic in the number n of records. If assumptions are made on the distribution of the multivariate data, faster heuristics can be devised: e.g., Solanas and DiPietro (2008) propose a linear-time multivariate microaggregation heuristic which works well if the data are (more or less) uniformly distributed. However, without such distributional assumptions, general-purpose multivariate microaggregation stays quadratic.

If n is very large, a sensible way to speed up microaggregation is to use blocking. Under this strategy, one or several categorical variables are taken as so-called blocking variables and records sharing a combination of values of the blocking variables are said to form a block; since blocks are disjoint from each other and they are smaller than the overall data set, they can be separately microaggregated in reasonable quadratic time or space. Of course, the blocking variables should be such that the resulting blocks are natural subsets of records; e.g. if *State* is used as a blocking variable, records in a block correspond to all respondents living in a particular state.

Instead of relying on the semantics of blocking variables to define blocks, some alternative blocking strategies attempt to use as blocks the natural clusters of multivariate data (see, e.g., Solanas *et al.* 2006).

We next describe the two microaggregation algorithms implemented in the μ-ARGUS package (Hundepool *et al.* 2008), and the approximation algorithm μ-approx.

Hansen–Mukherjee's optimal univariate microaggregation
In Hansen and Mukherjee (2003), a polynomial-time algorithm was proposed for *univariate* optimal microaggregation. The authors formulate the microaggregation problem as a shortest path problem on a graph. They first construct the graph and then show that the optimal microaggregation corresponds to the shortest path in this graph. Each arc of the graph corresponds to a possible group that may be part of an optimal k-partition. The arc label is the SSE that would result if that group were to be included in the k-partition. We next detail the graph construction.

Let $\mathbf{V} = \{v_1, \ldots, v_n\}$ be a vector consisting of n real numbers sorted into ascending order, so that v_1 is the smallest value and v_n the largest value. Let k be an integer group size such that $1 \leq k < n$. Now, a graph $\mathbf{G}_{n,k}$ is constructed as follows:

- For each value v_i in $\mathbf{V}$, create a node with label i. Create also an additional node with label 0.

- For each pair of graph nodes (i, j) such that $i + k \leq j < i + 2k$, create a directed arc (i, j) from node i to node j.

- Map each arc (i, j) to the group of values $C_{(i,j)} = \{v_h : i < h \leq j\}$. Let the length $L_{(i,j)}$ of the arc be the within-group sum of squares for $C_{(i,j)}$, that is,

$$L_{(i,j)} = \sum_{h=i+1}^{j} (v_h - \bar{v}_{(i,j)})^2,$$

where $\bar{v}_{(i,j)} = \frac{1}{j-i} \sum_{h=i+1}^{j} v_h$

It is proven in Hansen and Mukherjee (2003) that the optimal k-partition for $\mathbf{V}$ is found by taking as groups the $C_{(i,j)}$ corresponding to the arcs in the shortest path between nodes 0 and n. For minimal group size k and a data set of n real numbers sorted in ascending order, the complexity of this optimal univariate microaggregation is $O(k^2 n)$, that is, it is linear in the size of the data set.

The MDAV heuristic for multivariate microaggregation

The multivariate microaggregation heuristic implemented in μ-ARGUS is called MDAV. MDAV performs multivariate fixed group size microaggregation on unprojected data. MDAV is also described in Domingo-Ferrer and Torra (2005).

Algorithm 1 (MDAV)

1. *Compute the average record $\bar{x}$ of all records in the data set. Consider the most distant record x_r to the average record $\bar{x}$ (using the squared Euclidean distance).*

2. *Find the most distant record x_s from the record x_r considered in the previous step.*

3. *Form two groups around x_r and x_s, respectively. One group contains x_r and the $k-1$ records closest to x_r. The other group contains x_s and the $k-1$ records closest to x_s.*

4. *If there are at least $3k$ records which do not belong to any of the two groups formed in Step 3, go to Step 1 taking as new data set the previous data set minus the groups formed in the last instance of Step 3.*

5. *If there are between $3k-1$ and $2k$ records which do not belong to any of the two groups formed in Step 3:*

   (a) Compute the average record $\bar{x}$ of the remaining records.

   (b) Find the most distant record x_r from $\bar{x}$.

   (c) Form a group containing x_r and the $k-1$ records closest to x_r.

   (d) Form another group containing the rest of records. Exit the Algorithm.

6. *If there are less than $2k$ records which do not belong to the groups formed in Step 3, form a new group with those records and exit the Algorithm.*

The above algorithm can be applied independently to each group of variables resulting from partitioning the set of variables in the data set.

The μ-Approx approximation heuristic

μ-Approx (see Domingo-Ferrer *et al.* 2008) is an approximation heuristic for microaggregation, which finds a k-partition whose SSE is at most a fixed multiple of the minimum SSE that would be obtained with optimal microaggregation.

μ-Approx adapts an approximation heuristic to optimal suppression defined in Aggarwal *et al.* (2005). This heuristic is a graph-based algorithm that creates a directed forest (a set of directed trees) such that:

- Records are vertices.

- Each vertex has at most one outgoing edge.

- (u, v) is an edge only if v is one of the $k - 1$ nearest neighbours of u (according to a distance function).

- The size of every tree in the forest, i.e. the number of vertices in the tree, is between k and $\max(2k - 1, 3k - 5)$.

The Aggarwal *et al.* (2005) heuristic calls two procedures: (1) FOREST and (2) DECOMPOSE-COMPONENT. The first one creates edges in such a way that no cycle arises; also, the out-degree of each vertex is at most one and the size of the trees in the resulting forest is at least k. The second procedure decomposes trees with size greater than $\max(2k - 1, 3k - 5)$ into smaller component trees of size at least k.

Procedure 1 (FOREST)

1. *Start with an empty edge set so that each vertex is its own connected component.*

2. *Repeat until all components are of size at least k:*

   (a) *Pick any component T having size smaller than k.*

   (b) *Let u be a vertex in T without any outgoing edge. Since there are at most $k - 2$ other vertices in T, one of the $k - 1$ nearest neighbours of u, say v, must lie outside T. Add a directed edge (u, v) to the forest whose cost is the Euclidean distance $d(u, v)$ between u and v.*

Procedure 2 (DECOMPOSE-COMPONENT)

While *components of size $s > \max(2k - 1, 3k - 5)$ exist* do:

1. *Select one of those components.*

2. *Pick any vertex of the component as the candidate vertex u.*

3. *Root the tree at the candidate vertex u. Let U be the set of subtrees rooted at the children of u. Let ϕ be the size of the largest subtree of U, rooted at vertex v.*

4. *If $s - \phi \geq k - 1$, then*

   (a) *If $\phi \geq k$ and $s - \phi \geq k$ then partition the component into the largest subtree and the rest (by removing the edge between u and v). Clearly, the size of both resulting components is at least k.*

   (b) *If $s - \phi = k - 1$, partition the tree into a component containing the subtrees rooted at the children of v and another component containing the rest. To connect the children of v create a dummy vertex v' to replace v. Note that v' is only a Steiner vertex and does not contribute to the size of the first component. Clearly, the sizes of both components are at least k.*

   (c) *If $\phi = k - 1$, then partition into a component containing the subtree rooted at v along with the vertex u and another component containing the rest. In order to connect the children of u in the second component, create a Steiner vertex u'.*

(d) *Otherwise all subtrees have size at most $k - 2$. In this case, create an empty component and keep adding subtrees of u to it until the first time its size becomes at least $k - 1$. Clearly, at this point, its size is at most $2k - 4$. Put the remaining subtrees into a second component; this second component contains at least $k - 1$ vertices, since there are at least $s = \max(2k, 3k - 4)$ vertices in all (for $k \leq 4$, we have $s = 2k$ and a second component with at least 3 vertices; for $k > 4$, we have $s = 3k - 4$ and a second component with at least $k - 1$ vertices). Now, since $s \geq 2k$, at most one of the two components has size equal to $k - 1$. If such a component exists, add u to that component, else add u to the first component. In order to keep the partition not containing u connected, a Steiner vertex u' corresponding to u is placed in it.*

5. *Otherwise pick the root v of the largest subtree as the new candidate vertex and go to Step 3.*

End while

We can now state the μ-Approx approximation algorithm for optimal multivariate microaggregation. Given a multivariate numerical microdata set V, consider each microdata record as a vertex in a graph $\mathbf{G}_V = (V, \emptyset)$ with no edges and use the following algorithm for microaggregation:

Algorithm 2 (μ-Approx($\mathbf{G}_V, k$))

1. *Call procedure* FOREST.

2. *Call procedure* DECOMPOSE-COMPONENT. *Taking the resulting trees as groups yields a k-partition P, where groups have sizes between k and $\max(2k - 1, 3k - 5)$.*

3. *In a candidate optimal k-partition, no group size should be larger than $2k - 1$ (Domingo-Ferrer and Mateo-Sanz 2002). If $3k - 5 > 2k - 1$ (that is for $k \geq 5$), Step 2 can yield groups with more than $2k - 1$ records, which are partitioned as follows:*

   (a) *Compute the group centroid.*

   (b) *Take the record u that is farthest from the centroid.*

   (c) *Form a first group with u and its $k - 1$ nearest neighbours.*

   (d) *Take the rest of the records (at least k and at most $2k - 5$) as the second group.*

   This splitting step yields a k-partition P' where groups have sizes between k and $2k - 1$ (those groups in P with size $\leq 2k - 1$ are kept unchanged in P').

4. *Microaggregate using the groups in P'.*

Any bipartition procedure yielding two groups of size at least k can actually be used in Step 3. Whatever the bipartition, the resulting k-partition will always have a lower

SSE than the k-partition output by Step 2 (see Domingo-Ferrer and Mateo-Sanz 2002). The bipartition proposed above is the one used in the MDAV heuristic (see Domingo-Ferrer and Torra 2005 and Hundepool *et al.* 2008).

Example 3.7.2   *Algorithm μ-Approx was used to obtain a 3-anonymous version of the SME data set in Table 3.4. Like in Example 3.7.1, the company name was suppressed and microaggregation was carried out on the standardised versions of the quasi-identifier variables 'Surface' and 'No. employees'. Figure 3.2 depicts the 11 records projected on those two variables and grouped after the first two steps of Algorithm μ-Approx, that is, after procedure FOREST and after procedure DECOMPOSE-COMPONENT. The latter grouping corresponds to k-partition P, which in this example coincides with the k-partition P' output by the overall Algorithm μ-Approx. Figure 3.3 depicts the optimal k-partition, i.e., the one with the minimum SSE, whose computation is NP-hard in general (see Oganian and Domingo-Ferrer 2001).*

For k-partition P', the [0, 1]-bounded variability loss for the standardised versions of the two quasi-identifier variables is $SSE_{P'}/SST = 8.682/22 = 0.394$, quite close to $SSE_{opt}/SST = 0.34$ that is obtained for the optimal k-partition.

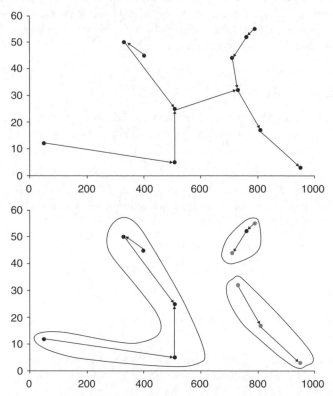

Figure 3.2   Top: tree obtained with FOREST on the SME data set. Bottom: 3-partition P obtained after DECOMPOSE-COMPONENT. In this example, P coincides with the 3-partition P' output by the overall Algorithm μ-Approx. Abscissae: surface; ordinates: number of employees.

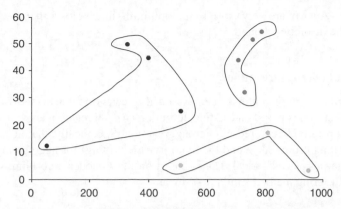

Figure 3.3   Optimal k-partition of the SME data set. Abscissae: surface; ordinates: number of employees.

Table 3.6 gives the 3-anonymous version of the SME data set output by Algorithm μ-Approx based on k-partition P'. In this case, the snooper looking for the turnover and net profit of a company K&K Sarl with crammed, small premises would hesitate between the 6th, the 7th, the 8th, the 9th or the 11th records of Table 3.6.

It is shown in Domingo-Ferrer *et al.* (2008) that the overall complexity of μ-Approx is dominated by $O(n^2)$, that is, it is quadratic in the number of records n, just like for MDAV. It is also shown that the SSE of the k-partition obtained with μ-Approx is upper bounded by $2(2k-1)[\max(2k-1, 3k-5)]^2 SSE_{opt}$, where SSE_{opt} is the minimum SSE attained by the optimal k-partition.

Empirical work in Domingo-Ferrer *et al.* (2008) shows that μ-Approx outperforms MDAV, because it finds k-partitions with lower SSE. Actually, the k-partitions

Table 3.6   Three-anonymous version of the SME data set after microaggregating quasi-identifier variables with Algorithm μ-Approx.

Surface (m²)	Number of employees	Turnover (Euros)	Net profit (Euros)
753.3	50	3 212 334	313 250
753.3	50	2 283 340	299 876
830	17	1 989 233	200 213
830	17	984 983	143 211
830	17	194 232	51 233
360	27	119 332	20 333
360	27	3 012 444	501 233
360	27	4 233 312	777 882
360	27	159 999	60 388
753.3	50	5 333 442	1 001 233
360	27	645 223	333 010

found by μ-Approx are very close to the optimum, much closer than the guarantee given by the theoretical bound.

Categorical microaggregation

Recently, microaggregation has been extended to categorical data (see Torra 2004). Such an extension is based on existing definitions for aggregation and clustering, the two basic operations required in microaggregation. Specifically, the median is used for aggregating ordinal data and the majority rule (voting) for aggregating nominal data. Clustering of categorical data is based on the k-modes algorithm, which is a partitive clustering method similar to c-means.

3.7.4   Data swapping and rank swapping

Data swapping was initially presented as an SDC method for databases containing only categorical variables (see Dalenius and Reiss 1978). The basic idea behind the method is to transform a database by exchanging values of confidential variables among individual records. Records are exchanged in such a way that low-order frequency counts or marginals are maintained.

Even though the original procedure was not very much used in practice (see Fienberg and McIntyre 2004), its basic idea had a clear influence in subsequent methods. In Reiss *et al.* (1982) and Reiss (1984) data swapping was introduced to protect continuous and categorical microdata, respectively. Another variant of data swapping for microdata is *rank swapping*. Although originally described only for ordinal variables in Greenberg (1987), rank swapping can also be used for any numerical variable (see Moore 1996). First, values of a variable X_i are ranked in ascending order, then each ranked value of X_i is swapped with another ranked value randomly chosen within a restricted range (e.g. the rank of two swapped values cannot differ by more than $p\%$ of the total number of records, where p is an input parameter). This algorithm is independently used on each variable in the original data set.

It is reasonable to expect that multivariate statistics computed from data swapped with this algorithm will be less distorted than those computed after an unconstrained swap. In earlier empirical work by these authors on continuous microdata protection Domingo-Ferrer and Torra (2001b), rank swapping has been identified as a particularly well-performing method in terms of the tradeoff between disclosure risk and information loss. Consequently, it is one of the techniques that have been implemented in the μ-ARGUS package Hundepool *et al.* (2008).

Yet data swapping and rank swapping do not prevent attribute disclosure, precisely because they only reorder data. For example, if the intruder knows which unit in the original data set has the highest income, she will simply have to look for the highest income in the swapped data set to get the exact true amount. It does not matter that this income is now attached to another unit. The intruder will also be able to refine her guess if she knows that the income of a certain unit lies within a defined quantile of the data.

Table 3.7 Example of rank swapping. Left: original file; right: rank-swapped file.

1	K	3.7	4.4	1	H	3.0	4.8
2	L	3.8	3.4	2	L	4.5	3.2
3	N	3.0	4.8	3	M	3.7	4.4
4	M	4.5	5.0	4	N	5.0	6.0
5	L	5.0	6.0	5	L	4.5	5.0
6	H	6.0	7.5	6	F	6.7	9.5
7	H	4.5	10.0	7	K	3.8	11.0
8	F	6.7	11.0	8	H	6.0	10.0
9	D	8.0	9.5	9	C	10.0	7.5
10	C	10.0	3.2	10	D	8.0	3.4

Example 3.7.3 *In Table 3.7, we can see an original microdata set on the left and its rank-swapped version on the right. There are four variables and 10 records in the original data set; the second variable is alphanumeric, and the standard alphabetic order has been used to rank it. A value of $p = 10\%$ has been used for all variables.*

3.7.5 Data shuffling

Data shuffling, proposed by Muralidhar and Sarathy (2006), is a special kind of data swapping for continuous or ordinal data which guarantees that marginal distributions in the shuffled data sets will be exactly the same as the marginal distributions in the original data. This is a very attractive property.

Like in data swapping and rank swapping, there is risk of attribute disclosure in shuffled data. On the other hand, data shuffling requires ranking of the whole data set. This can be computationally difficult for large data sets from censuses with millions of records.

The authors of Muralidhar and Sarathy (2006) obtained a US patent for their idea and thus the method cannot be implemented without an agreement from the authors. Therefore, we do not discuss this approach any further in this book.

3.7.6 Rounding

Rounding methods replace original values of variables with rounded values. For a given variable X_i, rounded values are chosen among a set of rounding points defining a *rounding set*. In a multivariate original data set, rounding is usually performed one variable at a time (*univariate* rounding). However, multivariate rounding is also possible (see, e.g., Willenborg and de Waal 2001). The operating principle of rounding makes it suitable for continuous data.

Example 3.7.4 *Assume a non-negative continuous variable X. We have to determine a set of rounding points $\{p_0, \ldots, p_r\}$. One possibility is to take rounding points as multiples of a base value b, that is, $p_i = bi$ for $i = 0, \ldots, r$. The set of attraction* *for each rounding point p_i is defined as the interval $[p_i - b/2, p_i + b/2)$, for $i = 1$ to $r - 1$; for p_0 and p_r, respectively, the sets of attraction are $[0, b/2)$ and $[p_r - b/2, X_{max}]$, where X_{max} is the largest possible value for variable X. Now an original value x of X is replaced with the rounding point corresponding to the set of attraction where x lies.*

3.7.7   Re-sampling

Originally proposed for protecting tabular data (Heer 1993 and Domingo-Ferrer and Mateo-Sanz 1999) re-sampling can also be used for microdata. Take t independent samples $S_1, \ldots, S_t$ of the values of an original variable X_i. The size of each sample must be equal to the number n of records in the original data set. Sort all samples using the same ranking criterion. Build the masked variable Z_i as $\bar{x}_1, \ldots, \bar{x}_n$, where n is the number of records and $\bar{x}_j$ is the average of the jth ranked values in $S_1, \ldots, S_t$.

3.7.8   PRAM

The post-randomisation method (PRAM, Gouweleeuw *et al.* 1997) is a disclosure control technique that can be applied to categorical data. Basically, it is a form of intended misclassification, using a known and predetermined probability mechanism. Applying PRAM means that, for each record in a microdata file, the score on one or more categorical variables is changed with a certain probability. This is done independently for each of the records.

PRAM is thus a perturbative method. Since PRAM uses a probability mechanism, the disclosure risk is directly influenced by this method. An intruder can never be certain that a record she thinks she has identified corresponds indeed to the identified person: with a certain probability this is a perturbed record.

Since the probability mechanism that is used when applying PRAM is known, characteristics of the (latent) true data can still be estimated from the perturbed data file. To that end, one can make use of correction methods similar to those used in case of misclassification and randomised response situations (see, e.g., Warner 1965, Chen 1979, Chaudhuri and Mukerjee 1988 and van den Hout and van der Heijden 2002).

PRAM was used in the 2001 UK Census to produce an end-user licence version of the Samples of Anonymised Records (SARs); see Gross *et al.* (2004) for a full description.

PRAM – the method

In this section, a short theoretical description of PRAM is given. For a detailed description of the method, see, e.g., Gouweleeuw *et al.* (1998a,1998b). For a discussion of several issues concerning the method and its consequences, see de Wolf *et al.* (1998).

Let ξ denote a categorical variable in the original file to which PRAM will be applied and let X denote the same variable in the perturbed file. Moreover, assume that ξ, and hence X as well, has K categories, labelled $1, \ldots, K$. The probabilities that define PRAM are denoted as

$$p_{kl} = P(X = l | \xi = k),$$

i.e., the probability that an original score $\xi = k$ is changed into the score $X = l$. These so-called transition probabilities are defined for all $k, l = 1, \ldots, K$.

Using these transition probabilities as entries of a $K \times K$ matrix, we obtain a Markov matrix that we will call the PRAM-matrix, denoted by $\mathbf{P}$.

Applying PRAM now means that, given the score $\xi = k$ for record r, the score X for that record is drawn from the probability distribution $p_{k1}, \ldots, p_{kK}$. For each record in the original file, this procedure is performed independently of the other records.

Example 3.7.5 *Let the variable ξ be gender, where $\xi = 1$ if male and $\xi = 2$ if female. We want to apply PRAM with PRAM-matrix*

$$\mathbf{P} = \begin{pmatrix} 0.9 & 0.1 \\ 0.1 & 0.9 \end{pmatrix}$$

to a microdata file with 110 males and 90 females. This would yield a perturbed microdata file with, in expectation, 108 males and 92 females. However, in expectation, 9 of these males were originally female, and similarly, 11 of the females were originally male.

Correcting analyses

The effect of PRAM on one-dimensional frequency tables is that

$$E(T_X | \xi) = \mathbf{P}^t T_\xi,$$

where $T_\xi = (T_\xi(1), \ldots, T_\xi(K))^t$ denotes the frequency table according to the original microdata file and T_X the frequency table according to the perturbed microdata file. A conditionally unbiased estimator of the frequency table in the original file is thus given by

$$\hat{T}_\xi = (\mathbf{P}^{-1})^t T_X.$$

Example 3.7.6 *Again consider the microdata file with 110 males and 90 females, i.e., with $T_\xi = (110, 90)^t$. Let the PRAM-matrix $\mathbf{P}$ be given by $p_{11} = p_{22} = 0.9$. Assume that a realisation of the perturbed data file has 107 males and 93 females, i.e., $T_X = (107, 93)^t$. Then*

$$\hat{T}_\xi = (\mathbf{P}^{-1})^t T_X = \begin{pmatrix} 1.125 & -0.125 \\ -0.125 & 1.125 \end{pmatrix} \begin{pmatrix} 107 \\ 93 \end{pmatrix} = \begin{pmatrix} 108.75 \\ 91.25 \end{pmatrix}$$

is an unbiased estimate of T_ξ.

This can be extended to n-dimensional frequency tables, by vectorising the frequency tables. In case PRAM is applied independently to each variable, the corresponding PRAM-matrix is then given by the Kronecker product of the PRAM-matrices of the individual dimensions.

Alternatively, in two dimensions, one could use the two-dimensional frequency tables[1] $\mathbf{T}_{\xi\nu}$ for the original data and $\mathbf{T}_{XY}$ for the perturbed data directly in matrix notation:

$$\hat{\mathbf{T}}_{\xi\nu} = (\mathbf{P}_X^{-1})^t \mathbf{T}_{XY} \mathbf{P}_Y^{-1},$$

where $\mathbf{P}_X$ denotes the PRAM-matrix corresponding to the categorical variable X and $\mathbf{P}_Y$ denotes the PRAM-matrix corresponding to the categorical variable Y.

An alternative way to correct for PRAM in calculating frequency tables, is to use so-called *calibration probabilities* as suggested in the misclassification literature (see, e.g., Kuha and Skinner 1997). Calibration probabilities are defined as

$$P(\xi = l \mid X = k) = \frac{p_{lk} P(\xi = l)}{\sum_{m=1}^{K} p_{mk} P(\xi = m)},$$

where p_{kl} are the entries of the PRAM-matrix $\mathbf{P}$. The calibration probabilities can be estimated by

$$\overleftarrow{p}_{kl} = \frac{p_{lk} T_\xi(l)}{\sum_{m=1}^{K} p_{mk} T_\xi(m)}.$$

Denoting the matrix with (estimated) calibration probabilities by $\overleftarrow{\mathbf{P}}$, one can show that

$$E\left(\overleftarrow{\mathbf{P}}^t T_X \mid \xi\right) = \overleftarrow{\mathbf{P}}^t \mathbf{P}^t T_\xi = T_\xi.$$

Hence, $\overleftarrow{\mathbf{P}}^t T_X$ is an unbiased estimator of T_ξ; see, e.g., van den Hout and Elamir (2006) for a discussion of the behaviour of this estimator.

For more information about correction methods for statistical analyses applied to data that have been protected with PRAM, we refer to, e.g., Gouweleeuw *et al.* (1998a), van den Hout (2000) and van den Hout (2004).

Choice of the PRAM matrix
The exact choice of transition probabilities influences both the amount of information loss as well as the amount of disclosure limitation. Moreover, in certain situations 'illogical' changes could occur, e.g. changing the gender of a female respondent with ovarian cancer to male or changing the marital status of a 5 year old to widowed. These kind of changes would attract the attention of a possible intruder and they should be avoided.

It is thus important to choose the transition probabilities in an appropriate way. Illogical changes could be avoided by appointing a probability of 0 to the anomalous combination of scores. In case of the ovarian cancer example, PRAM should not be applied to the variable gender individually, but to the crossing of the variables gender and diseases. In that case, each transition probability of changing a score into the score (male, ovarian cancer) should be set equal to 0. Note that, in general, the extension using the Kronecker product or the matrix notation mentioned directly

---

[1] When X has K categories and Y has L categories, the two-dimensional frequency table $\mathbf{T}_{XY}$ is a $K \times L$ matrix.

following Example 3.7.6 cannot be used to avoid anomalous changes as discussed above, since it is only valid in case PRAM is applied independently to each variable.

It is also possible to choose the transition probabilities in such a way that certain frequency tables calculated from the perturbed file are (in expectation) equal to the same frequency tables based on the original file:

$$E(T_X \mid \xi) = \mathbf{P}^t T_\xi = T_\xi, \tag{3.13}$$

i.e., the frequency table based on the original file is an eigenvector of the PRAM-matrix with eigenvalue 1. When PRAM is applied using a PRAM-matrix satisfying Equation (3.13), the method is called invariant PRAM.

Starting with a general PRAM-matrix $\mathbf{P}$ one can use the notion of calibration probabilities introduced previously, to construct a PRAM-matrix that satisfies Equation (3.13). Indeed, the matrix $\mathbf{R} = \mathbf{P}\overleftarrow{\mathbf{P}}$ is again a PRAM-matrix and satisfies Equation (3.13).

The choice of the transition probabilities in relation to the disclosure limitation and the information loss is more delicate. An empirical study on these effects is given in de Wolf and van Gelder (2004). A theoretical discussion on the possibility to choose the transition probabilities in an optimal way (in some sense) is given in Cator *et al.* (2005).

When to use PRAM

In certain situations methods like global recoding, local suppression and top coding would yield too much loss of detail in order to produce a safe microdata file. In these circumstances, PRAM is an alternative. Using PRAM, the amount of detail is preserved whereas the level of disclosure control is achieved by introducing uncertainty in the scores on identifying variables.

However, in order to make adequate inferences on a microdata file to which PRAM has been applied, the statistician needs to include sophisticated changes to the standard methods. This demands a good knowledge of both PRAM and the statistical analysis that is to be applied.

In case a researcher is willing to make use of a remote execution facility, PRAM might be used to produce a microdata file with the same structure as the original microdata file, but with some kind of synthetic data. Such microdata files might be used as a 'test' microdata file on which a researcher can try her scripts before sending these scripts to the remote execution facility. Since the results of the script are not used directly, the amount of distortion of the original microdata file can be chosen to be quite large. That way a safe microdata file is produced that still exhibits the same structure (and amount of detail) as the original microdata file.

In other situations, PRAM might produce a microdata file that is safe and leaves certain statistical characteristics of that file (more or less) unchanged. In that case, a researcher might perform his research on that microdata file in order to get an idea on the eventually needed research strategy. Once that strategy has been determined, the researcher might come to a Research Data Centre (RDC) facility in order to perform the analyses once more on the original microdata thereby reducing the amount of time that she has to be at the RDC facility.

3.7.9   MASSC

MASSC (Singh *et al.* 2004) is a masking method whose acronym summarises its four steps: (1) micro agglomeration, (2) substitution, (3) sub-sampling and (4) calibration. We briefly recall the purpose of those four steps:

1. Micro agglomeration is applied to partition the original data set into risk strata (groups of records which are at a similar risk of disclosure). These strata are formed using the key variables, i.e. the quasi-identifiers in the records. The idea is that those records with rarer combinations of key variables are at a higher risk.

2. Optimal probabilistic substitution is then used to perturb the original data (i.e. substitution is governed by a Markov matrix like in PRAM; see Singh *et al.* (2004) for details).

3. Optimal probabilistic sub-sampling is used to suppress some variables or even entire records (i.e. variables and/or records are suppressed with a certain set of probabilities as parameter).

4. Optimal sampling weight calibration is used to preserve estimates for outcome variables in the treated database whose accuracy is critical for the intended data use.

MASSC is interesting in that, to the best of our knowledge, it is the first attempt at designing a perturbative masking method in such a way that disclosure risk can be analytically quantified. Its main shortcoming is that its disclosure model simplifies reality by considering only disclosure resulting from linkage of key variables with external sources. Since key variables are typically categorical, the uniqueness approach can be used to analyse the risk of disclosure; however, doing so ignores the fact that continuous outcome variables can also be used for respondent re-identification. As an example, if respondents are companies and turnover is one outcome variable, everyone in a certain industrial sector knows which is the company with largest turnover. Thus, in practice, MASSC is a method only suited when continuous variables are not present.

3.8   Synthetic and hybrid data

Publication of synthetic – i.e. simulated – data is an alternative to masking for statistical disclosure control of microdata. The idea is to randomly generate data with the constraint that certain statistics or internal relationships of the original data set should be preserved.

Synthetic data can be used in at least three ways:

1. *Fully synthetic data*. No original data are released, only synthetic data.

2. *Partially synthetic data*. Only the most sensitive data (i.e. *some* variables and/or records) are replaced by synthetic versions before release. For the rest, original data are released.

3. *Hybrid data*. Original data and synthetic data are combined and the combination yields so-called hybrid data, which are the data subsequently released in place of original data. Depending on how the combination is computed, hybrid data can be closer to the original data or to the synthetic data.

We survey in this section methods for generating fully synthetic data, partially synthetic data and hybrid data. We largely follow Domingo-Ferrer *et al.* (2009).

3.8.1   Fully synthetic data

A forerunner: Data distortion by probability distribution

Data distortion by probability distribution was proposed by Liew *et al.* (1985) and is not usually included in the category of synthetic data generation methods. However, its operating principle is to obtain a protected data set by randomly drawing from the underlying distribution of the original data set. Thus, it can be regarded as a forerunner of synthetic methods.

This method is suitable for both categorical and continuous variables and consists of three steps:

1. Identify the density function underlying each of the confidential variables in the data set and estimate the parameters associated with that density function.

2. For each confidential variable, generate a protected series by randomly drawing from the estimated density function.

3. Map the confidential series to the protected series and publish the protected series instead of the confidential ones.

In the identification and estimation stage, the original series of the confidential variable (e.g. salary) is screened to determine which of a set of pre-determined density functions fits the data best. Goodness of fit can be tested by the Kolmogorov–Smirnov test. If several density functions are acceptable at a given significance level, selecting the one yielding the smallest value for the Kolmogorov–Smirnov statistics is recommended. If no density in the pre-determined set fits the data, the frequency imposed distortion method can be used. With the latter method, the original series is divided into several intervals (somewhere between 8 and 20). The frequencies within the interval are counted for the original series, and become a guideline to generate the distorted series. By using a uniform random number generating subroutine, a distorted series is generated until its frequencies become the same as the frequencies of the original series. If the frequencies in some intervals overflow, they are simply discarded.

Once the best-fit density function has been selected, the generation stage feeds its estimated parameters to a random value generating routine to produce the distorted series.

Finally, the mapping and replacement stage is only needed if the distorted variables are to be used jointly with other non-distorted variables. Mapping consists of ranking the distorted series and the original series in the same order and replacing each element of the original series with the corresponding distorted element.

It must be stressed here that the approach described in Liew *et al.* (1985) was for one variable at a time. One could imagine a generalisation of the method using multivariate density functions. However, such a generalisation is not trivial, because it requires multivariate ranking mapping, and can lead to very poor fitting.

Example 3.8.1 *A distribution fitting software (Crystal.Ball 2004) has been used on the original (ranked) data set 186, 693, 830, 1177, 1219, 1428, 1902, 1903, 2496 and 3406. Continuous distributions tried were normal, triangular, exponential, lognormal, Weibull, uniform, beta, gamma, logistic, Pareto and extreme value; discrete distributions tried were binomial, Poisson, geometric and hypergeometric. The software allowed for three fitting criteria to be used: Kolmogorov–Smirnov, χ^2 and Anderson–Darling. According to the first criterion, the best fit happened for the extreme value distribution with modal and scale parameters 1105.78 and 732.43, respectively; the Kolmogorov statistic for this fit was 0.1138. Using the fitted distribution, the following (ranked) data set was generated and used to replace the original one: 425.60, 660.97, 843.43, 855.76, 880.68, 895.73, 1086.25, 1102.57, 1485.37 and 2035.34.*

Fully synthetic data by multiple imputation

In 1993, Rubin (1993) suggested to create fully synthetic data sets based on the multiple imputation (MI) framework. His idea was to treat all units in the population that have not been selected in the sample as missing data, impute them according to the MI approach and draw simple random samples from these imputed populations for release to the public. Most surveys are conducted using complex sampling designs. Releasing simple random samples simplifies research for the potential user of the data, since the design does not have to be incorporated in the model. However, it is not necessary to release simple random samples. If a complex design is used, the analyst accounts for the design in the within variance u_i. For illustration, think of a data set of size n, sampled from a population of size N. Suppose further that the imputer has information about some variables X for the whole population, for example from census records, and only the information from the survey respondents for the remaining variables Y. Let Y_{inc} be the observed part of the population and Y_{exc} the non-sampled units of Y. For simplicity, assume that there are no item-missing data in the observed data set. The approach also applies if there are missing data. Details about generating synthetic data for data sets subject to item non-response are described in Reiter (2004). Now, the synthetic data sets can be generated in two steps:

1. Construct m imputed synthetic populations by drawing Y_{exc} m times independently from the posterior predictive distribution $f(Y_{exc}|X, Y_{inc})$ for the $N-n$ unobserved values of Y. If the released data should contain no real data for Y, all N values can be drawn from this distribution.

2. Make simple random draws from these populations and release them to the public.

The second step is necessary as it might not be feasible to release m whole populations for the simple matter of data size. In practice, it is not mandatory to generate complete populations. The imputer can make random draws from X in a first step and only impute values of Y for the drawn X. The analysis of the m simulated data sets follows the same lines as the analysis after MI for missing values in regular data sets (see Rubin 1987).

To understand the procedure of analysing fully synthetic data sets, think of an analyst interested in an unknown scalar parameter Q, where Q could be, e.g., the mean of a variable, the correlation coefficient between two variables, or a regression coefficient in a linear regression. Inferences for this parameter for data sets with no missing values usually are based on a point estimate q, an estimate for the variance of q, u and a normal or Student's t reference distribution. For analysis of the imputed data sets, let q_i and u_i for $i = 1, ..., m$ be the point and variance estimates for each of the m completed data sets. The following quantities are needed for inferences for scalar Q:

$$\bar{q}_m = \sum_{i=1}^{m} q_i/m, \tag{3.14}$$

$$b_m = \sum_{i=1}^{m} (q_i - \bar{q}_m)^2/(m-1), \tag{3.15}$$

$$\bar{u}_m = \sum_{i=1}^{m} u_i/m. \tag{3.16}$$

The analyst then can use $\bar{q}_m$ to estimate Q and

$$T_f = (1 + m^{-1})b_m - \bar{u}_m \tag{3.17}$$

to estimate the variance of $\bar{q}_m$. When n is large, inferences for scalar Q can be based on t distributions with degrees of freedom $v_f = (m-1)(1 - r_m^{-1})^2$, where $r_m = ((1 + m^{-1})b_m/\bar{u}_m)$. Derivations of these methods are presented in Raghunathan et al. (2003). Extensions for multivariate Q are presented in Reiter (2005b).

A disadvantage of this variance estimate is that it can become negative. For that reason, Reiter (2002) suggests a slightly modified variance estimator that is always positive: $T_f^* = \max(0, T_f) + \delta(\frac{n_{syn}}{n}\bar{u}_m)$, where $\delta = 1$ if $T_f < 0$, and $\delta = 0$ otherwise. Here, n_{syn} is the number of observations in the released data sets sampled from the synthetic population.

Theoretical properties

In general, the disclosure risk for the fully synthetic data is very low, since all values are synthetic values. However, it is not zero because, if the imputation model is too good and basically produces the same estimated values in all the synthetic data sets, it does not matter that the data are all synthetic. It might look like the data from a potential survey respondent an intruder was looking for. And once the intruder thinks that he has identified a single respondent and the estimates are reasonably close to the true values for that unit, it is no longer important that the data are all made up. The potential respondent will feel that her privacy is at risk. Still, this is very unlikely to occur since the imputation models would have to be almost perfect and the intruder faces the problem that he never knows if the imputed values are anywhere near the true values and if the target record is included in one of the different synthetic samples.

Regarding data utility, all values in the released fully synthetic data sets are generated from the distribution of Y given X. This means that in theory up to sampling error, the joint distribution of the original data does not differ from the joint distribution of the released data. Thus, any analysis performed on the released data will provide the same results as any analysis performed on the original data. However, the exact multivariate distribution of the data usually is unknown and any model for the joint distribution of the data will likely introduce some bias since the original distribution will seldom follow any multivariate standard distribution. For that reason, it is common to use an iterative algorithm called sequential regression MI (SRMI, Raghunathan et $al.$ 2001) that is based on the ideas of Gibbs sampling and avoids otherwise necessary assumptions about the joint distribution of the data. Imputations are generated variable by variable where the values for any variable Y_k are synthesised by drawing from the conditional distributions of $(Y_k|Y_{-k})$, where Y_{-k} represents all variables in the data set except Y_k. This allows for different imputation models for each variable. Continuous variables can be imputed with a linear model, binary variables can be imputed using a logit model, etc. Under some regularity assumptions, iterative draws from these conditional distributions will converge to draws from the joint multivariate distribution of the data. Further refinements for the imputation of heavily skewed variables based on kernel density estimation are given in Woodcock and Benedetto (2007).

As said above, the quality of the synthetic data sets will highly depend on the quality of the underlying model and for some variables, it will be very hard to define good models. Furthermore, specifying a model that considers all the skip patterns and constraints between the variables in a large data set can be cumbersome if not impossible. But if these variables do not contain any sensitive information or information that might help identify single respondents, why bother to find these models? Why bother to perturb these variables in the first place? Furthermore, the risk of biased imputations will increase with the number of variables that are imputed. For, if one of the variables is imputed based on a 'bad' model, the biased imputed values for that variable could be the basis for the imputation of another variable and this variable again could be used for the imputation of another one and so on. So, a small bias could increase to a really problematic bias over the imputation process. Partially synthetic data sets, described later, can be a helpful tool to overcome these drawbacks since only records that are actually at risk are synthesised.

Empirical examples
As of this writing no agency actually released fully synthetic data sets. However, Reiter (2005a) and Drechsler et $al.$ (2008b) evaluate this approach using real data sets. Reiter generated synthetic data sets for the US Current Population Survey. He computed more than 30 descriptive estimands for different sub-populations of the data and ran a linear and a logit regression on the data sets. He found that results

are encouraging for most estimates. The deviation between the true and the synthetic point estimates was seldom higher than 10% and for many estimands the synthetic 95% confidence interval covered the true estimate in more than 90% of the simulation runs. But Reiter also noted that some descriptive estimates were biased and especially the results for the logit regression were rather poor. He pointed out that the biased estimates were caused by the imputation model being different from the analysis model. He noted that, if the imputation model was modified to fit the analysis model, results from the synthetic data were similar to results from the original data. Hence, defining good imputation models is a very important step in the synthesis process. Reiter also shows that confidentiality would be guaranteed if the data would be considered for release.

Drechsler *et al.* (2008b) generate synthetic data sets for a German establishment survey, the IAB Establishment Panel. They compute several descriptive statistics and run a probit regression originally published in Zwick (2005) based on the original data of the Establishment Panel. The results from the synthetic data sets are very close to the results from the original data. A detailed disclosure risk evaluation in the paper shows that again confidentiality would be guaranteed if the synthetic data sets would be released.

Synthetic data by bootstrap

However attractive is MI, it requires complex imputation software. Therefore, other, simpler alternatives to generate synthetic data have been explored and are reviewed in what follows.

One of these alternatives, proposed in Fienberg (1994), consists of generating synthetic microdata by using bootstrap methods. Later, in Fienberg *et al.* (1998), this approach was used for categorical data.

The bootstrap approach bears some similarity to the data distortion by probability distribution and the MI methods described above. Given an original microdata set X with p variables, the data protector computes its empirical p-variate cumulative distribution function (c.d.f.) F. Now, rather than distorting the original data to obtain masked data, the data protector alters (or 'smoothes') the c.d.f. F to derive a similar c.d.f. F'. Finally, F' is sampled to obtain a synthetic microdata set Z.

Synthetic data by Latin Hypercube sampling

Latin Hypercube Sampling (LHS) appears in the literature as another method for generating multivariate synthetic   data sets. In Huntington and Lyrintzis (1998), the LHS-updated technique of Florian (1992) was improved, but the proposed scheme is still time intensive even for a moderate number of records. In Dandekar *et al.* (2002a), LHS is used along with a rank correlation refinement to reproduce both the univariate (i.e. mean and covariance) and multivariate structure (in the sense of rank correlation) of the original data set. In a nutshell, LHS-based methods rely on iterative refinement, are time intensive and their running time does not only depend on the number of values to be reproduced, but on the starting values as well.

Information Preserving Statistical Obfuscation

Information Preserving Statistical Obfuscation (IPSO), proposed by Burridge (2003), is a procedure allowing to create synthetic data which exactly preserve the mean vector and the covariance matrix of original data.

The basic form of IPSO will be called here IPSO-A. Informally, suppose two sets of variables X and Y, where the former are the confidential outcome variables and the latter are quasi-identifier variables. Then X are taken as independent and Y as dependent variables. A multiple regression of Y on X is computed and fitted Y'_A variables are computed. Finally, variables X and Y'_A are released by IPSO-A in place of X and Y.

In the above setting, conditional on the specific confidential variables x_i, the quasi-identifier variables Y_i are assumed to follow a multivariate normal distribution with covariance matrix $\Sigma = \{\sigma_{jk}\}$ and a mean vector $x_i B$, where B is the matrix of regression coefficients.

Let $\hat{B}$ and $\hat{\Sigma}$ be the maximum likelihood estimates of B and Σ derived from the complete data set (y, x). If a user fits a multiple regression model to (y'_A, x), she will get estimates $\hat{B}_A$ and $\hat{\Sigma}_A$ which, in general, are different from the estimates $\hat{B}$ and $\hat{\Sigma}$ obtained when fitting the model to the original data (y, x). The second IPSO method, IPSO-B, modifies y'_A into y'_B in such a way that the estimate $\hat{B}_B$ obtained by multiple linear regression from (y'_B, x) satisfies $\hat{B}_B = \hat{B}$.

A more ambitious goal is to come up with a data matrix y'_C such that, when a multivariate multiple regression model is fitted to (y'_C, x), *both* statistics $\hat{B}$ and $\hat{\Sigma}$, sufficient for the multivariate normal case, are preserved. This is done by the third IPSO method, IPSO-C.

In Mateo-Sanz *et al.* (2004a), a non-iterative method for generating continuous synthetic microdata based on the Cholesky decomposition is proposed. In a single step of computation, the method exactly reproduces the means and the covariance matrix of the original data set. The running time grows linearly with the number of records. This method can be regarded as a special case of the IPSO generator.

3.8.2 Partially synthetic data

Partially synthetic data by multiple imputation

In contrast to the creation of fully synthetic data sets, this approach replaces only observed values with synthetic values (see Little 1993) for variables that bear a high risk of disclosure (key variables) or for very sensitive variables. Masking these variables by replacing observed with imputed values prevents re-identification. The imputed values can be obtained by drawing from the posterior predictive distribution $f(Y|X)$, where Y indicates the variables that need to be modified to avoid disclosure and X are all variables that remain unchanged or variables that have been synthesised in an earlier step. Imputations are generated according to the MI framework but, compared to the fully synthetic data context, while the point estimate stays the same, the variance estimation differs slightly from the MI calculations for missing data. Yet, it differs from the estimation in the fully synthetic context as well: it is given by $T_p = b_m/m + \bar{u}_m$. Similar to the variance estimator for MI of missing data, b_m/m is the correction factor for the additional variance due to using a finite number of imputations. However, the additional b_m, necessary in the missing data context, is not needed here, since $\bar{u}$ already captures the variance of Q given the observed data. This is different in the missing data case, where $\bar{u}$ is the variance of Q given the completed data and $\bar{u} + b_m$ is the variance of Q given the observed data. Inferences for scalar Q can be based on t distributions with degrees of freedom $v_p = (m-1)(1 + r_m^{-1})^2$, where $r_m = (m^{-1}b_m/\bar{v}_m)$. Derivations of these methods are presented in Reiter (2003). Extensions for multivariate Q are presented in Reiter (2005b). The variance estimate T_p can never be negative, so no adjustments are necessary for partially synthetic data sets.

Theoretical properties

The disclosure risk is higher for partially synthetic data sets especially if the intruder knows that some unit participated in the survey, since true values remain in the data set and imputed values are generated only for the survey participants and not for the whole population. So, for partially synthetic data sets, assessing the risk of disclosure is an equally important evaluation step as assessing the data utility. It is essential that the agency identifies and synthesises all variables that bear a risk of disclosure. A conservative approach would be to also impute all variables that contain the most sensitive information. Once the synthetic data are generated, careful checks are necessary to evaluate the disclosure risk for these data sets. Only if the data sets prove to be satisfactory both in terms of data utility and in terms of disclosure risk, a release should be considered. Methods to calculate the disclosure risk for partially synthetic data sets are described in Reiter and Mitra (2009) and Drechsler and Reiter (2008).

Empirical examples

Some agencies in the United States have released partially synthetic data sets in the last years. For example, in 2007 the US Census Bureau released a partially synthetic, PUF for the Survey of Income and Program Participation (SIPP) that includes imputed values of social security benefits information and dozens of other highly sensitive variables (see SIPP 2007). A description of the project with detailed discussion of the data utility and disclosure risk is given in Abowd et al. (2006). The Census Bureau also created synthesised origin-destination matrices, i.e. where people live and work, available to the public as maps via the web (see Map 2007). In the Survey of Consumer Finances, the US Federal Reserve Board replaced monetary values at high disclosure risk with multiply-imputed values releasing a mixture of imputed values and the not replaced, collected values (Kennickell 1999). The Census Bureau also plans to protect the identities of people in group quarters (e.g. prisons and shelters) in the next release of PUFs of the American Communities Survey by replacing demographic data for people at high disclosure risk with imputations. Partially synthetic, public-use data sets are in the development stage in the US for the Longitudinal Business Database, the Longitudinal Employer-Household Dynamics survey and the American Communities Survey veterans and full sample data. Statistical agencies in Australia, Canada,

Germany and New Zealand (see Graham and Penny 2005) also are investigating the approach. Drechsler *et al.* (2008a) compare the partially and fully synthetic approach and give guidelines regarding which method agencies should pick for their data sets. Other applications of partially synthetic data are described in Abowd and Woodcock (2001, 2004), Abowd and Lane (2004), Little *et al.* (2004), Mitra and Reiter (2006), An and Little (2007) and Reiter and Drechsler (2010).

Model-based microdata

Multiply-imputed, partially synthetic data is not the only approach to mix original and synthetic data to obtain more plausible releasable data.

Model-based disclosure protection (see Franconi and Stander 2002 and Polettini *et al.* 2002) is an alternative approach, whose principle is to regress a set of confidential continuous outcome variables on a disjoint set non-confidential variables. The fitted values are then released for the confidential variables instead of the original values.

3.8.3 Hybrid data

Hybrid data by recordwise combination

In Dandekar *et al.* (2002b), it was proposed to create *hybrid data* by combining original and synthetic data. Such a combination allows better control than purely synthetic data over the individual characteristics of masked records. In Dandekar *et al.* (2002b), a rule is used to pair one original data record with one synthetic data record. This is done by going through all original data records and pairing each original record with the nearest synthetic record according to some distance. Once records have been paired, Dandekar *et al.* (2002b) suggest two possible ways for combining one original record X with one synthetic record X_s: additive combination and multiplicative combination. Additive combination yields

$$Z = \alpha X + (1 - \alpha)X_s$$

and multiplicative combination yields

$$Z = X^\alpha X_s^{(1-\alpha)},$$

where α is an input parameter in $[0, 1]$ and Z is the hybrid record. Dandekar *et al.* (2002b) present empirical results comparing the hybrid approach with rank swapping and microaggregation masking (the synthetic component of hybrid data is generated using LHS; see Dandekar *et al.* 2002a).

Another approach to combining original and synthetic microdata is proposed in Sebé *et al.* (2002). The idea here is to first mask an original data set using a masking method. Then a hill-climbing optimisation heuristic is run which seeks to modify the masked data to preserve the first and second-order moments of the original data set as much as possible without increasing the disclosure risk with respect to the initial masked data. The optimisation heuristic can be modified to preserve higher order moments, but this significantly increases computation. Also, the optimisation heuristic can use a random data set as an initial data set instead of a masked data set; in this case, the output data set is purely synthetic.

Sufficiency-based hybrid data

A method to generate sufficiency-based hybrid data was proposed by Muralidhar and Sarathy (2008):

- The resulting data are called hybrid because they are a mixture of original and synthetic data, like the hybrid data obtained with the method of Dandekar *et al.* (2002b).

- The resulting hybrid data are called sufficiency-based because they preserve the mean vector and the covariance matrix of the original data, like the synthetic data obtained with the above-described IPSO method by Burridge (2003).

We first describe the theoretical properties of the method and then empirical properties.

Theoretical properties

Let $\mathbf{X}(= X_1, \ldots, X_K)$ represent a set of K confidential variables, let $\mathbf{S}(= S_1, \ldots, S_L)$ represent a set of L non-confidential variables, and let $\mathbf{Y}(= Y_1, \ldots, Y_K)$ represent the set of K perturbed (i.e. hybrid) variables. Let n represent the number of records in the data set. Let $\Sigma_{\mathbf{XX}}$, $\Sigma_{\mathbf{SS}}$ and $\Sigma_{\mathbf{YY}}$ represent the covariance matrices of $\mathbf{X}$, $\mathbf{S}$ and $\mathbf{Y}$, respectively. Let $\Sigma_{\mathbf{XS}}$ and $\Sigma_{\mathbf{YS}}$ represent the covariance between $\mathbf{X}$ and $\mathbf{S}$, and $\mathbf{Y}$ and $\mathbf{S}$, respectively. Let $\bar{\mathbf{X}}$, $\bar{\mathbf{S}}$ and $\bar{\mathbf{Y}}$ be the mean vector of $\mathbf{X}$, $\mathbf{S}$ and $\mathbf{Y}$, respectively. Let α be a matrix of size $K \times K$ representing the multipliers of $\mathbf{X}$ and let β be a matrix of size $K \times L$ representing the multipliers of $\mathbf{S}$.

Using the above definition the hybrid values $\mathbf{y}_i$ are generated as:

$$\mathbf{y}_i = \gamma + \mathbf{x}_i \alpha^t + \mathbf{s}_i \beta^t + \mathbf{e}_i, \quad i = 1, \ldots, n. \tag{3.18}$$

The method is required to preserve variances, covariances and means, that is,

$$\Sigma_{\mathbf{YY}} = \Sigma_{\mathbf{XX}},$$
$$\Sigma_{\mathbf{YS}} = \Sigma_{\mathbf{XS}},$$
$$\bar{\mathbf{Y}} = \bar{\mathbf{X}}.$$

Based on the above preservation requirements, it turns out that:

$$\beta^t = \Sigma_{\mathbf{SS}}^{-1} \Sigma_{\mathbf{SX}}(\mathbf{I} - \alpha^t), \tag{3.19}$$

$$\gamma = (\mathbf{I} - \alpha)\bar{\mathbf{X}} - \beta \bar{\bar{\mathbf{S}}}, \tag{3.20}$$

$$\Sigma_{\mathbf{ee}} = (\Sigma_{\mathbf{XX}} - \Sigma_{\mathbf{XS}} \Sigma_{\mathbf{SS}}^{-1} \Sigma_{\mathbf{SX}}) - \alpha(\Sigma_{\mathbf{XX}} - \Sigma_{\mathbf{XS}} \Sigma_{\mathbf{SS}}^{-1} \Sigma_{\mathbf{SX}})\alpha^T, \tag{3.21}$$

where $\mathbf{I}$ is the identity matrix and $\Sigma_{\mathbf{ee}}$ is the covariance matrix of the noise terms $\mathbf{e}$.

Thus, α completely specifies the perturbation model shown in Equation (3.18). However, it must be checked that the selected α yields a positive definite covariance matrix for the error terms, that is, that the matrix obtained when that α is input to Expression (3.21) is positive definite.

Matrix α represents the extent to which the masked data is a function of the original data. There are three possible options for specifying matrix α:

1. α is a diagonal matrix and all values in the diagonal are equal. This would represent a situation where all the confidential variables are perturbed at the same level. In addition, the value of Y_i is a function only of X_i and does not depend on the value of X_j. Let α represent the value of the diagonal. In this case, it is easy to verify from Equation (3.21) that when $0 \leq \alpha \leq 1$, then Σ_{ee} will be positive definite.

2. α is a diagonal matrix and all values in the diagonal *are not equal*. This would represent a situation where the confidential variables are perturbed at different levels. As in the previous case, the perturbed values of a particular variable are a function of the original values of that particular confidential variable and not other confidential variables. However, in this case, after the specification of α, it is necessary to verify that the resulting Σ_{ee} is positive definite. If not, it may be necessary to re-specify α so that the resulting Σ_{ee} is positive definite.

3. α is not a diagonal matrix. In this case, the perturbed values for a particular variable are a function of the original values of that confidential variable as well as the original values of other confidential variables. This is the most general of the specifications and also the most complicated. In Muralidhar and Sarathy (2008), this third specification is considered as not entailing any advantages and is thereafter disregarded in the empirical work presented. This is only justified though, if only the mean vector and the covariance matrix should be preserved. For any other statistic between the confidential variables, the data utility will increase, if α is not a diagonal matrix.

Empirical examples

The theoretical properties above hold for a data set of any size, any underlying distribution and any noise distribution. While these results require no empirical evaluation, a few empirical examples are given in Muralidhar and Sarathy (2008) to illustrate the application of the approach proposed there. However, the examples given do not address the consequences of this perturbation approach for statistics other than the mean vector and the covariance matrix, which are preserved by definition.

Example 3.8.2   *As a first example, consider the case where the data provider requests that all variables be perturbed at the same level and specifies that $\alpha = 0.90$. Table 3 in Muralidhar Sarathy (2008) provides the hybrid variables (Y_1, Y_2) resulting from a data set consisting of 25 records with two non-confidential variables (S_1, S_2) and two confidential variables (X_1, X_2) with the following characteristics:*

$$\Sigma_{XX} = \begin{bmatrix} 1.0 & 0.4 \\ 0.4 & 1.0 \end{bmatrix},$$

$$\Sigma_{SS} = \begin{bmatrix} 1.0 & 0.6 \\ 0.6 & 1.0 \end{bmatrix},$$

$$\Sigma_{XS} = \begin{bmatrix} 0.2 & 0.4 \\ -0.3 & -0.2 \end{bmatrix}.$$

The mean vectors $\bar{\mathbf{X}}$ and $\bar{\mathbf{S}}$ were specified as $\mathbf{0}$ resulting in $\boldsymbol{\gamma} = \mathbf{0}$. Based on the data provider's request, $\boldsymbol{\alpha}$ is specified as:

$$\boldsymbol{\alpha} = \begin{bmatrix} 0.900 & 0.000 \\ 0.000 & 0.900 \end{bmatrix}. \tag{3.22}$$

In this case, one assumes that the data provider has requested that all variables be perturbed at the same level. Using Equation (3.19), one can compute $\boldsymbol{\beta}$ as:

$$\boldsymbol{\beta}^t = \begin{bmatrix} -0.006250 & 0.043750 \\ -0.028125 & -0.003125 \end{bmatrix}.$$

The resulting covariance of the noise term can be computed using Equation (3.21) as:

$$\Sigma_{ee} = \begin{bmatrix} 0.159125 & 0.089063 \\ 0.089063 & 0.172782 \end{bmatrix}.$$

It can be verified that Σ_{ee} above is positive definite. The specification of $\boldsymbol{\alpha}$ in Equation (3.22) implies that the perturbed values (Y_1, Y_2) of the confidential variables are heavily influenced by the original values (X_1, X_2) of the confidential variables. The non-confidential variables (S_1, S_2) play a very small role in the perturbation. The extent of the noise term is also relatively small, about 16% for the first and about 17% for the second confidential variable. The results show that the perturbed values $\mathbf{Y}$ have the same mean vector and covariance matrix as $\mathbf{X}$. It can easily be verified that for many traditional parametric statistical analyses (such as confidence intervals and hypothesis testing for the mean, analysis of variance, regression analysis, multivariate analysis of variance, multivariate multiple regression, etc.) using $(\mathbf{Y}, \mathbf{S})$ in place of $(\mathbf{X}, \mathbf{S})$ will yield exactly the same results.

As a variation of this first example, in Muralidhar and Sarathy (2008) it is assumed that the data provider wishes that the coefficient for the first variable should be 0.9 and that for the second variable should be 0.2. In this case, the data provider would like a much higher level of perturbation for variable X_2 than for variable X_1. From this specification

$$\boldsymbol{\alpha} = \begin{bmatrix} 0.900 & 0.000 \\ 0.000 & 0.200 \end{bmatrix}.$$

The resulting covariance matrix for the noise term is given by:

$$\Sigma_{ee} = \begin{bmatrix} 0.1591 & 0.3844 \\ 0.3844 & 0.8730 \end{bmatrix}.$$

It can be verified that the above covariance matrix is not positive definite and it would be necessary for the data provider to consider alternative specifications in this case. In order to maintain the sufficiency requirements, it is necessary that some restrictions be imposed on the selection of $\boldsymbol{\alpha}$. Extremely disparate specifications such as the one above are likely to create problems as illustrated above.

Example 3.8.3 *In the second example of Muralidhar and Sarathy (2008), one takes*

$$\boldsymbol{\alpha} = \begin{bmatrix} 0.800 & 0.000 \\ 0.000 & 0.300 \end{bmatrix}.$$

In this case, the two variables are perturbed at different levels, the first variable being perturbed less than the second variable. The resulting values for β and Σ_{ee} can be computed as follows:

$$\beta^t = \begin{bmatrix} -0.01250 & -0.19687 \\ 0.08750 & -0.02187 \end{bmatrix},$$

$$\Sigma_{ee} = \begin{bmatrix} 0.3015 & 0.3563 \\ 0.3563 & 0.8275 \end{bmatrix}.$$

It can be verified that Σ_{ee} is positive definite. The results of applying this specification on the original data set are also provided in Table 3 in Muralidhar and Sarathy (2008) (under Example 2 there). As before, it is easy to verify that the mean vector and covariance matrix of the masked data $(\mathbf{Y}, \mathbf{S})$ are exactly the same as that of the original data $(\mathbf{X}, \mathbf{S})$.

Consequently, for those types of statistical analyses for which the mean vector and covariance matrix are sufficient statistics, the results of the analysis using the masked data will yield the same results as the original data. However, any analysis that is not based on the first two moments of the distribution or considers sub-domains of the data that are not considered when creating the synthetic data will be distorted.

Example 3.8.4   *The third example given in Muralidhar and Sarathy (2008) shows the case where the perturbed values are generated as a function of only the non-confidential variables and the coefficients of the confidential variables are set to zero ($\alpha = 0$). This choice of α causes the method to become equivalent to the IPSO procedure Burridge (2003) described in Section 3.8.1: the method does no longer return sufficiency-based hybrid data, it now returns sufficiency-based synthetic data. The results of this example are displayed in Table 3 of Muralidhar and Sarathy (2008) (under their Example 3).*

The information loss resulting from the data is also presented in the table measured by the variance of the original and perturbed values. The measure clearly shows an increase in information loss measured using variance $(X - Y)$ as α approaches 0, because when $\alpha = 0$ the perturbed values are independent of the original values, which results in synthetic data. By contrast, as α approaches $\mathbf{I}$, the resulting information loss is very low. As observed earlier, the opposite is true for disclosure risk; as α approaches 0, disclosure risk decreases and as α approaches $\mathbf{I}$, the disclosure risk is maximal since all confidential values are released unchanged. Thus, the implementation of this procedure needs to be evaluated by considering the trade-off between information loss and disclosure risk.

Microaggregation-based hybrid data

In Domingo-Ferrer and González-Nicolás (2010), generic and specific procedures to generate hybrid data based on microaggregation are proposed.

A generic procedure

Let **V** be an original microdata set consisting of n records. On input of an integer parameter $k \in \{1, \ldots, n\}$, the generic procedure generates a hybrid microdata set **V'**. The greater k, the more synthetic is **V'**. Extreme cases are: $k = 1$, which yields **V'** = **V** (the output data are exactly the original input data) and $k = n$, which yields a completely synthetic output microdata set **V'**.

The procedure calls two algorithms:

1. A generic synthetic data generator $S(\mathbf{C}, \mathbf{C'}, parms)$, that is, an algorithm which, given an original data (sub)set **C**, generates a synthetic data (sub)set **C'** preserving the statistics or parameters or models of **C** specified in *parms*. For numeric data, this generator could be for example IPSO (see Section 3.8.1) or the sufficiency-based generator with $\alpha = \mathbf{0}$ (see Section 3.8.3); for categorical and numerical data, it could be based on MI (see Section 3.8.1).

2. A microaggregation heuristic (see Section 3.7.3), which, with input a set of n records and parameter k, partitions the set of records into clusters containing between k and $2k - 1$ records. Cluster creation attempts to maximise intra-cluster homogeneity. A generic version of MDAV (see Algorithm 1 in Section 3.7.3) can be used. Implementation of this MDAV-generic heuristic for a particular variable type requires specifying how the average record is computed and what distance is used. For numerical variables, computing averages using the arithmetic mean and using Euclidean distances are the natural choices. For ordinal and nominal data, the median and the modal value can be used, respectively; also, a number of distances for ordinal and for nominal data exist; see Domingo-Ferrer and Torra (2005) for details. A measure of variance to evaluate the homogeneity of clusters obtained on hierarchical nominal data is given in Domingo-Ferrer and Solanas (2008). If variables are all continuous, variable-size microaggregation such as μ-Approx (see Section 3.7.3 and Domingo-Ferrer *et al.* 2008) is an alternative to MDAV. In this type of microaggregation, the partition step returns clusters of size varying between k and $2k - 1$ records depending on the distribution of the data; such variable-size clusters are usually more homogeneous than fixed-size clusters, especially if the original data are naturally clustered.

Microaggregation can be defined operationally in terms of the following two steps:

Partition: The set of original records is partitioned into several clusters in such a way that records in the same cluster are *similar* to each other and so that the number of records in each cluster is at least k.

Aggregation: An aggregation operator (for example, the mean for continuous data or the median for categorical data) is computed for each cluster and is used to replace the original records. In other words, each record in a cluster is replaced by the cluster's prototype.

To create hybrid data, we will use *only the partition step* of microaggregation.

Procedure 3 *(microhybrid* $(\mathbf{V}, \mathbf{V}', parms, k))$

1. *Call microaggregation* $(\mathbf{V}, k)$. *Let* $C_1, \ldots C_\kappa$ *for some* κ *be the resulting clusters of records.*

2. *For* $i = 1, \ldots, \kappa$ *call* $\mathcal{S}(C_i, C_i', parms)$.

3. *Output a hybrid data set* $\mathbf{V}'$ *whose records are those in the clusters* $C_1', \ldots, C_\kappa'$.

At Step 1 of procedure *microhybrid* above, clusters containing between k and $2k - 1$ records are created. Then at Step 2, a synthetic version of each cluster is generated. At Step 3, the original records in each cluster are replaced by the records in the corresponding synthetic cluster (instead of replacing them with the average record of the cluster, as done in conventional microaggregation). The *microhybrid* procedure bears some resemblance to the condensation approach proposed in Aggarwal and Yu (2004). However, *microhybrid* is more general because:

- It can be applied to any data type (condensation is designed for numerical data only).

- Clusters do not need to be all of size k (their sizes can vary between k and $2k - 1$).

- Any synthetic data generator (chosen to preserve certain pre-selected statistics or models) can be used by *microhybrid*.

- Instead of using an *ad hoc* clustering heuristic like condensation, *microhybrid* can use any of the best microaggregation heuristics (e.g. those cited in Section 3.7.3) which should yield higher within-cluster homogeneity and thus less information loss.

We justify here the role of parameter k in *microhybrid*:

- If $k = 1$, and *parms* include preserving the mean of each variable in the original clusters, the output is the same original data set, because the procedure creates n clusters (as many as the number of original records). With $k = 1$, even variable-size heuristics will yield all clusters of size 1, because the maximum intra-cluster similarity is obtained when clusters consist all of a single record.

- If $k = n$, the output is a single synthetic cluster: the procedure is equivalent to calling the synthetic data generator $\mathcal{S}$ once for the entire data set.

- For intermediate values of k, several clusters are obtained at Step 1, whose parameters *parms* are preserved by the synthetic clusters generated at Step 2. As k decreases, the number of clusters (whose parameters are preserved in the data output at Step 3) increases, which causes the output data to look more and more like the original data. Each cluster can be regarded as a constraint on the synthetic data generation: the more constraints, the less freedom there is for generating synthetic data, and the output resembles more the original data. This is why the output data can be called hybrid.

It must be noted here that, depending on the synthetic generator used, there may be a lower bound for k *higher than* 1. For example, if using IPSO (see Section 3.8.1) with $|X|$ confidential variables and $|Y|$ non-confidential variables, it turns out that k must be at least $|Y| + 1$; the reason is that IPSO fits to the cluster data a multivariate linear regression model with the $|Y|$ non-confidential variables as independent variables, and it is well known that, in a linear regression, the sample size must be greater than the number of independent variables.

A specific procedure for numerical hybrid data

In the specific case where all variables in the data set **V** are numerical, we can use the *microhybrid* procedure with the following choices:

- Take one of the following two microaggregation procedures:
  - MDAV-generic with the arithmetic mean as average operator and the Euclidean distance, that is, the MDAV algorithm described in Section 3.7.3 and Domingo-Ferrer and Torra (2005) and implemented in the packages μ-ARGUS (see Hundepool *et al.* (2008)) and *sdcMicro* (see Templ 2008 and Templ and Petelin 2009);
  - The μ-Approx variable-size heuristic mentioned above.
- Take one of the following two synthetic data generators, whose preserved *parms* are the variable means and covariances:
  - IPSO (see Section 3.8.1).
  - The sufficiency-based generator with $\alpha = \mathbf{0}$ (see Section 3.8.3).

Let us call the resulting procedure $\mathbb{R}$-microhybrid.

Theoretical properties

It is shown in Domingo-Ferrer and González-Nicolás (2010) that, by construction, the hybrid data output by $\mathbb{R}$-*microhybrid* have *exactly the same means and covariances as the original data*.

It is also shown in that paper that:

- The better the microaggregation heuristic used (i.e. the lower the intra-cluster variance), the more accurately are third-order and fourth-order central moments preserved.

- There is approximate preservation *over random sub-domains* (any subset of the data) of means, variances, covariances and third and fourth-order central moments. The smaller k and the better the microaggregation heuristic, the more accurate is preservation over sub-domains.

Empirical examples

In Domingo-Ferrer and González-Nicolás (2010), the following experimental comparison of $\mathbb{R}$-*microhybrid* against sufficiency-based hybrid data and plain multivariate microaggregation is reported.

Two reference microdata sets (see Brand *et al.* 2002) proposed in the European project CASC were used:

1. The 'Census' data set which contains 1080 records with 13 numerical variables. Within this data set, as confidential variables X_1 and X_2 FICA (Social security retirement payroll deduction) and FEDTAX (Federal income tax liability) were selected, respectively. As non-confidential variables Y_1 and Y_2, INTVAL (amount of interest income) and POTHVAL (total other persons income) were selected, respectively.

2. The 'EIA' data set which contains 4092 records with 11 numerical variables (plus two additional categorical variables). If records are viewed as points in a multidimensional space, points in 'EIA' are less evenly distributed than points in 'Census'. In fact, points in 'EIA' tend to form natural clusters. As confidential variables X_1 and X_2, INDREVENUE (revenue from sales to industrial consumers) and INDSALES (sales to industrial consumers) were selected. As non-confidential variables Y_1 and Y_2, TOTREVENUE (revenue from sales to all consumers) and TOTSALES (sales to all consumers) were selected, respectively.

In order to conduct a fair comparison, values of k for $\mathbb{R}$-*microhybrid* and α for the sufficiency-based method were determined which yielded hybrid data with a similar disclosure risk. Disclosure risk was measured by using distance-based record linkage as described in Torra and Domingo-Ferrer (2003), between the hybrid and the original records. A hybrid record, actually a pair of values for the hybrid variables (X_1', X_2'), was linked to the original record whose values for (X_1, X_2) were at shortest Euclidean distance; if the pair of linked records shared the same values for the non-confidential variables (Y_1, Y_2), the match was considered correct. The percentage of correct matches was used as a measure of disclosure risk.

For the sufficiency-based method, the same values of α used in the empirical work presented in Muralidhar and Sarathy (2008) were taken, that is

$$\alpha_1 = \begin{pmatrix} 0.9 & 0.0 \\ 0.0 & 0.9 \end{pmatrix}; \ \alpha_2 = \begin{pmatrix} 0.8 & 0.0 \\ 0.0 & 0.3 \end{pmatrix}; \ \alpha_3 = \begin{pmatrix} 0.0 & 0.0 \\ 0.0 & 0.0 \end{pmatrix}.$$

For each value of α and each original data set, 10 hybrid data sets were generated and the average percentage of correct matches was computed. For $\mathbb{R}$-*microhybrid*, different values of k were tried. For each value of k and each original data set, 10 hybrid data sets were generated and the average percentage of correct matches was computed.

For the 'Census' and the 'EIA' data sets, Table 3.8 gives the average percentage of correct matches with the sufficiency-based method using parameters α_1, α_2 and α_3. For each data set, the table also lists those values of k for which $\mathbb{R}$-*microhybrid* was found to yield the most similar percentage of correct matches to the sufficiency-based

Table 3.8   Comparable values of α and k, i.e. yielding similar percentages of correct matches for the sufficiency-based method and $\mathbb{R}$-*microhybrid* with μ-Approx microaggregation, respectively. Data sets considered are 'Census' and 'EIA'.

Data set	α	% correct matches	k	% correct matches
'Census'	α_1	0.2	22	0.2
	α_2	0.1	23	0.1
	α_3	0.0	24	0.0
'EIA'	α_1	0.6	80	0.6
	α_2	0.4	90	0.4
	α_3	0.0	120	0.0

method with $\alpha_1, \alpha_2, \alpha_3$, respectively. For 'Census', the values of k comparable to $\alpha_1, \alpha_2, \alpha_3$ turned out to be $k = 22, 23, 24$, respectively. For 'EIA', they turned out to be $k = 80, 90, 120$, respectively. Note that as α changes from α_1 to α_2 and α_3, the resulting hybrid data are less influenced by the original data (in fact for α_3 they are purely synthetic). Therefore, it is natural that the corresponding values of k are also increasing (the higher k with $\mathbb{R}$-*microhybrid*, the less influenced are the resulting hybrid data by the original data).

After generating hybrid data with the sufficiency-based method and $\mathbb{R}$-*microhybrid* with, respectively, the values of α and k in Table 3.8, sampling was performed to capture utility loss when the user restricts her analysis to a subset of the data. The following steps were performed:

- For each data set and each parameter value, samples of size 10% of the data set were drawn 100 times by simple random sampling.

- Let Θ' be a statistic over the hybrid data and Θ the corresponding statistic over the original data. Let θ'_i and θ_i be, respectively, the values taken by Θ' and Θ in the ith sample of the hybrid data and the corresponding ith sample of the original data. Define the mean variation for the pair of statistics (Θ', Θ) over the 100 sample pairs as

$$\Delta(\Theta) = \frac{1}{100} \sum_{i=1}^{100} \frac{|\theta'_i - \theta_i|}{|\theta_i|}$$

Mean variations were computed for the following pairs of statistics: mean of X'_1 vs. X_1 (named $\Delta(m_1)$), mean of X'_2 vs. X_2 ($\Delta(m_2)$), variance of X'_1 vs. X_1 ($\Delta(\sigma_1^2)$), variance of X'_2 vs. X_2 ($\Delta(\sigma_2^2)$), covariance $Y_1 X_1$ vs. $Y_1 X'_1$ ($\Delta(\sigma_{11})$), covariance $Y_1 X_2$ vs. $Y_1 X'_2$ ($\Delta(\sigma_{12})$), covariance $Y_2 X_1$ vs. $Y_2 X'_1$ ($\Delta(\sigma_{21})$), covariance $Y_2 X_2$ vs. $Y_2 X'_2$ ($\Delta(\sigma_{22})$), third-order central moment of X'_1 vs. X_1 ($\Delta(m_1^3)$), third-order central moment of X'_2 vs. X_2 ($\Delta(m_2^3)$), fourth-order central moment of X'_1 vs. X_1 ($\Delta(m_1^4)$) and fourth-order central moment of X'_2 vs. X_2 ($\Delta(m_2^4)$).

Table 3.9   Average mean variation for several statistics on corresponding 10% samples of the hybrid data set and the original data set. Hybrid methods considered: sufficiency-based and $\mathbb{R}$-*microhybrid* with μ-Approx microaggregation, respectively. Correct matches computed for overall data set (no sampling). Data set: 'Census'.

Statistic	α_1	$k = 22$	α_2	$k = 23$	α_3	$k = 24$
% correct matches	0.60	0.60	0.40	0.40	0.00	0.00
$\Delta(m_1)$	0.0168	0.0029	0.0175	0.0036	0.0584	0.0048
$\Delta(m_2)$	0.0203	0.0048	0.0215	0.0056	0.0699	0.0066
$\Delta(\sigma_1^2)$	0.0694	0.0131	0.0747	0.0155	0.1253	0.0199
$\Delta(\sigma_2^2)$	0.0625	0.0200	0.0770	0.0246	0.1188	0.0347
$\Delta(\sigma_{11})$	0.2097	0.0293	0.2584	0.0333	0.5267	0.0443
$\Delta(\sigma_{12})$	0.1722	0.0346	0.2522	0.0524	0.4723	0.0488
$\Delta(\sigma_{21})$	0.3600	0.0595	0.4364	0.0870	1.2375	0.1619
$\Delta(\sigma_{22})$	0.2437	0.0736	0.3223	0.2937	1.2618	0.1637
$\Delta(m_1^3)$	3.4807	0.5501	5.1872	0.8345	62.6002	0.5509
$\Delta(m_2^3)$	0.1990	0.0309	0.2303	0.0345	0.5332	0.0427
$\Delta(m_1^4)$	0.5597	0.1192	0.6987	0.2022	1.9927	0.2412
$\Delta(m_2^4)$	0.1830	0.0600	0.2458	0.0901	0.4939	0.1165

Tables 3.9 and 3.10 report, respectively for the 'Census' and the 'EIA' data sets, the above mean variations. The following can be seen:

- For 'Census', $\mathbb{R}$-*microhybrid* *clearly* outperforms the sufficiency-based method for all parameter values and for all considered statistics. 'Clear outperformance' means that the mean variation using the sufficiency-based method is between 5 and 10 times greater than the mean variation using $\mathbb{R}$-*microhybrid* for all statistics, except for $\Delta(m_1^3)$ when α_3 and $k = 24$ are taken, in which case it is more than 100 times greater.

- For 'EIA', $\mathbb{R}$-*microhybrid* with $k = 100$ *slightly* outperforms the sufficiency-based method for all statistics with α_3, but is *slightly* outperformed by the sufficiency-based method for all statistics with the other two parameter choices (α_1, $k = 80$ and α_2, $k = 90$, respectively). 'Slight outperformance' means that the ratio between mean variations is between 1 and 3 in either sense.

An interesting feature of $\mathbb{R}$-*microhybrid* method is that the trade-off between disclosure risk and information loss depends only on parameter k. Tables 3.11 and 3.12 show, respectively, for 'Census' and 'EIA', how taking smaller values of k increases the disclosure risk (percentage of correct matches) and reduces information loss (mean variation for the considered statistics). Furthermore, the tables show that the disclosure risk incurred by $\mathbb{R}$-*microhybrid* is lower than the one incurred by plain multivariate microaggregation for the same k.

Table 3.10   Average mean variation for several statistics on corresponding 10% samples of the hybrid data set and the original data set. Hybrid methods considered: MS and $\mathbb{R}$-*microhybrid* with μ-Approx microaggregation, respectively. Correct matches computed for the overall data set (no sampling). Data set: 'EIA'.

Statistic	α_1	$k = 80$	α_2	$k = 90$	α_3	$k = 100$
% correct matches	0.20	0.20	0.10	0.10	0.00	0.00
$\Delta(m_1)$	0.0129	0.0147	0.0150	0.0150	0.0444	0.0165
$\Delta(m_2)$	0.0118	0.0113	0.0231	0.0123	0.0378	0.0155
$\Delta(\sigma_1^2)$	0.0195	0.0705	0.0276	0.0732	0.0919	0.0849
$\Delta(\sigma_2^2)$	0.0172	0.0435	0.0164	0.0559	0.0812	0.0665
$\Delta(\sigma_{11})$	0.0128	0.0363	0.0204	0.0424	0.0670	0.0509
$\Delta(\sigma_{12})$	0.0132	0.0344	0.0204	0.0387	0.0641	0.0443
$\Delta(\sigma_{21})$	0.0116	0.0303	0.0300	0.0378	0.0608	0.0442
$\Delta(\sigma_{22})$	0.0112	0.0297	0.0289	0.0330	0.0599	0.0391
$\Delta(m_1^3)$	0.0691	0.1838	0.0860	0.1964	0.3225	0.2170
$\Delta(m_2^3)$	0.0807	0.2854	0.0949	0.3379	0.3493	0.3671
$\Delta(m_1^4)$	0.0532	0.1259	0.0695	0.1381	0.2482	0.1520
$\Delta(m_2^4)$	0.0552	0.1978	0.0695	0.2272	0.2668	0.2390

Table 3.11   Average mean variation for several statistics on corresponding 10% samples of the hybrid data set and the original data set. Hybrid method: $\mathbb{R}$-*microhybrid* with μ-Approx microaggregation and several values of k. Correct matches computed for the overall data set (no sampling); between parentheses, percentage of correct matches if plain multivariate microaggregation was used. Data set: 'Census'.

Statistic	$k = 7$	$k = 10$	$k = 15$	$k = 20$	$k = 22$	$k = 23$	$k = 24$
% correct matches	3.30 (11.11)	2.00 (7.69)	1.00 (5.00)	0.40 (3.70)	0.20 (3.52)	0.10 (3.43)	0.00 (2.87)
$\Delta(m_1)$	0.0012	0.0016	0.0023	0.0035	0.0029	0.0036	0.0048
$\Delta(m_2)$	0.0017	0.0024	0.0033	0.0052	0.0048	0.0056	0.0066
$\Delta(\sigma_1^2)$	0.0042	0.0074	0.0101	0.0151	0.0131	0.0155	0.0199
$\Delta(\sigma_2^2)$	0.0063	0.0110	0.0151	0.0230	0.0200	0.0246	0.0347
$\Delta(\sigma_{11})$	0.0084	0.0126	0.0206	0.0464	0.0293	0.0333	0.0443
$\Delta(\sigma_{12})$	0.0070	0.0137	0.0219	0.0507	0.0346	0.0524	0.0488
$\Delta(\sigma_{21})$	0.0144	0.0247	0.0488	0.0584	0.0595	0.0870	0.1619
$\Delta(\sigma_{22})$	0.0220	0.0261	0.0407	0.0513	0.0736	0.2937	0.1637
$\Delta(m_1^3)$	0.1940	0.2377	0.3057	0.6841	0.5501	0.8345	0.5509
$\Delta(m_2^3)$	0.0108	0.0200	0.0236	0.0376	0.0309	0.0345	0.0427
$\Delta(m_1^4)$	0.0404	0.0853	0.1059	0.1066	0.1192	0.2022	0.0901
$\Delta(m_2^4)$	0.0202	0.0412	0.0473	0.0517	0.0600	0.0901	0.1165

Table 3.12 Average mean variation for several statistics on corresponding 10% samples of the hybrid data set and the original data set. Hybrid method: $\mathbb{R}$-*microhybrid* with μ-Approx microaggregation and several values of k. Correct matches computed for the overall data set (no sampling); between parentheses, percentage of correct matches if plain multivariate microaggregation was used. Data set: 'EIA'.

Statistic	$k = 10$	$k = 15$	$k = 20$	$k = 80$	$k = 90$	$k = 120$
% correct matches	7.80 (7.50)	4.50 (5.06)	3.30 (3.76)	0.60 (0.93)	0.40 (0.83)	0.00 (0.64)
$\Delta(m_1)$	0.0047	0.0079	0.0080	0.0147	0.0150	0.0165
$\Delta(m_2)$	0.0035	0.0060	0.0060	0.0113	0.0123	0.0155
$\Delta(\sigma_1^2)$	0.0311	0.0408	0.0426	0.0705	0.0732	0.0849
$\Delta(\sigma_2^2)$	0.0154	0.0262	0.0297	0.0435	0.0559	0.0665
$\Delta(\sigma_{11})$	0.0174	0.0232	0.0238	0.0363	0.0424	0.0509
$\Delta(\sigma_{12})$	0.0116	0.0159	0.0177	0.0344	0.0387	0.0443
$\Delta(\sigma_{21})$	0.0131	0.0190	0.0201	0.0303	0.0378	0.0442
$\Delta(\sigma_{22})$	0.0090	0.0135	0.0159	0.0297	0.0330	0.0391
$\Delta(m_1^3)$	0.0929	0.0975	0.1200	0.1838	0.1964	0.2170
$\Delta(m_2^3)$	0.1683	0.1604	0.2126	0.2854	0.3379	0.3671
$\Delta(m_1^4)$	0.0363	0.0627	0.0728	0.1259	0.1381	0.1520
$\Delta(m_2^4)$	0.0570	0.1005	0.1145	0.1978	0.2272	0.2390

3.8.4   Pros and cons of synthetic and hybrid data

Synthetic data are appealing in that, at a first glance, they seem to circumvent the re-identification problem: since published records are invented and do not derive from any original record, it might be concluded that no individual can complain from having been re-identified. At a closer look, this advantage is less clear. If, by chance, a published synthetic record matches a particular citizen's non-confidential variables (age, marital status, place of residence, etc.) and confidential variables (salary, mortgage, etc.), re-identification using the non-confidential variables is easy and that citizen may feel that his confidential variables have been unduly revealed. In that case, the citizen is unlikely to be happy with or even understand the explanation that the record was synthetically generated.

Here lies one of the important advantages of the synthetic data approaches based on MI compared to IPSO or sufficiency-based synthetic data. While the latter are only applicable to continuous (numeric) data, the MI approach is also applicable to categorical variables. So, if the data releasing agency wants to prevent any form of disclosure (attribute or identity disclosure), the MI approach allows synthesising some categorical variables, too. For fully synthetic data sets the actual disclosure risk is further reduced, since the synthetic data is generated for new samples from the population and the intruder never knows if a unit in the released data was actually included in the original data.

Partially synthetic data sets on the other hand have the advantage that the synthesis can be tailored specifically to the records at risk. For some data sets, it might only be necessary to synthesise certain subsets of the data set, e.g. the income for survey participants with an income above 100 000 Euro. Obviously, the decision on which records will remain unchanged is a delicate task and a careful disclosure risk evaluation is necessary in this context.

On the other hand, limited data utility is a problem of synthetic data. Only the statistical properties explicitly captured by the model used by the data protector are preserved. A logical question at this point is why not directly publish the statistics one wants to preserve or simply the parameters of the imputation model rather than release a synthetic microdata set. Possible defences against this argument are:

- Synthetic data are normally generated by using more information on the original data than is specified in the model whose preservation is guaranteed by the data protector releasing the synthetic data.

- As a consequence of the above description, synthetic data may offer utility beyond the models they exactly preserve.

- It is impossible to anticipate all possible statistics an analyst might be interested in. So access to some microdata set should be granted.

- Not all users of a PUF will have a sound background in statistics. Some of the users might only be interested in some descriptive statistics and they would not be able to generate the results if only imputation parameters were released.

- The imputation models in most applications can be very complex, because different models are fitted for every variable and often for different subsets of the data set. This might lead to hundreds of parameters just for one variable. Thus, it is much more convenient even for the skilled user of the data to have the synthesised data set available.

- The most important reason for not releasing the imputation parameters is that the parameters themselves could be disclosive in some occasions. For that reason, only some general statements about the generation of the PUF should be released. For example, these general statements could provide information as to which variables where included in the imputation model, but not the exact parameters. The goal is that the user can judge if her analysis would be covered by the imputation model, but she should not be able to use the parameters to disclose any confidential information.

Anyway, since there is some truth in the limited utility of synthetic data, hybrid data obtained by mixing the original data and synthetic data are an interesting option combining the strengths of masked and synthetic data. Indeed, as discussed above, hybrid data (e.g. sufficiency-based and microaggregation-based) exactly preserve a set of pre-selected statistics, approximately preserve other statistics and offer a low disclosure risk.

3.9 Information loss in microdata

A strict evaluation of information loss must be based on the data uses to be supported by the protected data. The greater the differences between the results obtained on original and protected data for those uses, the higher the loss of information. However, very often microdata protection cannot be performed in a data use specific manner, for the following reasons:

- Potential data uses are very diverse and it may be even hard to identify them all at the moment of data release by the data protector.

- Even if all data uses can be identified, issuing several versions of the same original data set so that the ith version has an information loss optimised for the ith data use may result in unexpected disclosure.

Since data often must be protected with no specific data use in mind, generic information loss measures are desirable to guide the data protector in assessing how much utility harm is being inflicted by a particular SDC technique.

Defining what a generic information loss measure is can be a tricky issue. Roughly speaking, it should capture the amount of information loss for a reasonable range of data uses. We will say there is little information loss if the protected data set is analytically valid and interesting according to the following definitions by Winkler (1999):

- A protected microdata set is *analytically valid* if it approximately preserves the following with respect to the original data (some conditions apply only to continuous variables):

  - Means and covariances on a small set of sub-domains (subsets of records and/or variables).

  - Marginal values for a few tabulations of the data.

  - At least one distributional characteristic.

- A microdata set is *analytically interesting* if six variables on important sub-domains are provided that can be validly analysed.

More precise conditions of analytical validity and analytical interest cannot be stated without taking specific data uses into account. As imprecise as they may be, the above definitions suggest some possible measures:

- Compare raw records in the original and the protected data sets. The more similar the SDC method to the identity function, the lower the information loss (but the higher the disclosure risk). This requires pairing records in the original data set and records in the protected data set. For masking methods, each record in the protected data set is naturally paired to the record in the original data set it originates from. For synthetic protected data sets, pairing is more artificial. In Dandekar *et al.* (2002b), it was proposed to pair a synthetic record to the nearest original record according to some distance.

- Compare some statistics computed on the original and the protected data sets. The above definitions list some statistics which should be preserved as much as possible by an SDC method.

3.9.1   Information loss measures for continuous data

We will discuss in what follows two sets of measures:

1. *Unbounded measures.* These are based on measuring the mean square error, the mean absolute error or the mean variation caused by an SDC technique on several statistics. These measures are intuitive but, by definition, they are not bounded within any pre-defined range.

2. *Bounded measures.* These are based on probability and they are bounded within $[0, 1]$. Boundedness of information loss is convenient to compare and trade off against disclosure risk, which is naturally bounded within $[0, 1]$.

Unbounded information loss measures

Assume a microdata set with n individuals (records) $I_1, I_2, \ldots, I_n$ and p continuous variables $Z_1, Z_2, \ldots, Z_p$. Let X be the matrix representing the original microdata set (rows are records and columns are variables). Let X' be the matrix representing the protected microdata set. The following tools are useful to characterise the information contained in the data set:

- Covariance matrices V (on X) and V' (on X').

- Correlation matrices R and R'.

- Correlation matrices RF and RF' between the p variables and the p principal components $PC_1, \ldots, PC_p$ obtained through principal components analysis.

- Communality between each of the p variables and the first-principal component PC_1 (or other principal components PC_i's). Communality is the percentage of each variable that is explained by PC_1 (or PC_i). Let C be the vector of communalities for X and C' the corresponding vector for X'.

- Matrices F and F' containing the loadings of each variable in X, respectively, in X', on each principal component. The ith variable in X can be expressed as a linear combination of the principal components plus a residual variation, where the jth principal component is multiplied by the loading in F relating the ith variable and the jth principal component (Chatfield and Collins 1980). The same holds if X and F are replaced by X' and F', respectively.

In Domingo-Ferrer *et al.* (2001) and Domingo-Ferrer and Torra (2001a) it was proposed to measure information loss through the discrepancies between matrices X, V, R, RF, C and F obtained on the original data and the corresponding X', V', R', RF', C' and F' obtained on the protected data set. In particular, discrepancy between correlations is related to the information loss for data uses such as regressions and cross tabulations.

Matrix discrepancy can be measured in at least three ways:

Mean square error: Sum of squared componentwise differences between pairs of matrices, divided by the number of cells in either matrix.

Mean absolute error: Sum of absolute componentwise differences between pairs of matrices, divided by the number of cells in either matrix.

Mean variation: Sum of absolute percent variation of components in the matrix computed on protected data with respect to components in the matrix computed on original data, divided by the number of cells in either matrix. This approach has the advantage of not being affected by scale changes of variables.

Table 3.13 summarises the measures proposed in Domingo-Ferrer *et al.* (2001) and Domingo-Ferrer and Torra (2001a). In this table, p is the number of variables, n the number of records, and components of matrices are represented by the corresponding lowercase letters (e.g. x_{ij} is a component of matrix X). Regarding $X - X'$ measures, it makes also sense to compute them on the averages of variables rather than on all data (call this variant $\bar{X} - \bar{X}'$). Similarly, for $V - V'$ measures, it would also be sensible to use them to compare only the variances of the variables, i.e. to compare the diagonals of the covariance matrices rather than the whole matrices (call this variant $S - S'$).

Table 3.13   Unbounded information loss measures for continuous microdata.

	Mean square error	Mean absolute error	Mean variation
$X - X'$	$\dfrac{\sum_{j=1}^{p}\sum_{i=1}^{n}(x_{ij}-x'_{ij})^2}{np}$	$\dfrac{\sum_{j=1}^{p}\sum_{i=1}^{n}|x_{ij}-x'_{ij}|}{np}$	$\dfrac{\sum_{j=1}^{p}\sum_{i=1}^{n}\frac{|x_{ij}-x'_{ij}|}{|x_{ij}|}}{np}$
$V - V'$	$\dfrac{\sum_{j=1}^{p}\sum_{1\le i\le j}(v_{ij}-v'_{ij})^2}{\frac{p(p+1)}{2}}$	$\dfrac{\sum_{j=1}^{p}\sum_{1\le i\le j}|v_{ij}-v'_{ij}|}{\frac{p(p+1)}{2}}$	$\dfrac{\sum_{j=1}^{p}\sum_{1\le i\le j}\frac{|v_{ij}-v'_{ij}|}{|v_{ij}|}}{\frac{p(p+1)}{2}}$
$R - R'$	$\dfrac{\sum_{j=1}^{p}\sum_{1\le i<j}(r_{ij}-r'_{ij})^2}{\frac{p(p-1)}{2}}$	$\dfrac{\sum_{j=1}^{p}\sum_{1\le i<j}|r_{ij}-r'_{ij}|}{\frac{p(p-1)}{2}}$	$\dfrac{\sum_{j=1}^{p}\sum_{1\le i<j}\frac{|r_{ij}-r'_{ij}|}{|r_{ij}|}}{\frac{p(p-1)}{2}}$
$RF - RF'$	$\dfrac{\sum_{j=1}^{p}w_j\sum_{i=1}^{p}(rf_{ij}-rf'_{ij})^2}{p^2}$	$\dfrac{\sum_{j=1}^{p}w_j\sum_{i=1}^{p}|rf_{ij}-rf'_{ij}|}{p^2}$	$\dfrac{\sum_{j=1}^{p}w_j\sum_{i=1}^{p}\frac{|rf_{ij}-rf'_{ij}|}{|rf_{ij}|}}{p^2}$
$C - C'$	$\dfrac{\sum_{i=1}^{p}(c_i-c'_i)^2}{p}$	$\dfrac{\sum_{i=1}^{p}|c_i-c'_i|}{p}$	$\dfrac{\sum_{i=1}^{p}\frac{|c_i-c'_i|}{|c_i|}}{p}$
$F - F'$	$\dfrac{\sum_{j=1}^{p}w_j\sum_{i=1}^{p}(f_{ij}-f'_{ij})^2}{p^2}$	$\dfrac{\sum_{j=1}^{p}w_j\sum_{i=1}^{p}|f_{ij}-f'_{ij}|}{p^2}$	$\dfrac{\sum_{j=1}^{p}w_j\sum_{i=1}^{p}\frac{|f_{ij}-f'_{ij}|}{|f_{ij}|}}{p^2}$

In Yancey *et al.* (2002), it was observed that dividing by x_{ij} causes the $X - X'$ mean variation to rise sharply when the original value x_{ij} is close to 0. This dependency on the particular original value being undesirable in an information loss measure, the authors of Yancey *et al.* (2002) proposed to replace the mean variation of $X - X'$ by the more stable measure

$$\frac{1}{np} \sum_{j=1}^{p} \sum_{i=1}^{n} \frac{|x_{ij} - x'_{ij}|}{\sqrt{2}S_j}, \tag{3.23}$$

where S_j is the standard deviation of the jth variable in the original data set.

Bounded information loss measures

Trading off information loss and disclosure risk

There is a broad choice of methods for continuous microdata protection (see Sections 3.6 and 3.7). To increase the *embarras du choix*, most of such methods are parametric (e.g. in microaggregation, one parameter is the minimal number of records in a cluster), so the user must go through two choices rather than one: a primary choice to select a method and a secondary choice to select parameters for the method to be used.

The optimal method and parameterisation will be the ones yielding an optimal tradeoff between information loss and disclosure risk. Thus, we need to be able to combine measures of information loss and measures of disclosure risk. Two approaches to do this are conceivable:

Explicit: A score (formula) is adopted which combines information loss and disclosure risk measures. This is the approach adopted in Domingo-Ferrer and Torra (2001b). Using a score permits to regard the selection of a masking method and its parameters as an optimisation problem. This was exploited in Sebé *et al.* (2002): a masking method was applied to the original data file and then a post-masking optimisation procedure was applied to decrease the score obtained.

Implicit: No specific score can do justice to all methods for all data uses and all disclosure scenarios. Thus, another possibility is for the data protector to separately compute several information loss and disclosure risk measures and choose the most appropriate method based on a combination of the most relevant measures for the specific data use/disclosure scenario. This implicit approach was adopted in Yancey *et al.* (2002), for example.

Whether explicit or implicit, a combination of information loss and disclosure risk measures is best performed if both types of measures can be bounded within the same range. Unfortunately, while disclosure risk measures are bounded, the above information loss measures for continuous data are not:

- Disclosure risk measures can normally be regarded as probabilities or proportions bounded between 0 and 1, e.g. the probability that a certain respondent is re-identified or the proportion of correctly re-identified records after a record linkage attack.

- However, being mean square errors, mean absolute errors and mean variations, the information loss measures of Table 3.13 are unbounded[2]. Moreover, mean variations may become huge when measured on magnitudes close to 0.

In Trottini (2003), the above mismatch was detected and a solution consisting of enforcing upper bounds on information loss measures was proposed. In practice, the proposal in Trottini (2003) was to limit those measures in Table 3.13 based on the mean variation to a pre-defined maximum value. Such an *unnatural* truncation clearly damages the accuracy of the resulting measures.

Probabilistic information loss measures

In Mateo-Sanz *et al.* (2005), probabilistic measures were proposed which *naturally* yield bounded information loss measures for continuous variables. These are described as follows.

The original data set X is viewed as a population with n records and the protected data set X' as a sample with n' records.

Given a population parameter θ on X, we can compute the corresponding sample statistic $\hat{\Theta}$ on X'. Let us assume that $\hat{\theta}$ is the value taken by $\hat{\Theta}$ for a specific sample. The more different is $\hat{\theta}$ from θ, the more information is lost when publishing the sample X' instead of the population X. Mateo-Sanz *et al.* (2005) express that loss of information through probability as follows.

If the sample size n' is large, the distribution of $\hat{\Theta}$ tends to normality with mean θ and variance $\text{Var}(\hat{\Theta})$. According to Stuart and Ord (1994), values of n' greater than 100 are often large enough for normality of all sample statistics to be acceptable. Fortunately, most protected data sets released in official statistics consist of $n' > 100$ records, so that assuming normality is safe. Thus, the standardised sample discrepancy

$$Z = \frac{\hat{\Theta} - \theta}{\sqrt{\text{Var}(\hat{\Theta})}}$$

can be assumed to follow a $N(0, 1)$ distribution.

Therefore, a probabilistic information loss measure $pil(\theta)$ referred to parameter θ is the probability that the absolute value of the discrepancy Z is less than or equal to the actual discrepancy in the specific sample X', that is

$$pil(\hat{\Theta}) = 2P(0 \leq Z \leq \frac{|\hat{\theta} - \theta|}{\sqrt{\text{Var}(\hat{\Theta})}}). \tag{3.24}$$

From Expression (3.24), it can be seen that little information loss translates to $pil(\hat{\Theta})$ close to 0: the sample discrepancy between $\hat{\theta}$ and θ is so small that it is hard for Z to stay below that discrepancy. Conversely, large information loss translates to $pil(\hat{\Theta})$ close to 1: the sample discrepancy is so large that Z can hardly surpass it.

Mateo-Sanz *et al.* (2005) identify and denote several population parameters θ and corresponding sample statistics $\hat{\Theta}$ which can be relevant to measure information

---

[2] Note that unboundedness is only a problem for continuous variables. For categorical variables, information loss measures are naturally bounded, because of the finite range of such variables.

loss. They start with population parameters (on X) and then continue with sample statistics (on X').

Let the rth moment about zero of the jth variable of X be denoted by:

$$\mu_r^0(j) = \frac{\sum_{i=1}^n x_{ij}^r}{n}.$$

The rth central moment of the jth variable of X is expressed as:

$$\mu_r(j) = \frac{\sum_{i=1}^n (x_{ij} - \mu_1^0(j))^r}{n}.$$

The (r, s)th central moment of the jth and j'th variables of X can be computed as:

$$\mu_{rs}(j, j') = \frac{\sum_{i=1}^n (x_{ij} - \mu_1^0(j))^r (x_{ij'} - \mu_1^0(j'))^s}{n}.$$

If $r = s = 1$, we get the covariance $\mu_{11}(j, j')$ between variables j and j'. In this way, the correlation coefficient can be expressed as

$$\rho(j, j') = \frac{\mu_{11}(j, j')}{(\mu_{02}(j, j')\mu_{20}(j, j'))^{1/2}}.$$

We now turn to the moments on a protected data set X' with n' records corresponding to the original data set X. The moments on X' are regarded as statistics. We denote the rth moment about zero of the jth variable of X' by:

$$m_r^0(j) = \frac{\sum_{i=1}^{n'} (x_{ij}')^r}{n'}.$$

The rth central moment of the jth variable of X' is expressed as:

$$m_r(j) = \frac{\sum_{i=1}^{n'} (x_{ij}' - m_1^0(j))^r}{n'}.$$

The (r, s)th central moment of the jth and j'th variables of X' can be computed as:

$$m_{rs}(j, j') = \frac{\sum_{i=1}^{n'} (x_{ij}' - m_1^0(j))^r (x_{ij'}' - m_1^0(j'))^s}{n'}.$$

The correlation coefficient between two variables in X' can be expressed as:

$$r(j, j') = \frac{m_{11}(j, j')}{(m_{02}(j, j')m_{20}(j, j'))^{1/2}}.$$

Expression (3.24) can be used to derive an information loss measure for any particular statistic. For the sake of concreteness and comparability, we will consider here the same statistics as Domingo-Ferrer and Torra (2001b) and Domingo-Ferrer *et al.* (2001), with the slight adaptation that direct comparison of data is replaced with quantile comparison. Given two variables j and j', this yields the following measures:

- $pil(m_1^0(j))$ for the mean.
- $pil(m_2(j))$ for the variance.

- $pil(m_{11}(j, j'))$ for the covariance.

- $pil(r(j, j'))$ for Pearson's correlation.

- $pil(Q_q(j))$ for quantiles.

Being probabilities, all the above measures are naturally bounded within the $[0, 1]$ interval. In order to get data set-wide measures, we must average over the various variables (as in Table 3.13). This yields:

$$PIL(m_1^0) = \frac{\sum_{j=1}^{p} pil(m_1^0(j))}{p}, \tag{3.25}$$

$$PIL(m_2) = \frac{\sum_{j=1}^{p} pil(m_2(j))}{p}, \tag{3.26}$$

$$PIL(m_{11}) = \frac{\sum_{1 \le j < j' \le p} pil(m_{11}(j, j'))}{p(p-1)/2}, \tag{3.27}$$

$$PIL(r) = \frac{\sum_{1 \le j < j' \le p} pil(r(j, j'))}{p(p-1)/2}. \tag{3.28}$$

Some remarks follow:

- The normality assumption for the $r(j, j')$ statistic between two variables j and j' only holds when the population correlation $\rho(j, j')$ is sufficiently centred within the interval $[-1, 1]$. For values of $\rho(j, j')$ close to -1 or 1, computing $pil(r(j, j'))$ using Expression (3.24) with a standard normal Z yields an over-pessimistic information loss measure ($\text{Var}(r(j, j'))$ is very small). Still, when one takes the average $PIL(r)$ over all pairs of variables, the result is usually coherent with the average $PIL(m_{11})$ for covariances, as one would expect.

- Using the fact that the correlation coefficient is bounded in $[-1, 1]$, one might think of using as a non-parametric (and non-probabilistic) alternative to $PIL(r)$ the following one:

$$\frac{\sum_{1 \le j < j' \le p} |r(j, j') - \rho(j, j')|}{p(p-1)}. \tag{3.29}$$

Expression (3.29) is half the mean absolute error given in Table 3.13, which is bounded between 0 and 1. However, being non-probabilistic, this measure takes values often incoherent with the average information loss for covariances $PIL(m_{11})$. For example, one can easily get a $PIL(m_{11})$ in $[0.75, 1]$ and an 'optimistic' Expression (3.29) in $[0, 0.25]$. This lack of coherence is clearly undesirable for statistics as related as the covariance and Pearson's correlation.

- The information loss measure $PIL(Q_q)$ for quantiles bears some resemblance to the information loss measure in Agrawal and Aggarwal (2001) consisting of half the expected value of the L_1-norm between the densities of the original and protected variable: both measures are bounded in the $[0, 1]$ interval, but they are not equivalent because the measure in Agrawal and Aggarwal (2001) is not a probability.

In order to use Expression (3.24) to construct $pil(m_1^0(j))$, $pil(m_2(j))$, $pil(m_{11}(j, j'))$, $pil(r(j, j'))$ and $pil(Q_q(j))$, we need the variance of each statistic or at least an approximation to it. This is a technical but unavoidable issue.

Since we take the original data set X as the population and the masked data set X' as the sample, our sampling method is the particular masking method used to obtain X' from X. Therefore, to be strict, the variance of each sample statistic depends on the masking method. However, deriving the expression of the statistic variance for each masking method whose information loss is to be measured is a cumbersome and hardly feasible task. Our primary goal is to obtain information loss measures which can be easily applied to any masking method. In that spirit, we suggest to sacrifice accuracy to applicability and compute variances as if the sampling method were simple random sampling. Of course, some masking methods may substantially differ from simple random sampling: for example, if masking consists of replacing original values by their overall mean, one has zero variance for the sample mean and so on. However, the fact that the new measures using those simplified variances are highly correlated with previous information loss measures in the literature (see experiments in Mateo-Sanz *et al.* 2005), shows that the above is a reasonable approximation.

We will drop indices j, j' of variables in the remainder of this section to improve readability. Following Chapter 10 of Stuart and Ord (1994), we have that, under simple random sampling, the variance of the sample mean is

$$\text{Var}(m_1^0) = \frac{\mu_2}{n'}.$$

The variance of the sample variance is

$$\text{Var}(m_2) = \frac{\mu_4 - \mu_2^2}{n'}.$$

The variance of the sample covariance is

$$\text{Var}(m_{11}) = \frac{\mu_{22} - \mu_{11}^2}{n'}.$$

The variance of the sample Pearson's correlation coefficient is

$$\text{Var}(r) = \frac{\rho^2}{n'} \left\{ \frac{\mu_{22}}{\mu_{11}^2} + \frac{1}{4} \left(\frac{\mu_{40}}{\mu_{20}^2} + \frac{\mu_{04}}{\mu_{02}^2} + \frac{2\mu_{22}}{\mu_{20}\mu_{02}} \right) - \left(\frac{\mu_{31}}{\mu_{11}\mu_{20}} + \frac{\mu_{13}}{\mu_{11}\mu_{02}} \right) \right\}.$$

Finally, if $q \in [0, 1]$, the variance of the sample q-quantile Q_q is

$$\text{Var}(Q_q) = \frac{q(1 - q)}{n' f_{Q_q}^2},$$

where f_{Q_q} is the value of the variable's density function for the abscissa Q_q. If we take the data set X as our population, it is unlikely that we know the analytical expression of the variables' density functions. A simple method to estimate f_{Q_q} is to approximate it by counting the proportion of records included in an interval around Q_q for the specific variable being considered and then dividing by the interval width. It remains to decide what interval should be taken. A possible (and arbitrary) option is $(Q_q - \epsilon, Q_q + \epsilon)$, where ϵ is the range of the variable divided by 1000.

Kernel methods (see e.g. Rosenblatt 1956, Parzen 1962 or Silverman 1982) are an alternative for density estimation based on histogram smoothing. They may yield better density estimates than the simple approach sketched above, but they usually require more computation; see Härdle (1991) for a comprehensive discussion on kernel density estimation.

Empirical application of the proposed probabilistic measures is reported in Mateo-Sanz *et al.* (2005). In particular, the proposed measures seem to be highly correlated to the corresponding measures in Table 3.13:

- $PIL(Q)$ is correlated to the mean variation of $X - X'$ in Table 3.13.

- $PIL(m_1^0)$ is correlated to the mean variation of means of variables.

- $PIL(m_2)$ is correlated to the mean variation of variances of variables.

- $PIL(m_{11})$ is correlated to the mean variation of covariances of variables $(V - V')$.

- $PIL(r)$ is correlated to the mean absolute error of Pearson's correlations $(R - R')$.

3.9.2   Information loss measures for categorical data

Straightforward computation of measures in Table 3.13 on categorical data is not possible. The following alternatives are considered in Domingo-Ferrer and Torra (2001a):

- Direct comparison of categorical values.

- Comparison of contingency tables.

- Entropy-based measures.

Direct comparison of categorical values

Comparison of matrices X and X' for categorical data requires the definition of a distance for categorical variables. Definitions consider only the distances between pairs of categories that can appear when comparing an original record and its protected version (see discussion above on pairing original and protected records).

For a nominal variable V (a categorical variable taking values over an unordered set), the only permitted operation is comparison for equality. This leads to the following distance definition:

$$d_V(c, c') = \begin{cases} 0 & \text{if } c = c' \\ 1 & \text{if } c \neq c', \end{cases}$$

where c is a category in an original record and c' is the category which has replaced c in the corresponding protected record.

For an ordinal variable V (a categorical variable taking values over a totally ordered set), let $\leq_V$ be the total order operator over the range $D(V)$ of V. Define the distance between categories c and c' as the number of categories between the minimum and the maximum of c and c' divided by the cardinality of the range:

$$d_V(c, c') = \frac{|c'' : \min(c, c') \leq c'' < \max(c, c')|}{|D(V)|},$$

where $| \cdot |$ stands for the cardinality operator.

Comparison of contingency tables

An alternative to directly comparing the values of categorical variables is to compare their contingency tables. Given two data sets X and X' (the original and the protected set, respectively) and their corresponding t-dimensional contingency tables for $t \leq K$, we can define a contingency table-based information loss measure *CTBIL* for a subset W of variables as follows:

$$CTBIL(X, X'; W, K) = \sum_{\substack{\{V_{j_1}\cdots V_{j_t}\} \subseteq W \\ |\{V_{j_1}\cdots V_{j_t}\}| \leq K}} \sum_{i_1 \cdots i_t} |x_{i_1 \cdots i_t} - x'_{i_1 \cdots i_t}|, \qquad (3.30)$$

where $x_{subscripts}$, respectively, $x'_{subscripts}$, is the entry of the contingency table of the original data, respectively, protected data, at position given by *subscripts*.

Because the number of contingency tables to be considered depends on the number of variables $|W|$, the number of categories for each variable, and the dimension K, a normalised version of Expression (3.30) may be desirable. This can be obtained by dividing Expression (3.30) by the total number of cells in all considered tables.

Distance between contingency tables generalises some of the information loss measures used in the literature. For example, μ-ARGUS (Hundepool *et al.* 2008) measures information loss for local suppression by counting the number of suppressions. The distance between two contingency tables of dimension one returns twice the number of suppressions. This is because, when category A is suppressed for one record, two entries of the contingency table are changed: the count of records with category A decreases and the count of records with the 'missing' category increases.

Entropy-based measures

In de Waal and Willenborg (1999), Kooiman *et al.* (1998) and Willenborg and de Waal (2001), the use of Shannon's entropy to measure information loss is discussed for the following microdata masking methods (described in Sections 3.6 and 3.7): local suppression, global recoding and PRAM. Entropy is an information-theoretic measure, but it can be used in SDC if the protection process is modelled as the noise that would be added to the original data set in the event of it being transmitted over a noisy channel.

PRAM, described in Section 3.7.8, is a method that generalises noise addition, suppression and recoding methods. Therefore, our description of the use of entropy will be limited to PRAM.

PRAM operates by randomly changing the values of a categorical variable according to a prescribed Markov matrix. Let V be a variable in the original data set and V' be the corresponding variable in the PRAM-protected data set. Let $\mathbf{P}_V$ be the PRAM Markov matrix with entries $\{p_{ij}\} = \{P(V' = j | V = i)\}$.

The conditional uncertainty of V given that $V' = j$ is:

$$H(V | V' = j) = -\sum_{i=1}^{n} P(V = i | V' = j) \log P(V = i | V' = j), \qquad (3.31)$$

where n is the number of scores variable V can attain. The probabilities in Expression (3.31) can be derived from $\mathbf{P}_V$ using Bayes's formula. Finally, the entropy-based information loss measure *EBIL* is obtained by accumulating Expression (3.31) for all individuals r in the protected data set X'

$$EBIL(\mathbf{P}_V, X') = \sum_{r \in X'} H(V | V' = j_r),$$

where j_r is the value taken by V' in record r.

The above measure can be generalised for multivariate data sets if V and V' are taken as being multidimensional variables (i.e. representing several one-dimensional variables).

While using entropy to measure information loss is attractive from a theoretical point of view, its interpretation in terms of data utility loss is much less obvious than for the previously discussed measures.

3.10   Release of multiple files from the same microdata set

The instances of the production of multiple files from the same data set are growing very fast as international institutes or EU or world based projects urge the need to develop customised files that could be compared at international level: recent examples are the Generation and gender project (http://www.unece.org/pau/ggp/Welcome.html) or the IPUMS project (https://international.ipums.org/international/). The problem encountered in such situation is a simple one: the file required by international institutions is generally not a problematic one in itself, but it might differ for some classifications from other files already released at national or EU level. For example, nowadays an international organisation could require for a certain survey a level of geography not extremely detailed but, at the same time, it would need indications on the socio-demographic characteristics of the municipality. Such requirements could then be in contrast with previously released files with more detailed geography where information on the size of the town or its rural/urban nature were not present. This type of problem is the microdata counterpart of the linked tables problem and, like for the latter, an optimal solution can be found only when the different data to be released are anonymised at the same time. Therefore, to be optimal, the anonymisation of different types of microdata files should be planned at the same time. At a national level, the multiple release problem has already been encountered as the production of different files for different users is becoming more common nowadays. However, despite the need of data anonymisation procedures targeted to the different data users, the problem of releasing different files is still at an embryonic stage: see Trottini *et al.* (2006) for a dissemination strategy proposal for the household expenditure survey in Italy, Abowd and Lane (2004) for an implementation proposal and Ichim and Franconi (2010) for a proposal to treat multiple releases in different dimensions. What is most commonly applied in most agencies adopting a dual dissemination (PUF and MFR) is the mere adoption of more aggregated classifications for the categorical variables and various forms of top and bottom coding as well as the introduction of bands for the continuous variables when developing the PUF. This causes the required drastic decrease of the risk of disclosure but presents, as a side effect, a severe drop in the information content of the microdata file. However, the free availability of the PUF should not be the synonym of the production of files being of very limited interest and analytical validity for the final users. Targeted utility-based perturbation methods or, more recently, synthetic data generation methods can be used to release perturbed data that still present interesting level of information content. Certainly PUF and MFR must be hierarchically designed in terms of information content (see Trottini *et al.* 2006). This means that all the information in the PUF should also be contained in the corresponding MFR. The hierarchical structure of the two data sets greatly simplifies assessment of the disclosure risk and information loss associated with the anonymisation procedure. Because of the hierarchy, in fact, there is no gain for a user having access to the MFR, to access the PUF. The hierarchy requires coherence in the choice of the variables to be included in both files and on the corresponding levels of detail. The inclusion of a variable in the PUF implies its inclusion in the MFR; non-nested classifications for the same variable should not be allowed, and so on. The proposal by Casciano *et al.* (2011) shows further steps towards a real integrated approach to the issue of releasing PUF and MFR from the same survey.

3.11   Software

3.11.1 μ-ARGUS

The μ-ARGUS software (see Hundepool *et al.* 2008) has been developed to facilitate to statisticians, mainly those in NSIs, to apply the SDC methods described earlier to create safe microdata files. It is a tool to apply the SDC methodology, not a black-box that will create a safe file without knowing the background of the SDC methodology. The development of μ-ARGUS has started at Statistics Netherlands by implementing the Dutch methods and rules (see also Section 3.12.1). With this software as a starting point, many other methods have been added. Several of these methods have been developed and/or actually implemented during the CASC project.

In this section, we will give a short overview of μ-ARGUS. For a full description, we refer to the extensive manual (see Hundepool *et al.* 2008).

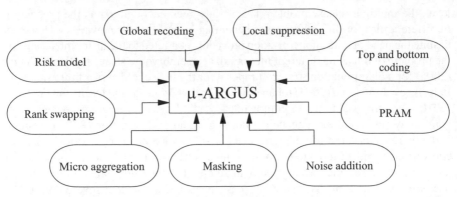

Figure 3.4   Overview of μ-ARGUS.

The starting point of μ-ARGUS has been implementation of the threshold rules for identifying unsafe records and procedures for global recoding and local suppression. μ-ARGUS can both protect fixed and free format ASCII files. A recent extension has been to support the SPSS file format.

Many of the methods described previously can be applied to a data set using μ-ARGUS. Figure 3.4 gives an overview of the currently implemented methods. It is planned to include other methods as well in the near future, if time permits.

μ-ARGUS is a flexible interactive program that will guide the user through the process of data protection. In a typical μ-ARGUS run the user will go through the following steps, given that the microdata set is available:

1. Metadata: μ-ARGUS needs to know the structure of the data set. Not only the general aspects but also additional SDC-specific information. As there was no suitable flexible standard for metadata allowing specification of the SDC-specific parts of the metadata, one relies on μ-ARGUS own metadata format. This can be prepared (partially) externally or it can be specified interactively during a μ-ARGUS session.

2. Threshold-rule/risk models: selection and computation of frequency tables on which several SDC methods (like risk models, threshold rule) are based.

3. Global recoding: selection of possible recodings and inspection of the results.

4. Selection and application of other protection methods like:

   (a) Microaggregation

   (b) PRAM

   (c) Rounding

   (d) Top and bottom coding

   (e) Rank swapping

   (f) Noise addition

5. Risk model: selection of the risk level.

6. Generation of the safe microdata file: during this process all data transformations specified above will be applied to the actual microdata. This is also the moment that all remaining unsafe combinations will be protected by local suppressions. Also an extensive report will be generated.

When the above scheme has been followed a safe microdata file has been generated. μ-ARGUS is capable of handling very large data sets. Only during the first phase, when the data file is explored and the frequency tables are computed, some heavy computations are performed. This might take some time depending on the size of the data file. However, all the real SDC work (global recoding and the other methods named under 4 and 5 above) are done at the level of the information prepared during this first phase. This will be done very quickly. In the final phase, when the protected data file is made, the operation might be time consuming depending on the size of the data file. This architecture of μ-ARGUS has the advantage that all real SDC work, that will be done interactively, will have a very quick response time. Inspecting the results of various recodings is easy and simple.

3.11.2   sdcMicro

sdcMicro (see Templ 2008; Templ and Petelin 2009), is a free and open software for the generation of anonymised microdata written in the **R** language. **R** is a highly extendable system for statistical computing and graphics, distributed over the net. *sdcMicro* contains almost all popular methods for the anonymisation of both categorical and continuous variables. Furthermore, several new methods, some of them not present in μ-ARGUS, have been implemented. The package can also be used for the comparison of methods and for measuring the information loss and disclosure risk of the masked data.

sdcMicro can be downloaded from the Comprehensive **R** Archive Network (see CRAN 2011).

sdcMicro architecture

sdcMicro exploits the object orientation of the **R** language. In **R** everything is an object and every object belongs to a specific class. The class of an object determines how it will be treated, and generic functions perform either a task or an action on its arguments according to the class of the argument itself. The class mechanism offers to the programmers a facility for designing and writing special-purpose functions which is extensively used in *sdcMicro*. Nearly all functions, e.g. the one for the individual risk methodology or the frequency calculation, produce objects from certain classes. Different *print*, *summary* and *plot* methods are provided for each of these objects depending on their class. `plot(ir1)` produces a completely different result than `plot(fc1)` assuming that the objects `ir1` and `fc1` are objects from different classes, i.e. resulting from different functions in package *sdcMicro*.

This object-oriented approach allows simple use of the package to any user, independently of her proficiency in **R**. Furthermore, users can try out methods with several different parameters and they can easily compare the methods with the implemented summary and plot methods.

No real metadata management needs to be done by the user, not even after importing the data into **R**. The methods can be directly applied by the user on her data sets or on objects from certain classes. The only metadata task that has to be done is to determine which of the variables should be considered as the key variables, weight vector and confidential numerical variables.

An online documentation is included in the package containing all explanations on all input and output parameters of every function. Furthermore, various examples are included for each of the functions. All the examples can be easily executed by the users. Since *sdcMicro* is checked automatically every day from the CRAN server, it is guaranteed that these examples work. Furthermore, additional checks are executed, e.g. the consistency of the documentation and the functions can be guaranteed.

To be able to deal with large data sets, intensive calculation steps are implemented in C++ and included in **R** via the **R**/C++ interface. The estimation of the frequency counts, for the population is an example of such intensive calculation steps. When using loops in **R** for the calculation of the frequency counts, the computation time is too long. Only when using functions in **R** which are implemented in C or Fortran is it possible to calculate the frequency counts with up to six key variables in an adequate time. The developed **R**/C++ interface can also handle a large number of key variables in a reasonable time, even when doing the calculation on much larger data sets than the CASC reference data sets (see Brand *et al*. 2002).

Implemented methods

After reading the data (in a certain format) into **R**, the user is able to protect the categorical key variables and/or protect the continuous (numerical) variables. In addition to that, several functions are provided for the generation of synthetic data and for the measurement of the disclosure risk and the data utility for continuous data.

SDC methods for continuous variables
The anonymisation of continuous variables is often necessary to avoid a successful attack by record linkage methods. Many different methods are implemented in *sdcMicro* for protecting continuous variables. A non-exhaustive list is:

- More than 10 methods for microaggregation, including multivariate microaggregation (MDAV) and individual ranking microaggregation (microaggregating each variable separately).

- Five methods for adding noise, including simple additive noise and correlated noise.

- Rank swapping.

- Synthetic data generation methods.

Methods for masking categorical variables

Function `freqCalc()` calculates the frequency count for each observation as well as the frequency counts with respect to the sampling weights. Function `freqCalc()` returns an object of class *freqCalc*. For objects of this class, a print and a summary method is provided. Objects of class *freqCalc* can be used as input parameters to the functions `globalRecode()` and `indivRisk().globalRecode()` provides some functionality for recoding variables. `indivRisk()` estimates the individual risk for re-identification as implemented in μ-ARGUS and produces objects of class *indivRisk* for which a plot and a print method are implemented. Concerning the risk calculation, the uncertainty on the frequency counts of the population is accounted for in a Bayesian fashion by assuming that the distribution of the population frequency given the sample frequency is negative binomial distributed.

The objects of class *indivRisk* can be used as an input to the implemented local suppression functions. The local suppression function `localSupp2()` searches automatically for a quasi-optimal solution to find a minimum amount of suppression. Additionally, the user may set weights for each variable to indicate which variables are less critical for suppression or which ones are more critical (the weight can be set to 0 for those variables which should not suffer any suppressions and set to a low value for those which should suffer only a few suppressions). `localSupp2wrapper()` ensures *k*-anonymity and a low risk of re-identification.

The functions `freqCalc()`, `indivRisk()` and `globalRecode()` are recommended to be used in an exploratory and sometimes in an iterative way. It takes just a look to find out how many suppressions are necessary to achieve *k*-anonymity and/or low individual risk for different global recoding settings. In addition to that, one can simply reproduce any previous step of the anonymisation. So, one can try out different global recodings of various variables and evaluate their impact on information loss. Moreover, the frequency count calculation procedure can deal with missing values in the data.

Also, function `pram()` is available which implements the PRAM method and produces objects of class *pram*. A print method and a summary method are provided for objects of this class.

3.11.3   *IVEware*

IVEware (IVEware 2011) is a package for MI (see Section 3.8.1) developed by researchers at the Survey Methodology Program, Survey Research Center, Institute for Social Research and University of Michigan. It performs:

- Single or MIs of missing values using the sequential regression MI method (SRMI) described in Raghunathan *et al.* (2001) and in Section 3.8.1.

- A variety of descriptive and model-based analyses accounting for complex design features like clustering, stratification and weighting.

- MI analyses for both descriptive and model-based survey statistics.

- Creation of partially or fully synthetic data sets using SRMI to protect confidentiality and limit statistical disclosure.

- Combination of information from multiple sources by vertical concatenation of data sets and MI of the missing portions to create a larger rectangular data set.

The above functions are performed by the six modules of IVEware: IMPUTE, DESCRIBE, REGRESS, SASMOD, SYNTHESIZE and COMBINE. Specifically, *SYNTHESIZE is the module that implements SRMI for statistical disclosure control;* see Reiter (2002), Raghunathan *et al.* (2003) and Little *et al.* (2004) for details.

3.12   Case studies

In this section, we provide examples based on real surveys to describe a possible process of microdata anonymisation in practice or rules applied by statistical agencies to release their microdata. The surveys analysed are one from the social domain, the European Labour Force Survey (LFS), and one on business data, the European Structure of Earnings Survey (SES). Additionally, the PUF resulting from a mortality follow-up study at the American National Center for Health Statistics (NCHS) is discussed. Notice how the complexity of the reasoning on business data may rise sharply as compared to social microdata.

3.12.1   Microdata files at Statistics Netherlands

As has been shown in previous sections, there are many sophisticated ways of making a safe protected microdata set. And it is far from a simple straightforward task to select the most appropriate method for the disclosure protection of a microdata set. This requires a solid knowledge of the survey in question as well as a good overview of all the methods described in the previous sections.

However, as an introduction, we will describe here the method/set of rules currently applied at Statistics Netherlands for making both microdata files for researchers as well as PUFs. This approach can be easily applied, as it is readily available in μ-ARGUS (see Hundepool *et al.* 2008). These rules are based on the ARGUS threshold rule (see Section 3.5.1) in combination with global recoding and local suppression (see Sections 3.6.2 and 3.6.4). This rule only concentrates on the identifying variables or key variables, as these are the starting point for an intrusion. The rules have primarily been developed for microdata about persons.

Microdata for researchers
For the microdata for researchers Statistics Netherlands uses the following set of rules:

1. Direct identifiers should not be released and, therefore, should be removed from the microdata set.

2. The quasi identifiers are sub-divided into extremely identifying variables, very identifying variables and identifying variables. Only direct regional variables are considered to be extremely identifying. Very identifying variables are very visible variables like gender, ethnicity, etc. Each combination of values of an extremely identifying variable, a very identifying variable and an identifying variable should occur at least 100 times in the population.

3. The maximum level of detail for occupation, firm and level of education is determined by the most detailed direct regional variable. This rule does not replace rule 2, but is instead a practical extension of that rule.

4. A region that can be distinguished in the microdata should contain at least 10 000 inhabitants.

5. If the microdata concern panel data, direct regional data should not be released. This rule prevents the disclosure of individual information by using the panel character of the microdata.

If these rules are violated, global recoding and local suppression are applied to achieve a safe file. Both global recoding and local suppression lead to information loss, because either less detailed information is provided or some information is not given at all. A balance between global recoding and local suppression should always be found in order to make the information loss due to the statistical disclosure control measures as low as possible. It is recommended to start by recoding some variables globally until the number of unsafe combinations that has to be protected is sufficiently low. Then the remaining unsafe combinations have to be protected by local suppressions.

For business microdata, these rules are not appropriate. Opposed to personal microdata, business data tend to be much more skewed. Each business is much more visible in a microdata set. This makes it very hard to make a safe business microdata set.

Public use files

The software package μ-ARGUS (see Hundepool *et al*. 2008) is also of help in producing public use microdata files. For the public use microdata files, Statistics Netherlands uses the following set of rules:

1. The microdata must be at least 1 year old before they may be released.

2. Direct identifiers should not be released. Also direct regional variables, nationality, country of birth and ethnicity should not be released.

3. Only one kind of indirect regional variables (e.g. the size class of the place of residence) may be released. The combinations of values of the indirect regional variables should be sufficiently scattered, i.e. each area that can be distinguished should contain at least 200 000 persons in the target population and, moreover, should consist of municipalities from at least six of the 12 provinces in the Netherlands. The number of inhabitants of a municipality in

an area that can be distinguished should be less than 50% of the total number of inhabitants in that area.

4. The number of identifying variables in the microdata is at most 15.

5. Sensitive variables should not be released.

6. It should be impossible to derive additional identifying information from the sampling weights.

7. At least 200 000 persons in the population should score on each value of an identifying variable.

8. At least 1000 persons in the population should score on each value of the crossing of two identifying variables.

9. For each household from which more than one person participated in the survey we demand that the total number of households that correspond to any particular combination of values of household variables is at least five in the microdata.

10. The records of the microdata should be released in random order.

According to this set of rules, the PUFs are protected much more severely than the microdata for research. Note that for the microdata for research, it is necessary to check certain trivariate combinations of values of identifying variables and for the PUFs, it is sufficient to check bivariate combinations, but the thresholds are much higher. However, for PUFs, it is not allowed to release direct regional variables. When no direct regional variable is released in a microdata set for research, then only some bivariate combinations of values of identifying variables should be checked according to the SDC rules. For the corresponding PUFs, all the bivariate combinations of values of identifying variables should be checked.

3.12.2   The European Labour Force Survey microdata for research purposes

The European Union Labour Force Survey (EU LFS) is conducted in the 27 Member States of the European Union, 3 candidate countries and 3 countries of the European Free Trade Association (EFTA) in accordance with Council Regulation (EC) No. 577/98. The EU LFS is a large household sample survey providing quarterly results on labour participation of people aged 15 and over as well as on persons outside the labour force.

The surveys are conducted by the national statistical institutes across Europe and are centrally processed by Eurostat using the same concepts, definitions and common classifications (NACE, ISCO, ISCED and NUTS), following International Labour Organisation guidelines, recording the same set of characteristics in each country. The quality reports annually delivered by Eurostat (see Eurostat 2011c) provide an overview of the complexity of a survey carried out in more than 30 countries with differences in population sizes, sampling design, stratification, rotation schemes and methods to calculate weights.

In 2011, the quarterly LFS sample size across the EU was about 1.5 million individuals. More information on the survey can be found at the Eurostat web page (see Eurostat 2011b).

Now we go through the SDC process outlined in Section 3.2.

Action 1.1 *Decision on the need for confidentiality protection for releasing the file.*

The survey collects information on individuals inside households. Some of the main observed variables are:

1. *Demographic variables.* Gender, year of birth, marital status, relationship to reference person and place of residence.

2. *Labour status.* Labour status during the reference week, etc.

3. *Employment characteristics of the main job.* Professional status, economic activity of local unit, country of place of work, etc.

4. Education and training.

5. Income.

6. Technical item relating to the interview.

This is a classical example of social survey on individuals in households where confidentiality protection is needed.

Action 2.1 *Identification of data structure.*

Many countries deliver data on households with information on each individual belonging to them. So a hierarchical structure is present in these files. This should be addressed by the risk-assessment method as well as by the statistical disclosure limitation procedure.

Action 2.2 *Priorities from the users' point of view.*

In the quality reports (see Eurostat 2011c) a brief analysis of users' needs is presented. Apart from the obvious variables related to work, gender and education, the regional detail seems important as well as the indication of the employer's economic activity (according to the NACE classification). The International Standard Classification of Occupations (ISCO) is mandatory for Member States at level three, while the most detailed level of the geographical classification NUTS is again three. Academic users would need detailed information on education and geography and well as country of birth.

Action 2.3 *Decision on types of release.*

As for the current situation only MFR purposes are implemented at EU level. PUF, defined in the Reg. (EC) 223/2009 on European statistics, are still under study.

As the microdata to be released is for scientific research purposes, the quality of the released data is extremely important. Also in some EU countries, statistical agencies are not allowed to change the values of the answers provided by respondents.

To this end, EU countries decided not to perturb the microdata and therefore only global recoding of quasi-identifiers was carried out together with, in some cases, sub-sampling as a way to reduce risk.

Action 2.4 *Constraints from the data producer side.*
List of aggregates published through different channels to be maintained.

Action 3.1 *Definitions of disclosure scenarios (i.e. how a breach of confidentiality could be carried out) and disclosure risk.*
A possible scenario could be spontaneous identification (see Section 3.3.1), where an intruder has a direct knowledge of some statistical units belonging to the sample and whether such units assume extreme values for some variables or for some combinations. In the Labour Force data set there are several variables that may lead to a spontaneous identification. Some of these variables are: professional status, number of persons working at local unit, income, economic activity, etc. To avoid a possible spontaneous identification such variables are usually checked to see whether there are unusual patterns or very rare keys and, if necessary, some recoding may be suggested. Also, the external register scenario could be considered for the LFS data. The two external archives taken as example in the Italian study are: the Electoral roll for the individual archive scenario, and the population register for the household archive scenario. The electoral roll is a register containing information on people having electoral rights. It contains mainly demographic variables (gender, age, place of residence and birth) and variables such as, marital status, professional and/or educational information. The quasi-identifiers considered as reliable for re-identification attempts under this scenario are: gender, age and region of residence. Exact place of birth is removed from the microdata file. Other variables have not been considered as their quality was not deemed sufficient for re-identification purposes. The population register is a public register containing demographic information at individual and household level. Particularly, the set of quasi-identifiers considered for the household archive scenario comprises the variables gender, age, region of residence and marital status as individual information, and the household size and parental relationship as household information.

Action 3.2 *Identification of methods to estimate or measure disclosure risk.*
Different countries decided to apply different risk-assessment methods. In countries where a complex multi-staged stratified random sample design was carried out with calibration procedures at detailed level of the NUTS classification, the individual risk described in Section 3.5.2 might be used, especially when hierarchical information on the household needs to be released.

Action 4.1 *Identification of disclosure limitation methods to be applied, choice of parameters involved, etc.*
The decision to avoid perturbation methods to protect the data from possible disclosure implies the use of non-perturbative methods (see Section 3.6) such as global recoding, local suppression, top and bottom coding and sampling.

The use of local suppression for the continuous variables makes it impossible for the user to replicate some already published tables. As the users are researchers, this is deemed neither acceptable nor useful. So global recoding is mostly used. When individuals at risk are concentrated on specific categories of a quasi-identifier then global recoding for such specific categories is applied. This is the case, for example, of the categories *widowed*, *divorced* or *separated* that are combined in a single category on variable *marital status*. Or, for the variable professional status, the categories *self-employed with employees* and *without employees* are combined in a single category.

As the variable occupation is deemed crucial, the ISCO is delivered at the maximum level of detail: level three (ISCO4D is not delivered at all).

To protect the variable income, income deciles are introduced.

Eight countries (Bulgaria, Germany, Spain, France, the Netherlands, Finland, the United Kingdom and Norway) use a sub-sample to survey all or some of the 39 structural variables of the survey. The sub-sample coincides in general with one rotation panel in total sample, except for Germany, where the sub-sample is about a randomly selected 10% of the total yearly sample. These countries release the sub-sample.

3.12.3   The European Structure of Earnings Survey microdata for research purposes

The European Union SES is conducted every fourth year in the 27 Member States of the European Union and 2 countries of the European Free Trade Association (EFTA); as of this writing, the latest available wave was carried out in 2006.

The objective of the survey is to provide accurate and harmonised data on earnings for policy-making and research purposes. The SES is a large enterprise sample survey providing detailed and comparable information on relationships between the level of remuneration, individual characteristics of employees (sex, age, occupation, length of service, highest educational level attained, etc.) and their employer (economic activity, size and location of the enterprise).

A tailored questionnaire is carried out, although in some countries existing surveys, administrative sources or a combination of such sources are used. A sample of employees is drawn from a stratified sample of local units/enterprises. Generally, a two-stage sampling design is applied. In a first stage, a stratified sample of local units/enterprises is selected. In a second stage, a systematic sample of employees is selected. Large enterprises are included in the sample with probability one.

By means of calibration procedures, the total number of enterprises and employees are generally maintained. The estimation domains are derived from the economic activity, the territoriality and size of enterprises expressed in terms of number of employees.

The main information about the local unit/enterprise to which the sampled employees are attached refers to the geographical location (*region*), the size of the enterprise to which the local unit belongs (*size*) and the principal economic activity of the local unit (*NACE*).

The main information collected on individual characteristics of each employee in the sample relates to gender, age, occupation, level of education and training, contractual working time, etc. Also, for each employee in the sample, information on working periods (e.g. number of hours paid during the reference month or number of overtime hours paid in the reference month) is collected, as well as information on earnings such as gross annual and monthly earnings, annual bonuses, earnings related to overtime and average gross hourly earnings.

Now, we go through the first four stages of the SDC process outlined in Section 3.2. We outline different approaches adopted by different countries.

Action 1.1 *Decision on the need for confidentiality protection for releasing the file.*

As individuals as well as enterprises are involved in the survey, confidentiality protection is needed for both types of units.

Action 2.1 *Identification of data structure.*

The SES survey has a clear hierarchical structure enterprise/local unit vs employee that needs to be maintained. In coherence to such hierarchical structure, both levels (enterprise/local unit and employee) need to be addressed by the SDC process. Two disclosure scenarios need to be adopted, taking into account the particular observed phenomenon: one for enterprise re-identification and one for employee re-identification. Consequently, risk definition and assessment as well as SDC procedures need to be developed for each of these types of units.

The variable identifying the enterprise is removed. The exact number of employees in the local unit is removed. The variable *age* is recoded in classes: 14–19, 20–29, 30–39, 40–49, 50–59 and 60+.

Action 2.2 *Priorities from the users' point of view. This includes priorities for variables to be included, types of statistical analyses on data, types of release for microdata.*

The SES is a classical example of Linked Employer-Employee data set. The users most interested in the microdata are researchers. Information on structural variables for the enterprises (NACE, region and size) is necessary for any analysis. Demographic variables for the employees, such as age and gender, are essential as well as level of education. Finally, all the variables related to earnings, the core of the survey, are of paramount importance.

The statistical analyses carried out on the data include gender pay gap, statistical models, etc; an overview of the type of analyses carried out on the SES data is available as part of the ESSnet project on Common tools and harmonised methodology for SDC in the European Statistical System (see the bibliographic reference ESSnet on Common tools and harmonised methodology for SDC in the ESS 2012).

In order to maintain the gender pay gap of the original microdata, the anonymised microdata should maintain the difference between average gross earnings of male paid employees and of female paid employees divided by the earnings of male paid employees. The gender pay gap is usually estimated at domain level like economic branch (1-digit NACE code), education and age groups.

Possible domains that should be maintained to improve the utility of the released data are the mean earnings according to several breakdowns, e.g. by NACE 1-digit level and gender, by full or part-time employment and gender, by age groups and gender, by working duration and age groups and by region.

Action 2.3 *Decision on types of release.*

As researchers are the main users for such data, files for research purposes are implemented at EU level.

Action 2.4 *Constraints from the data producer side.*

European publications are released at quite aggregated a level: national level, 1-digit NACE and size class, 1-digit ISCO. For transparency reasons, the disclosure limitation methodology applied to the microdata should maintain weighted totals in the above mentioned domain for the published variables. Some countries take into consideration more detailed levels of the variables.

Action 3.1 *Definitions of disclosure scenarios (i.e. how a breach of confidentiality could be carried out) and disclosure risk.*

We now analyse disclosure scenarios and disclosure risk for each type of units.

Disclosure scenario: enterprises

SES microdata contain enterprise variables related to economic activity, region and size. These are structural variables generally publicly available that could be used to identify an enterprise. Therefore, these are the quasi-identifiers for the unit enterprise. As a business register was used as a sampling frame, if an enterprise is present in the sample, it is also included in the business register.

As the typical user of the file is a researcher, it may be assumed that he would not perform a complete record linkage for re-identification purposes. Nonetheless, the researcher may be particularly interested in some units, maybe important or visible. For example, he might know in advance the most famous/dominant enterprises. Alternatively, some units may be highlighted during the analysis because of their particular score. In other words, it may be supposed that structural information might be used for the re-identification of particular enterprises. Public business registers report general information on name, address, number of employees (size), principal economic activity of an enterprise (NACE classification), region (NUTS), etc. Among these variables, the ones that are present in the 2006 SES are the principal economic activity (NACE), the region (NUTS) and the size of the enterprise (size). Hence, NACE code, region and size of the enterprise are the quasi-identifiers. A more detailed description of the disclosure scenario can be found in Ichim and Franconi (2007). All the quasi-identifiers chosen according to the scenario are categorical.

Disclosure risk: enterprises

As discussed in Section 3.4.1, various approaches can be followed to define whether a unit, in this case an enterprise, is at risk of disclosure. Here, we mention three different strategies applied by different countries: Strategy A is used by the Italian sample of the SES (see Ichim and Franconi 2007), Strategy B is the one proposed by Eurostat (see Eurostat 2010), and Strategy C is used for the German sample (see Hafner and Lenz 2006).

Strategy A: Risk based on the population

If information is available, the disclosure risk can be based on the population. In the case of SES, the statistical agency has the statistical business register that can be assimilated to the population of interest and that contains the quasi-identifiers. With respect to the assumed disclosure scenario, the enterprises at risk are the sampled enterprises showing a combination of quasi-identifiers that has a frequency in the population below a given threshold. Generally, the threshold value is equal to 3. Hence, an enterprise/local unit in the sample is considered at risk when it presents a combination of quasi-identifiers whose frequency in the population is 1 or 2. Again more information can be found in Ichim and Franconi (2007).

Strategy B: Risk based on the sample

A strategy based on information available on the sole sample has been adopted by Eurostat (2010). A threshold on the percentage of records having a unique combination of quasi-identifiers is proposed to define the enterprises/local units at risk. In particular, if an enterprise shows a combination of quasi-identifiers for which less than three enterprises exist in the sample then the combination is defined as sensitive. According to the proposed strategy, a sample of a certain country is considered at risk of disclosure if the ratio between the number of sensitive combinations NACE-NUTS-SIZE and the total number of combinations in the SES is higher than the threshold of 10%.

Strategy C: Risk based on record linkage

In Hafner and Lenz (2006) the risk of disclosure for the enterprises present in the German sample of the SES are defined by means of matching experiments. In Germany the Markus database is the largest commercial register that includes information on, among others, name, address, legal form, branch of economic activity, number of employees, etc. According to this strategy an enterprise is at risk of disclosure if a successful match can be carried out between the record relative to an enterprise in the SES microdata and the information in the Markus database.

Disclosure scenario: employees

Strategy A: Spontaneous identification scenario

With regard to which variable the identification of employees could be attempted, several considerations hold concerning availability of variables. As the file is released to researchers for research purposes a *nosy colleague* scenario is not deemed possible. Therefore, variables such as paid hours or absence days, that could be known only by a nosy colleague, are not considered in the scenario as quasi-identifiers, and therefore not perturbed. Moreover, the identification of an employee with respect to his/her number of absence days or number of working days would be quite difficult. Therefore, it is considered that identification of employees would be possible only by means of some previous 'ideas' on ranges of earnings variables, especially visible high-range salaries. In the SES microdata, there are two such variables: (1) Monthly Earnings and (2) Annual Earnings. Considering that the microdata file to be released is intended for research purposes, the annual earnings is more adequate for usage for identification purposes. This choice is mainly due to the fact that annual earnings generally include the 'management' bonuses based on productivity results.

Moreover, variables on earnings if combined only with the observed demographic variables (Gender, Age) cannot be used for employee identification, as such demographic variables are not at all discriminating (high frequencies). In other words, it is believed that an intruder cannot identify an employee only by means of her or his gender, age, number of paid hours and earnings variables.

On the other hand, enterprise information could be used to identify an employee. In this approach, the hypothesis is that variables related to information on the enterprise can be combined with variables Gender and Age to identify an employee. About the information content of the identification, it is assumed that an intruder could be interested only in extremely high earnings. 'Small-medium' earnings are not judged at risk also because many occupational categories are subject to some kind of national contract, at least as a common base. It is supposed that only high earnings corresponding to large (and generally well-known) enterprises could present some interest from an intruder's point of view. Given the structure of the economy, it is believed that small-medium size enterprises are hardly identifiable because their number is far too high. In conclusion, the spontaneous identification scenario assumes for the SES that the employee identification is possible by means of the following quasi-identifiers:

* Information on the enterprise (NACE x Nuts x Size).

* Demographic variables (Gender x Age (in classes)).

* Extremely high earnings related to large enterprises, i.e. 250 or more employees.

According to this strategy, the only variable that will be perturbed is the Annual Earnings; the variable Monthly Earnings is modified in order to maintain the original relationship between such variables. Again more information on this approach can be found in Ichim and Franconi (2007).

Strategy B: All records at risk
The strategy adopted by Eurostat (2010) is to consider all records possibly at risk. All the continuous variables will undergo some form of perturbation in order to avoid employee identification.

Disclosure risk: employees

Strategy A: Selective risk
According to the disclosure scenario described in Strategy A, only a small selection of employees, i.e. those working for large enterprises and showing high salaries are at risk of disclosure when the file is released to researchers.

Strategy B: No proper risk assessment
According to this strategy, no proper risk assessment is carried out as all records are considered at risk; as a consequence all records from the employees will be subject to a SDC method.

Action 3.2 *Identification of methods to estimate or measure disclosure risk.*

Estimate disclosure risk: enterprises

Strategy A: Risk based on the population
In order to estimate the risk according to this strategy, it is sufficient to evaluate the frequency of the combinations of NACE, region (NUTS) and size in the register of the statistical agency. Whenever both sample and register frequencies for a certain combination are equal to one or two, then the combination is at risk, i.e. the enterprises showing such combination are at risk.

Strategy B: Risk based on the sample
If the ratio between the number of sensitive combinations NACE-NUTS-SIZE for a certain country and the total number of combinations in the SES is higher then the threshold of 10% the enterprises are at risk and SDC procedures need to be applied.

Strategy C: Risk based on record linkage
The records that are correctly matched are at risk.

Estimate disclosure risk: employees

Strategy A: Selective risk
First of all a definition of employees with high annual earnings is necessary. A threshold $T_{earnings}$ on the variable Annual Earnings if chosen. Records exceeding such threshold are considered as extremely high earnings, hence subject to spontaneous identification. The threshold value should be the same for all combinations of key variables. Taking into account the fact that the earnings probability distribution is very skewed, the value could be a high quintile such as for example, the 99% quantile of the distribution. Then, we search for employees with high earnings that present unique combinations on the quasi-identifiers: NACE, Region, Size (250 or more), Gender, Age (in classes) and the variable Annual Earnings of employees in classes. If an employee with such characteristics is present, then it is considered at risk of identification. Note that, in this way, the number of employees at risk of identification is not a priori defined. More details on this approach can be found in Ichim and Franconi (2007).

Strategy B: No proper risk assessment
No proper risk estimate/measure is carried out under this strategy.

Strategy C: Risk based on association
The strategy in Hafner and Lenz (2006) permits to measure dependencies between employer and employee data, to evaluate whether these dependencies have an impact

on the re-identification risk of the employer and, if necessary, to anonymise the data of the employees in such a manner that the confidentiality of the employer is guaranteed.

Action 4.1 *Identification of disclosure limitation methods to be applied, choice of parameters involved, etc.*

SDC methods: enterprise
As all identifying variables are categorical, a way to reduce the risk of identification consists in recoding too detailed variables by applying global recoding. The use of local suppression was not deemed useful as the structural variables for the employer are deemed crucial for any analysis.

SDC methods: employees

Strategy A: Model-based SDC
In this approach, the perturbation is applied only to those records which are deemed at risk of identification according to the definition given in Strategy A above. As protection should be applied taking into account also the possible usages of the microdata file, an imputation of the values at risk by mean of the corresponding fitted values of a constrained regression is adopted. The constraints imposed to the regression are to maintain published totals, gender pay gap, etc. The response variable of the regression must be the one for whom perturbation is required (e.g. annual earnings). If necessary, i.e. for goodness-of-fit reasons, suitable variable transformations could be investigated. For the Italian SES, Annual Earnings is modelled as a linear combination of Size, Gender, Age, Management Position, Occupation, FtPt, Length of service, montly Earnings and Paid Hours Monthly, respectively. Such a model was used for each combination of NACE and NUTS. Such a model can be modified and a robust version of it could be introduced. The importance of such a model is proportional to the number of records to which this model is applied, i.e. to the number of employees at risk for whom the Annual Earnings is substituted by the fitted values of the model.

The weighted least squares method is applied to estimate the parameters, for each combination of NACE and NUTS and Gender. The least squares weights are the sampling weights of the employees. More details as well as the **R** code to perform this approach are available at the ESSnet project mentioned above (see in the bibliography "ESSnet on Common tools and harmonised methodology for SDC in the ESS 2012"). The main drawback of the standard least squares method is that any data utility constraints are not guaranteed. Consequently, to ensure the coherence with the already published totals and gender pay gap, a constrained minimisation problem was solved, by using dummy variables to represent the categorical variables in the regression. The main constraint was given in terms of published totals: for each combination of NACE and NUTS, the relative difference between the original and perturbed value should not be higher that 50%, as required by the survey experts. Actually, this constraint was transposed in terms of weighted totals of units at risk of identification, since these are the only records to be modified. Each perturbed value can be restricted to belong to a pre-defined interval. These constraints actually control the perturbation introduced in each record.

For the employees considered at risk of identification, only the values of the two earnings variables (annual and monthly) involved in the disclosure scenario are perturbed. The values of the variable Annual Earnings of the employees at risk of identification are substituted by the corresponding fitted values. Then, the related Monthly Earnings value is proportionally modified. By leaving unchanged all the values corresponding to employees not at risk of identification, the information loss is limited.

Strategy B: Microaggregation
Microaggregation, in the form of individual ranking, is applied to all continuous variables. Various different constraints for microaggregation have been proposed. The crucial point here is to maintain consistency among variables: for example, the number of hours actually paid during the reference month should be consistent with the Monthly Earnings or the number of overtime hours paid in the reference month should be consistent with overtime earnings. So, careful implementation is needed to apply a generic method such as microaggregation.

The other important point is to maintain the chosen statistics. The approach to reach this aim can be either to constrain the microaggregation procedure in order to force the wanted results (constrained microaggregation by NACE, size and region) or to allow the classical individual ranking and then check a posteriori if the needed result was reached. If this is not the case, corrections need to be put in place.

Action 4.2 *Identification of methods to measure information loss.*
To test the utility and to demonstrate the comparability of the data as well as to ensure that information loss was limited, comparison of various statistics have been developed.

3.12.4   NHIS-linked mortality data public use file, USA

The National Center for Health Statistics (NCHS) completed a mortality follow-up study for the 1986–2000 National Health Interview Survey (NHIS) years through a probabilistic record linkage with the National Death Index (NDI). The NHIS is a cross-sectional household survey of US population on a broad range of health topics and socio-demographic information. NCHS allows access to the NHIS-linked mortality data through its Research Data Center (RDC); this file is called the restricted access file as many detailed information are available to researchers. In 2007, a PUF has been developed from the restricted access version; information on the survey and the PUF are available at National Center for Health Statistics (2011).

We now go through various stages of the SDC process described in Section 3.2.

Action 1.1 *Decision on the need for confidentiality protection for releasing the file.*
The variables gathered on individuals range from health to socio-demographic information. Therefore, confidentiality protection is necessary.

Action 2.1 *Identification of data structure.*
The restricted access files include mortality status, exact date of death (month, day, year), and underlying and contributing cause-of-death codes, as reported on the

death certificate. It was felt that such information was too identifying for release and, therefore, the PUFs include mortality status, approximate follow-up time (quarter and year of death), information on causes of death to a regrouped code and a variable indicating whether diabetes, hypertension or hip fractures were reported in the contributing cause-of-death codes; for some records, these information have been imputed. Also detailed informations on age at interview in years (not top-coded), date of birth (month, day, year) and age at death have been recoded.

Action 2.2 *Priorities from the users' point of view. This includes priorities for variables to be included, types of statistical analyses on data and types of release for microdata.*

The description of the study, Lochner *et al.* (2008), states that the aim of the PUF is to maximise the amount of information on mortality included. Lochner *et al.* (2008) also present an overview of types of analyses carried out in the RDC showing the point of view of the users on the microdata.

Action 2.3 *Decision on types of release.*

Having developed already a file for researchers in the RDC, NCHS decided to release a more open type of microdata: the PUF.

Action 2.4 *Constraints from the data producer side. This includes constraints derived from publication of the same data through different channels.*

Rather than considering the ability to reproduce published tables or other forms of release, NCHS focuses the attention on maintaining high utility in the data by checking that the results of the analyses carried out using the PUF were very similar to those obtained using the restricted access file.

Action 3.1 *Definitions of disclosure scenarios (i.e. how a breach of confidentiality could be carried out) and disclosure risk.*

The disclosure scenario for a PUF often relies on all available information on the web. As a consequence, all NHIS decedent 'unique' records which can be correctly matched to existing publicly available data sources by mean of socio-demographic variables and mortality information are defined at risk of being re-identified.

Action 3.2 *Identification of methods to estimate or measure disclosure risk.*

As the risk is defined by means of matching experiments, the way to 'measure' it is to actually carry out extensive record linkage experiments to identify the units at risk.

Action 4.1 *Identification of disclosure limitation methods to be applied, choice of parameters involved, etc.*

All cases considered at risk were subject to data perturbation and were randomly assigned to have either date of death or underlying cause-of-death perturbed. To further reduce re-identification risk, an additional random sample of decedents was subjected to perturbation. Information regarding vital status was not perturbed.

Cases requiring date of death perturbation had either the quarter or year randomly perturbed, and in some cases both fields were perturbed. For those cases requiring underlying-cause-of-death perturbation, a hot-deck method was implemented and imputed the 113 grouped underlying cause-of-death recodings by replacing the original value with a value from a decedent with similar characteristics. The perturbed cases are not identified on the PUFs.

Action 4.2 *Identification of methods to measure information loss.*

To test the utility and to demonstrate the comparability of the data as well as to ensure that information loss was limited, analyses conducted on the restricted-use files were replicated on the PUFs. The details of the comparative analysis as well as the results can be found in Lochner *et al.* (2008).

3.12.5   Other real case instances

Other worked examples of the process that have been brought to the decision of releasing microdata can be found in the following:

- Franconi and Ichim (2007) on the development of a MFR for the European Community Innovation Survey, a business survey devoted to the investigation of innovation in enterprises.

- Franconi *et al.* (2011) on issues of the creation of a MFR for the European Farm Structure Survey.

- Heldal (2011) on an attempt to release event history data for researchers.

4

Magnitude tabular data

4.1 Introduction

In this section, we will introduce the main concepts and terminology of 'tabular data' in Section 4.1.1 and the more complex case of hierarchical and linked tables in Section 4.1.2.

The remaining subsections of the introduction are meant as a quick tour through the main contents of the current chapter and as a reference to its respective sections. We mention the main disclosure risk concepts, protection methods and information loss concepts. After introducing these basic concepts, we address more practical issues, like software and how to use it efficiently.

4.1.1 Magnitude tabular data: Basic terminology

Statistical magnitude tables display sums of observations of a quantitative variable where each sum relates to a group of observations defined by categorical variables observed for a set of respondents.

Respondents are typically companies but can also be individuals or households, etc. The categorical *grouping variables* (also referred to as *classification* or *spanning* variables – we use those terms synonymously in this chapter) typically give information on geography or economic activity or size, etc. of the respondents. The *cells* of a table are defined by cross-combinations of the grouping variables.

Each *table cell* presents a sum of a quantitative *response* variable such as income, turnover, expenditure, sales, number of employees, number of animals owned by farms, etc. These sums are the *cell values* (sometimes also referred to as *cell totals*) of a magnitude table. The individual observations of the variable (for each individual respondent) are the *contributions* to the cell value.

Statistical Disclosure Control, First Edition. Anco Hundepool, Josep Domingo-Ferrer,
Luisa Franconi, Sarah Giessing, Eric Schulte Nordholt, Keith Spicer and Peter-Paul de Wolf.
© 2012 John Wiley & Sons, Ltd. Published 2012 by John Wiley & Sons, Ltd.

Table 4.1   First example.

	Industry A	Industry B	...	Total
Region 1 turnover	540	231	...	
(respondents)	(12)	(15)		
Region 2 turnover	48	125	...	
(respondents)	(2)	(8)		
...	...	...	...	
Total turnover				
(total respondents)				

The *dimension* of a table is given by the number of grouping variables used to specify the table. We say that a table contains *margins* or *marginal cells*, if not all cells of a table are specified by the same number of grouping variables. The smaller the number of grouping variables, the higher the *level* of a marginal cell. A two-dimensional table of some business survey may, for instance, provide sums of observations grouped by economic activity and company size classes. At the same time, it may also display the sums of observations grouped by economic activity or size class only. These are then margins/marginal cells of this table. If a sum across all observations is provided, we refer to it as the *total* or *overall total*. Table 4.1 is an example of a 2-dimensional table with row margins ('Total turnover') and column margins ('Total').

4.1.2   Complex tabular data structures: Hierarchical and linked tables

This section presents a brief introduction and some terminology of hierarchical and linked tables. For mathematical formulation of those structures, see, for example, de Wolf (2007) and de Wolf and Giessing (2008).

Data collected within government statistical systems must meet the requirements of many users, who differ widely in the particular interest they take in the data. Some may need community level data, while others need detailed data on a particular branch of the economy but no regional detail. As statisticians, we try to cope with this range of interest in our data by providing the data at several levels of detail.

When constructing tables for a publication presenting the results of a survey, statisticians usually combine classification variables in multiple ways. If two tables presenting data on the same response variable share some categories of at least one spanning variable, there will be cells which are presented in both tables – those tables are said to be *linked* by the cells they have in common, i.e., by cells that are logically identical. In order to offer a range of statistical detail, statisticians use elaborate classification schemes to categorise respondents. A respondent will often belong to various categories of the same classification scheme – for instance, a particular community within a particular county within a particular state – and may thus fall into three categories of the regional classification.

The structure between the categories of hierarchical variables also implies sub-structure for the table. When, in this chapter, we talk about sub-tables without sub-structure, we mean a table constructed in the following way.

For any classification variable we pick one particular non-bottom-level category (the 'food-production sector' for instance). Then we construct a 'sub-variable'. This sub-variable consists only of the category picked in the first step and those categories of the level immediately below belonging to this category (bakers, butchers, etc.). After doing that for each classification variable, the table specified through a set of these sub-variables is free from sub-structure then, and is a sub-table of the original one. Any cell within the sub-table does also belong to the original table. Many cells of the original table will appear in more than one sub-table: the sub-tables are linked. Example 4.1.1 provides a simple instance of a one-dimensional table with hierarchical structure.

Example 4.1.1 *Assume the categories A, AA, AB, AA1, AA2 and AA3 of some variable EXP resemble a hierarchical structure as depicted in Figure 4.1.*

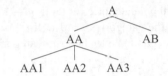

Figure 4.1   Tree view of a hierarchy.

Let the full table present turnover by the categories of EXP. Then the two sub-tables of this table are given in Figure 4.2.

EXP	Turnover		EXP	Turnover
A			**AA**	
AA			**AA1**	
AB			**AA2**	
			AA3	

Figure 4.2   Two sub-tables derived from Figure 4.1.

Note that cell AA is contained in both sub-tables.

As we see later in this chapter, for tabular data protection obtaining information on the structure of spanning variables is very important. There are three well-established models to collect this kind of meta-information:

1. A simple list of the variable categories, explaining which one is the 'total', will do if the spanning variable only defines a *simple, non-hierarchical* breakdown.

A
@AA
@@AA1
@@AA2
@@AA3
@AB

Figure 4.3   Indented code list of hierarchy of Example 4.1.1.

2. Some classifications provide the required information through a *specific coding scheme*, where level i of the hierarchy is defined by the first n_i characters of the code. In Example 4.1.1, level 1 is defined by the first, level 2 by the first two, and level 3 by all three characters. In such a case, the sequence $(n_i)_{i=1,...,L}$ defines the hierarchical structure (L denoting the maximum level).

3. A more general approach to describe a hierarchy is an *indented code list*. The order of the categories in that list is important: if a category x at level m (indented by $m - 1$ indentation characters) splits into n categories x_1 to x_n at level $m + 1$, then the n categories at level $m + 1$ that follow after category x in the list must be the categories x_1 to x_n. In this way, all categories in the hierarchy should be expanded into their sub-categories. Figure 4.3 presents an indented code list for the hierarchy of Example 4.1.1 (using '@' as indentation character).

4.1.3   Risk concepts

At first sight, one might find it difficult to understand how the kind of summary information published in magnitude tables presents a disclosure risk at all. However, it often occurs that cells of a table relate to a single or to only a few respondents. The number of this kind of small cells in a table will increase, the more grouping variables are used to specify the table, the higher the amount of detail provided by the grouping variables, or the more uneven the distributions of respondents are over the categories of the grouping variables.

If a table cell relates to a small group of respondents (or even only one), then publication of that cell value may imply a disclosure risk. This is, e.g., the case, if an intruder could identify these respondents using information displayed in the table.

Example 4.1.2 *Consider a table cell displaying the turnover of companies in the mining sector for a particular region X. Let us assume that company A is the only mining company in this region. This is a fact that will be known to certain intruders (like, e.g., another mining company (company B) in a neighbouring region Y). So, if that table cell is published, company B would be able to disclose the turnover of company A.*

In order to establish if a disclosure risk is connected to the publication of a cell value in a table, and in order to protect against this risk, data providers (like, e.g., national statistical institutes) should apply tabular data protection methods. In many countries, this is a legal obligation to official statistical agencies. It may also be regarded as a necessary requirement in order to maintain the trust of respondents. After all, if in Example 4.1.2, company A realises that company B might, by looking into the published table, disclose the value of turnover it has reported, and if it considers this value as confidential information, it may refuse to respond to that survey in the next period, or (if the survey is compulsory) it may choose to provide incorrect or inaccurate information.

Sections 4.2 and 4.3 provide an introduction into disclosure risk assessment concepts of magnitude tabular data protection. In Section 4.2, we present the most common methods for disclosure risk assessment for each individual cell of a table (and for special combinations of individual cells). These methods are called *primary* disclosure control methods. Table cells for which a disclosure risk has been established are called *sensitive*, *primary sensitive*, *confidential*, *risky* or *unsafe* cells. Primary disclosure risk is usually assessed by applying certain sensitivity rules. Section 4.2 illustrates the concept of sensitivity rules and the most commonly practised rules.

While detailed tabulated summary information on small groups of statistical objects might be of interest to certain user groups, it is also a responsibility of official statistics to provide summary information at a high aggregate level by producing summary statistics on large groups of a population. For example, summary information on companies by sub-groups of sub-sectors at the district level could be of interest to these local companies and the household income at neighbourhood level could be of interest to people who want to move to a certain neighbourhood. On the other hand, official statistics should provide information about *all* companies of an economic sector or the household income at national level as well. Because of this, it is not enough to have methodologies to only assess *primary* disclosure risks and protection methodologies of individual cells. It implies a need for the so-called *secondary* tabular data risk assessment and protection methodologies. This becomes evident considering the following example.

Example 4.1.3 *Assume that a table displays the sum of a variable 'production' by three sub-sectors of an economic sector. Assume that this sum is sensitive for one of the sub-sectors and that the table is protected by cell suppression, meaning that the confidential cell value is suppressed.*

With respect to the total production for this sector, we distinguish between two cases: (1) either it is foreseen to be published – we then consider it as a cell of the table (i.e. the 'total') – or (2) it is not foreseen. In Table 4.2, the grey column displays these two cases: (1) either the total 56 600 is published or (2) it is not published.

If the cell values of the two non-sensitive sub-sectors and the 'total' are displayed, users of the publication can disclose the cell value for the sensitive sub-sector by taking the difference between the 'total' and the sub-sector values for the two non-sensitive sectors. In Table 4.2, e.g., the value of the production in sub-sector I is calculated as $56\,600 - 47\,600 - 8002 = 998$.

Table 4.2   Production (in million Euros).

Sector X	Sub-sector I	Sub-sector II	Sub-sector III
56 600		47 600	8 002
	(sensitive)	(non-sensitive)	(non-sensitive)

This kind of disclosure risk is often referred to as *disclosure by differencing*. This can happen even when the sector total is not published directly. For example, if in another table, the sector results are published in a breakdown by geography.

Example 4.1.4 *Consider the same table on production for sector X as in Example 4.1.3, but this time broken down by region, see Table 4.3. Again the grey column depicts the two cases where the state total is either published or not.*

Table 4.3   Production in sector X (in million Euros).

State	Region A	Region B	Region C	Region D
56 600	20 300	10 100	5 100	21 100
	(non-sensitive)	(non-sensitive)	(non-sensitive)	(non-sensitive)

Even if the state total is not published directly, the total can then be computed by adding the regional results: $20\,300 + 10\,100 + 5100 + 21\,100 = 56\,600$. *This corresponds to the total of the sector X of Table 4.2, and hence the confidential sub-sector result in that table can be computed.*

Note that disclosure by differencing is typically not a risk when tables present ratios or indexes. In those situations, there are neither additive structures between the cells of such a table, nor between a cell value and the values of the contributing units. Think, for instance, of mean wages. Then the cell values would be a sum of wages divided by a sum of the number of employees of several companies. There may, of course, be exceptions. If, for instance, it is realistic to assume that the denominators of a ratio can be estimated quite closely, it might indeed be adequate to apply disclosure control methods based on the enumerator variable. That is, if in the case of the mean wages, the number of employees can be estimated quite closely both on the company level and on the cell level, it might indeed be adequate to apply disclosure control methods based on the wages themselves.

In order to avoid disclosure by differencing, data disseminators usually apply some kind of *secondary protection* to tables containing primary sensitive cells. In the case of Example 4.1.3, one would typically select one of the two non-sensitive sub-sector cells and suppress it as well. This would be called a *secondary suppression*. An alternative option could be to perturb the data. One might choose to perturb in a balanced way, for example by increasing the cell value of the sensitive cell and decreasing the cell values of the two non-sensitive cells by about the same amount to achieve that the sector X total result remains more or less unchanged.

From a general perspective, one could say that the purpose of secondary protection is chiefly to *preserve data utility* under the constraint that disclosure by differencing risks are eliminated to the degree the disseminator considers necessary. While some 'small', primary confidential cells within detailed tables may have to be protected, sums for larger groups, like, e.g., the margins of those detailed tables, should be preserved to the extent possible. Secondary protection should strike a balance between preserving information – especially that of higher level aggregates – and ensuring that primary protection of sensitive cells cannot be undone by some differencing, i.e. avoiding what we call *secondary disclosure risk*. Section 4.3 introduces the concepts of secondary disclosure risk assessment.

4.1.4   Protection concepts

Basically, there are two different classes of protection methods for tabular data: (1) pre-tabular and (2) post-tabular methods. Pre-tabular methods manipulate the microdata before they are summed up for tabulation and hence do not depend on any particular tabulation. Post-tabular methods are applied after tabulation.

Basically, any of the microdata protection methods of Chapter 3 could be considered as a pre-tabular protection method for tables. In the present chapter, we only consider methods that have been proposed mainly in the context of magnitude tabular data, are suitable to protect strongly skewed business data and are able to take into account both the preservation of data utility and the limitation of disclosure risk.

As in the case of microdata, another way of categorising protection methods is of course to classify them as either non-perturbative or perturbative. Table 4.4 shows one prominent example for each of the four possible cross-combinations of both classifications. In the present chapter, we will focus on these four protection methods which will be presented in Sections 4.4 and 4.5.

4.1.5   Information loss concepts

As pointed out in Willenborg and de Waal (2001), information loss concepts for tabular data are different from those for microdata. This is due to the fact that tables are often seen as final products, and not so much as a starting point for further analysis. This conception is probably even more prominent in case of magnitude tables.

In order to find a good balance between protection of individual response data and provision of information, it is necessary to somehow rate the information loss

Table 4.4   Classification of protection methods for magnitude tabular data.

Type of method	Perturbative	Non-perturbative
Pre-tabular	Multiplicative noise	Global recoding
Post-tabular	Controlled tabular adjustment	Cell suppression

connected to a cell that is suppressed or perturbed. By doing so, we can try to control the protection process in order to achieve 'optimal' behaviour of the algorithms. In Sections 4.6.1 and 4.6.2, we will explain the concept of cell costs with respect to rating information loss. On the other hand, especially in case of perturbative methods, it is essential to measure data utility *after* protection and to compare and evaluate the results of the applied methods. This issue is addressed in Section 4.6.3.

4.1.6   Implementation: Software, guidelines and case study

In order to come from theory to practice, first of all one needs software. Section 4.7 discusses the software issue, with special attention to the τ-ARGUS software package (Hundepool *et al.* 2011) which is freely available and covers the majority of methods described in this chapter.

Section 4.8 provides some guidance and discussion regarding commonly practised strategies for modelling a set of tables planned to be released for a statistic in a systematic way. This should be a starting point for the development of a specific disclosure control concept for that statistic. Central to a disclosure control concept for a statistic is the composition of one or more sets of tables designed in such a way that it can be handled by the software tools efficiently. All three aspects 'risk limitation', 'preservation of data utility' and 'effort' (in terms of computing time and in terms of implementing the respective concept in a production process) should be addressed when considering different possibilities.

Finally, Section 4.9 rounds the discussion up with a case study as an example for how to come to a disclosure control concept using suggestions of the guidelines.

4.2   Disclosure risk assessment I: Primary sensitive cells

We begin this section by describing intruder scenarios typically considered by statistical agencies in the context of disclosure control for magnitude tables. Considering these intruder scenarios, statistical agencies have developed some 'safety rules' as measures to assess disclosure risks. This section will also introduce the most popular rules using some illustrative examples. Finally, we will compare the different rules and give guidance on making a decision between alternative rules.

4.2.1   Intruder scenarios

If a table cell relates to a small group of respondents (or even only one), publication of that cell value may imply a disclosure risk. This is, for example, the case if these respondents could be identified by an intruder using information displayed in the table.

Let us revisit Example 4.1.2 of Section 4.1.3. In that example, the intruder (company B) only needs to know that the cell value reports the turnover of mining companies in region X, since company B is assumed to be able to identify company A as the only mining company in region X. Hence, publication of the cell value implies a disclosure risk: if company B looks into the publication, it will be able to disclose the turnover of company A.

But what if a cell value does not relate to one, but to two or more respondents?

Example 4.2.1 *Let us assume this time that both companies A and B are located in region X, and are the only mining companies there. Let us further assume that they both are aware of this fact. Then again publication of the cell value implies a disclosure risk, this time to both companies: if any of the two companies looks into the publication and subtracts its own contribution (i.e. the turnover it reported) from the cell value, it will be able to disclose the turnover of the other company.*

Example 4.2.2 *Assume now that the table cell relates to more than two respondents. Imagine this time that four companies (A, B, C and D) are located in region X. Then, theoretically, three of them (B, C and D, say) could exchange information on each other's contribution to disclose the turnover of company A. Companies B, C and D are then said to form a coalition. Such a coalition might be a rather theoretical construct. An equivalent but perhaps more likely scenario could be that of another party who knows the contributions of companies B, C and D (perhaps a financial advisor working for all three companies) who would then be able to disclose also the turnover of company A by subtracting the contributions of B, C and D from the cell value.*

The examples above are based on the intruder scenario typical for business data: it is usually assumed, that the 'intruders', those who might be interested in disclosing individual respondent data, may be 'other players in the field'. For example, competitors of the respondent or other parties who are generally well informed on the situation in the part of the economy to which the particular cell relates. Such intruder scenarios make sense, because, unlike microdata files for researchers, tabular data released by official statistics are accessible to everybody – which means they are accessible in particular to those well-informed parties.

In the scenarios of the previous examples, there is a risk that magnitude information is disclosed exactly. But how about approximate disclosure?

Example 4.2.3 *Let us reconsider Example 4.1.2 once more. Assume this time that in region X, there are 51 companies that belong to the mining sector: one large company A and 50 very small companies S1–S50. Assume further that 99% of the turnover in mining in region X is contributed by company A. In that scenario, the cell value corresponding to the total turnover in the mining sector for region X, is a pretty close approximation of the turnover of company A. And even though a potential intruder (e.g. mining company B of neighbouring region Y) may not be able to identify all 51 mining companies of region X, it is very likely that she will know that there is one very big company in region X and which company that is.*

The presumption of the sensitivity of a variable often matters in the choice of a particular protection method. For example, especially in the case of tables presenting business magnitude information, many agencies decide that, because of the

Table 4.5   Sensitivity rules.

Rule	Definition
	A cell is considered *unsafe*, when ...
Minimum frequency rule	the number of contributors to the cell is less than a pre-specified minimum frequency n (the common choice is $n = 3$).
(n, k)-dominance rule	the sum of the n largest contributions exceeds $k\%$ of the cell total, i.e., $x_1 + \cdots + x_n > k/100\,X$.
$p\%$ rule	the cell total minus the two largest contributions x_1 and x_2 is less than $p\%$ of the largest contribution, i.e., $X - x_2 - x_1 < p/100\,x_1$.[a]

[a]For the case of coalitions of $n - 1$ respondents $(n > 2)$: $X - x_n - \cdots - x_2 - x_1 < p/100\,x_1$.

sensitivity, this kind of information must be protected against the approximate disclosure illustrated by Example 4.2.3 above.

Considering the above explained intruder scenarios, statistical agencies have developed some 'safety rules' (also referred to as 'sensitivity rules' or 'sensitivity measures'). Safety rules are measures to assess disclosure risks. In Section 4.2.2, we will introduce the most popular safety rules, starting with an overview. After that, the rules (or rather, classes of rules) will be discussed in detail. We explain in which situations it may make sense to use those rules, using simple examples for illustration where necessary.

4.2.2   Sensitivity rules

Throughout this section, we denote the decreasingly ordered contributions to a cell with N respondents by $x_1 \geq x_2 \geq ... \geq x_N$. The corresponding cell total, or cell value, is denoted by $X = \sum_{i=1}^{N} x_i$.

Table 4.5 briefly presents the most commonly used sensitivity rules. Note that both the dominance rule and the $p\%$ rule are meaningful only when all contributions to a cell are non-negative.[1] Moreover, when $k = 100$ or $p = 0$, respectively, the dominance rule and the $p\%$ rule do not make sense either.

In case the dominance rule is used, it can be shown that an (n, k)-dominance rule implies a minimum frequency rule with threshold $100n/k$. That is, the total number of contributors in a cell should be larger than or equal to $100n/k$ to have a cell that is possibly safe according to an (n, k)-dominance rule. To see this, consider a cell with a total of N contributors where $N < 100n/k$, i.e., a cell that is unsafe according to a minimum frequency rule with threshold $100n/k$. Then that cell is also unsafe

---

[1] When all contributors are non-positive, a simple transformation would make them all non-negative and these rules would be meaningful as well.

according to an (n, k)-dominance rule:

$$
\begin{aligned}
x_1 + \cdots + x_n &= \frac{k}{100}(x_1 + \cdots + x_n) + \left(1 - \frac{k}{100}\right)(x_1 + \cdots + x_n), \\
&\geq \frac{k}{100}(x_1 + \cdots + x_n) + \left(1 - \frac{k}{100}\right)nx_n && (4.1) \\
&= \frac{k}{100}(x_1 + \cdots + x_n) + \frac{k}{100}\left(\frac{100n}{k} - n\right)x_n \\
&> \frac{k}{100}(x_1 + \cdots + x_n) + \frac{k}{100}(N - n)x_n && (4.2) \\
&\geq \frac{k}{100}(x_1 + \cdots + x_n) + \frac{k}{100}(x_{n+1} + \cdots + x_N) && (4.3) \\
&= \frac{k}{100}X.
\end{aligned}
$$

In Equation (4.1), we used that $x_i \geq x_n$ for $i = 1, \ldots, n$, in Equation (4.2) we used the assumption $N < 100n/k$ and in Equation (4.3) we used that $x_i \leq x_n$ for $i = n + 1, \ldots, N$.

For a $p\%$ rule (without coalition), it is easy to see that less than three contributors to a cell would always result in an unsafe cell for any value of p. That is, a $p\%$ rule without coalitions implies a minimum frequency rule with threshold 3.

Choice of a particular sensitivity rule is usually based on certain intruder scenarios involving assumptions about additional knowledge available in public or to particular users of the data, and on some (intuitive) notion on the sensitivity of the variable involved.

Minimum frequency rule

When the disseminating agency thinks it is enough to prevent exact disclosure, all cells with at least as many respondents as a certain, fixed minimum frequency n are considered safe. Example 4.1.2 of Section 4.1.3 (cell value referring to one company in the mining sector of a region) illustrates the disclosure risk for cells with frequency 1. Example 4.2.1 on two mining companies in a single cell shows that there is a similar risk for cells with frequency 2.

The mentioned examples explain the rule of thumb that normally the minimum frequency n will be set equal to 3. An exception is the case when for some n_0 larger than 3, the agency thinks it is realistic to assume that a coalition of $n_0 - 2$ respondents contributing to the same cell may pool their data to disclose the contribution of another respondent. In such a case, we set n equal to n_0. Indeed, Example 4.2.2 provides an instance for this case with $n_0 = 5$: the intruder knows the pooled data of $5 - 2 = 3$ companies (B, C and D).

Note that sensitivity rules are usually applied to all cells in a table equally. Therefore, one should keep in mind that assumed coalitions should be equally likely to all cells in the table under consideration.

Concentration rules

In case of only non-negative contributions, a published cell total is of course always an upper bound for each individual contribution to that cell. This bound is the closer to an individual contribution, the larger the size of the contribution. This fact is the mathematical foundation of the well-known concentration rules. Concentration rules like the dominance and $p\%$ rule make sense only if it is assumed specifically that the intruders are able to identify the largest contributors to a cell. The commonly applied sensitivity rules differ in the particular kind and precision of additional knowledge assumed to be around.

When a particular variable is deemed strongly confidential, preventing only exact disclosure may be judged inadequate. In such a case, a concentration rule should be specified. For reasons that will be explained later, we recommend use of the so-called $p\%$ rule. Another, well-known concentration rule is the 'n respondent, k percent' dominance rule.

Traditionally, some agencies use a combination of a minimum frequency rule together with a $(1, k)$-dominance rule. However, this approach is inadequate. It ignores the problem that in some cases, the contributor with the second largest contribution to a cell which is non-sensitive according to this rule is able to derive a close upper estimate for the contribution of the largest one by subtracting its own contribution from the aggregate total. Example 4.2.4 provides an instance.

Example 4.2.4 *Consider the $(1, 90)$-dominance rule. Let the total value of a table cell be $X = 100\,000$ and let the largest contributions be $x_1 = 50\,000$ and $x_2 = 49\,000$. Since $50\,000 < (90/100) \cdot 100\,000$ the cell is safe according to the $(1, 90)$-dominance rule. Hence, according to that rule, there seems to be no risk of disclosure. However, the second largest contributor is able to derive an upper estimate $\hat{x}_1 = 100\,000 - 49\,000 = 51\,000$ for the largest contribution which overestimates the true value of 50 000 by 2% only: quite a good estimate.*

Unlike the $(1, k)$-dominance rule, both the $(2, k)$-dominance rule and $p\%$ rule take the additional knowledge of the second largest contributor into account properly.

It should be noted that when applying a concentration rule to a particular table, the used values of the corresponding parameters should be kept confidential. Giving the exact value of the parameters could be used to derive additional information of a confidential cell, as is shown in Example 4.2.5 in case of a $p\%$ rule. A similar example could be given in case of an (n, k)-dominance rule.

Example 4.2.5 *Assume that a table is published along with the information that a $p\%$ rule is used with $p = 10$. That is, any cell in that table that is unsafe according to this $p\%$ rule is not published. Let C be a cell that is not published because it violates the $p\%$ rule.*

Assume that company A contributes 5 million Euros to cell C and knows that she is the largest contributor to this cell. Denote the ordered contributions to cell C by $x_1 \geq x_2 \geq \cdots \geq x_N$. Even without knowing the value of N but using the fact that cell

C is unsafe according to the p% rule with p = 10, company A can calculate that

$$\sum_{i=3}^{N} x_i < \frac{10}{100} 5 \text{ mln}$$

by the formula in Table 4.5. Hence, she can derive that the cell total satisfies

$$X = x_1 + x_2 + \sum_{i=3}^{N} x_i < (5 + 5 + 0.5) \text{ mln},$$

i.e., the cell total cannot exceed 10.5 million Euros.

In general, a cell total that is unsafe according to a p% rule cannot exceed $(2 + p/100)x_1$, where x_1 is the largest contributor to that cell.

Comparing the p% rule and the dominance-rule

We show in the following that according to both types of concentration rules, an aggregate total (e.g. cell value) X is considered sensitive, if it provides an upper estimate for one of the individual contributions that is relatively close to the true value of this contribution.

Assume that there are no coalitions of respondents, i.e., intruders only know their own contribution. Moreover, assume that all contributions to the cell are non-negative. Then the upper estimate with the smallest relative error of any contribution can be obtained by the second largest contributor $\hat{x}_1 = X - x_2$, i.e., by subtracting its own contribution from the cell total. All other scenarios of a contributor subtracting its own value from the cell total in order to estimate any other contribution, result in an upper estimate with the same or larger relative error.

In the, rather unlikely, scenario that $n - 1$ respondents (with $n > 2$) pool their data in order to disclose the contribution of another, the upper estimate with the smallest relative error can be obtained by the coalition of the $n - 1$ largest respondents $x_2, x_3, \ldots, x_n$, using the estimate $\hat{x}_1 = X - x_2 - x_3 - \cdots - x_n$.

How to determine whether such an estimate is 'relatively close'?

When we adapt the relation in Table 4.5 for the (n, k) rule to the case of $n = 2$ and subtract both sides from X, the result is

$$(X - x_2) - x_1 < (100 - k)/100 \, X. \tag{4.4}$$

In this formulation, the $(2, k)$ rule looks very similar to the formulation of the $p\%$ rule given in Table 4.5. Both rules define an aggregate to be sensitive, when the estimate $\hat{x}_1 = X - x_2$ does not overestimate the true value of x_1 'sufficiently': an aggregate total X is sensitive if

$$\hat{x}_1 - x_1 < p/100 \, x_1 \qquad \text{in case of a } p\% \text{ rule}$$
$$\hat{x}_1 - x_1 < (100 - k)/100 \, X \quad \text{in case of an } (n, k) \text{ rule.}$$

The difference between both rules is thus in how they determine this 'sufficiency'. According to the $p\%$ rule, it is expressed as a rate (i.e. $p\%$) of the true value of the largest contribution x_1, while according to the $(2, k)$ rule, it is expressed as a rate (i.e.

$(100 - k)\%)$ of the aggregate total X. Considering this, the concept of the $p\%$ rule seems to be more natural than that of the $(2, k)$ rule. Therefore, we recommend use of the $p\%$ rule instead of a $(2, k)$-dominance rule.

Comparing the two concentration rules, one could pose the following three questions:

Given a table T and a value of k,

1. For which values of p does it hold that all cells that are sensitive according to a $p\%$ rule are also sensitive according to a $(2, k)$-dominance rule?

2. For which values of p does it hold that all cells that are sensitive according to a $(2, k)$-dominance rule are also sensitive according to a $p\%$ rule?

3. For which values of p does it hold that all cells that are sensitive according to a $p\%$ rule are also sensitive according to a $(2, k)$-dominance rule *and vice versa*?

The formulation of both rules for sensitive cells as given in Table 4.5 can be rewritten into

$$\sum_{i=3}^{N} x_i < p/100 \, x_1$$

and

$$\sum_{i=3}^{N} x_i < \frac{100 - k}{k}(x_1 + x_2).$$

From this we can derive that the first question is answered by

$$p \leq 100 \frac{100 - k}{k}\left(1 + \frac{x_2}{x_1}\right) \tag{4.5}$$

whereas the second question is answered by

$$p \geq 100 \frac{100 - k}{k}\left(1 + \frac{x_2}{x_1}\right). \tag{4.6}$$

We then see immediately that the third question cannot be answered in general: it is not possible to fulfil both Equations (4.5) and (4.6) with a single p value for all cells at the same time, since the bounds depend on the ratio of the largest and second largest value of a cell.

How to obtain the p parameter?
Looking at Equations (4.5) and (4.6), it seems reasonable to choose p equal to $100(100 - k)/k$ when replacing a $(2, k)$-dominance rule by a $p\%$ rule. Thus, replacing a $(2, 80)$-dominance rule by a 25% rule or a $(2, 95)$-dominance rule by a 5.26% rule.

Using this particular choice for the parameter p, the set of cells that is unsafe according to the $p\%$ rule will usually be a subset of the set of cells that is unsafe according to the corresponding $(2, k)$-dominance rule. In Example 4.2.6, a situation is described in which a cell is sensitive according to a $(2, k)$-dominance rule and at the same time non-sensitive according to a $p\%$ rule with $p = 100(100 - k)/k$.

Example 4.2.6 *Consider a table cell with total value $X = 110\,000$ and the largest contributions x_1 and x_2 equal to 52000 and 50000, respectively. Using a $(2, k)$-dominance rule with $k = 90$, this cell is sensitive:*

$$(x_1 + x_2) = 102\,000 > 99\,000 = 0.9\,X.$$

However, setting $p = 100(100 - k)/k \approx 11$, this cell is not sensitive according to the $p\%$ rule:

$$(X - x_2) - x_1 = 8000 > 5720 = 0.11\,x_1.$$

If we derive p from the formula $p = 100(100 - k)/k$ when replacing a $(1, k)$-dominance rule, we will obtain a much larger number of sensitive cells. Denoting the set of unsafe cells according to a $(1, k)$-dominance rule by $\mathcal{U}_{(1,k)}$, according to a $(2, k)$-dominance rule by $\mathcal{U}_{(2,k)}$ and according to a $p\%$ rule with $p = 100(100 - k)/k$ by $\mathcal{U}_p$, we have that $\mathcal{U}_{(1,k)} \subseteq \mathcal{U}_p \subseteq \mathcal{U}_{(2,k)}$ (see Figure 4.4). To see this, note that it is not possible to have a cell being sensitive according to the $(1, k)$-dominance rule and at the same time not sensitive according to the $p\%$ rule with $p = 100(100 - k)/k$. Indeed, sensitivity by the $(1, k)$-dominance rule yields (rewriting the formula of Table 4.5)

$$(100 - k)x_1 > kx_2 + k \sum_{i=3}^{N} x_i$$

whereas non-sensitivity by the $p\%$ rule with $p = 100(100 - k)/k$ yields

$$(100 - k)x_1 \le k \sum_{i=3}^{N} x_i.$$

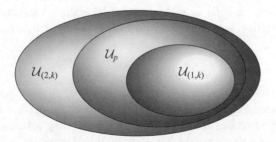

Figure 4.4   Comparing sets of sensitive cells according to $p\%$ rule ($\mathcal{U}_p$), $(1, k)$-dominance rule ($\mathcal{U}_{(1,k)}$) and $(2, k)$-dominance rule ($\mathcal{U}_{(2,k)}$) when $p = 100\,(100-k)/k$.

Table 4.6   Number of sensitive cells according to $(2, k)$-dominance rule and $p\%$ rule.

	Table A	Table B	Table C	Table D	Table E	Table F
$k = 70$	3 634	664	12 796	9 657	1 589	7 812
$p = 43$	3 303	600	12 250	8 948	1 469	6 829
$p = 64$	3 616	662	12 719	9 657	1 573	8 021
$p = 86$	3 954	715	13 074	10 266	1 670	9 048
$k = 80$	3 172	557	12 082	8 678	1 420	6 233
$p = 25$	2 964	518	11 753	8 199	1 340	5 568
$p = 38$	3 217	576	12 123	8 769	1 437	6 513
$p = 50$	3 413	632	12 412	9 196	1 505	7 251
$k = 90$	2 752	470	11 346	7 620	1 222	4 626
$p = 11$	2 660	455	11 154	7 356	1 175	4 296
$p = 17$	2 800	485	11 445	7 795	1 256	4 893
$p = 22$	2 892	504	11 625	8 015	1 310	5 357

Obviously, these two inequalities cannot hold at the same time since x_2 and k are assumed to be positive.

Similarly, sensitivity by to the $p\%$ rule with $p = 100 (100 - k)/k$ yields

$$(100 - k)x_1 > k \sum_{i=3}^{N} x_i,$$

and non-sensitivity by the $(2, k)$-dominance rule yields

$$(100 - k)x_1 \le k \sum_{i=3}^{N} x_i - (100 - k)x_2.$$

Again these two inequalities cannot hold at the same time ($k < 100$), and hence $U_p \subseteq U_{(2,k)}$ for this choice of p.

To get a feeling about the parameters p and k, Table 4.6 shows some examples of number of sensitive cells in tables, typical for national statistical institutes. To put this in perspective, some general information on the considered tables is given in Table 4.7.

The (p, q) rule

A well-known extension of the $p\%$ rule is the so-called prior-posterior or (p, q) rule. With this extended rule, one can formally account for general knowledge about individual contributions assumed to be around *prior* to publication. In particular, the information that the second largest contributor knows the contribution of the other respondents up to $q\%$. That is, she has two values X_R^L and X_R^U and knows that $X_R^L \le (100 - q)/100 \, X_R$ and $X_R^U \ge (100 + q)/100 \, X_R$, where $X_R = \sum_{i=3}^{N} x_i$ is the contribution of the smaller respondents. Using these bounds, the second largest

Table 4.7   General info on tables discussed in Table 4.6.

	Dimensions	Total cells	Empty cells
Table A	3	14 400	4 442
Table B	3	2 016	514
Table C	3	79 856	63 796
Table D	3	38 304	17 262
Table E	2	31 330	28 799
Table F	2	28 084	2 540

contributor can bound the value of the largest contribution by

$$X - x_2 - X_R^U \le x_1 \le X - x_2 - X_R^L. \tag{4.7}$$

Obviously, the best a priori information in this scenario is when $X_R^L = (100 - q)/100 \; X_R$ and $X_R^U = (100 + q)/100 \; X_R$. Hence, an aggregate is considered to be sensitive according to a (p, q) rule when either

$$(X - x_2 - (100 - q)/100 \; X_R) - x_1 < p/100 \; x_1 \tag{4.8}$$

or

$$x_1 - (X - x_2 - (100 + q)/100 \; X_R) < p/100 \; x_1. \tag{4.9}$$

This is equivalent to saying that a cell is sensitive according to a (p, q) rule whenever $X_R < p/q \; x_1$. Note that the knowledge of an intruder can only increase when it is assumed that $p < q$.

Moreover, it follows that the sensitivity of the cells is determined by the ratio p/q only. Therefore, in determining the sensitivity of cells, any (p, q) rule with $q < 100$ can also be expressed as a (p^*, q^*) rule, with $q^* = 100$ and

$$p^* = 100p/q. \tag{4.10}$$

Note that a (p, q) rule with $q = 100$ is equivalent to a $p\%$ rule.

Of course, we can also adapt the (n, k)-dominance rule to account for $q\%$ relative a priori bounds. Let, e.g., $n = 2$. According to Equation (4.4) above, an aggregate should then be considered unsafe when the second largest respondent could estimate the largest contribution x_1 to within $(100 - k)$ percent of X. That is, a cell is considered to be sensitive when either

$$(X - x_2 - (100 - q)/100 \; X_R) - x_1 < (100 - k)/100 \; X \tag{4.11}$$

or

$$x_1 - (X - x_2 - (100 + q)/100 \; X_R) < (100 - k)/100 \; X. \tag{4.12}$$

Just as in the case of the $p\%$ rule, we see that the aggregate is sensitive, when $X_R < (100 - k)/q \; X$. Again this only depends on the ratio $(100 - k)/q$. For given parameters k and q, the $(2, k)$-dominance rule with $q\%$ a priori knowledge can be expressed as a $(2, k^*)$-dominance rule with q^* a priori knowledge, with $q^* = 100$ and

$$(100 - k^*)/100 = (100 - k)/q. \tag{4.13}$$

For a more analytical discussion of sensitivity rules the interested reader is referred to Cox (2001) and for more generalised formulations considering coalitions to Loeve (2001).

Sensitivity rules for special cases

Negative Contributions
When disclosure risk has to be assessed for a variable that can take not only positive, but also negative values, we suggest to reduce the value of p in a $p\%$ rule or increase the value of k in a dominance rule, relative to the situation of positive values only. It may even be adequate to take that reduction to the extent of replacing a concentration rule by a minimum frequency rule.[2]

This recommendation is motivated by the following consideration. We have already explained that the $p\%$ rule is equivalent to a (p, q) rule with $q = 100$. When contributions may take negative as well as positive values, it makes sense to assume that the a priori bound $(100 - q)/100 \, X_R$ can be negative as well. This can be expressed by using a value $\tilde{q} = 100 f$, with $f > 1$. According to Equation (4.10) this $(p, \tilde{q})$ rule is equivalent to a $\tilde{p}\%$ rule with $\tilde{p} = 100 p/\tilde{q}$ and

$$\tilde{p} = 100p/\tilde{q} = 100p/(100 f) = p/f < p.$$

In case of the dominance rule, because of Equation (4.13), $\tilde{k}$ can be determined by $(100 - \tilde{k})/100 = (100 - k)/(100 f)$. Again, since $f > 1$, this means that $\tilde{k} > k$.

Waivers
Sometimes, respondents authorise publication of an aggregate even if this publication might cause a risk of disclosure for their contributions. Such authorisations are also referred to as *waivers*. In case waivers are present in a cell, sensitivity rules need to be adjusted.

Denote the largest contribution from the respondents *without* waivers by x_s and the largest contribution of all other respondents (with and without waivers) by x_r. Since for all i and j with $i \neq j$ it holds that $x_i + x_j \leq x_r + x_s$, the worst case is that respondent r tries to estimate the largest contribution without waiver, i.e., x_s.

The sensitivity rules can then, e.g. be reformulated as: a cell is sensitive if

$$(X - x_r) - x_s < p/100 \, x_s \tag{4.14}$$

in case of a $p\%$ rule or

$$x_r + x_s > k/100 \, X \tag{4.15}$$

in case of a $(2, k)$-dominance rule.

---

[2] Ideas for how to adapt secondary protection (see Section 4.3) to the case of negative contributions are presented in Giessing (2008).

Foreign Trade Rule

In foreign trade statistics, traditionally a special rule is applied. Only for those enterprises that have actively *asked* for protection of their information, special sensitivity rules are applied. This implies that if the largest contributor to a cell asks for protection and contributes over a certain percentage to the cell total, that cell is considered a primarily unsafe cell.

Given the level of detail in the foreign trade tables, application of the standard sensitivity rules would imply that a very large proportion of the table ought to be suppressed. This is not considered a desirable situation and for a long time there have not been serious complaints about not applying the standard rules. This has led to this special foreign trade rule. A legal basis for this rule is laid down in the European Regulation 638/2004. In the software τ-ARGUS (see Hundepool *et al.* 2011) a special option called *the request rule* has been introduced to apply this rule. The secondary cell suppression (see Section 4.4.2) will be done in the standard way.

Holdings

In many data sets, especially economic data sets, the reporting unit is different from the entities we want to protect. Often companies have several branches in various regions. So, the marginal cell may have several contributions but from one company only. Therefore, it might be a mistake to think that such a cell is safe if we look only at the two largest contributors, while the three largest ones might belong to only one company.

We need to group the contributions from one company together into one contribution before we apply the sensitivity rules. The following example shows the effect of this action.

Example 4.2.7 *Consider the row of Table 4.8 on turnover (in million Euros) by type of industry and region and assume that company Q has a branch in Region A as well as in Region B with turnover $x_{QA} = 800$ and $x_{QB} = 300$ million Euros, respectively. Moreover, assume that the second largest contribution x_P to the turnover in Region A is equal to 350 million Euros. Regarding both branches of company Q as individual contributions to the row total, that total is non-sensitive according to a*

Table 4.8   Turnover in million Euros by type of industry and region.

	Region A	Region B	Region C	Total
⋮	⋮	⋮	⋮	⋮
Type X	1200	370	30	1600
⋮	⋮	⋮	⋮	⋮

p% rule with p = 15:

$$\sum_{i=3}^{N} x_i = 1600 - x_{QA} - x_P = 450 \leq 120 = p/100\,x_{QA}.$$

However, if both branches are combined to a single contributor to the row total (i.e. with contribution $x_Q = x_{QA} + x_{QB}$), that cell would be sensitive according to the same p% rule:

$$\sum_{i=3}^{N} x_i = 1600 - x_Q - x_P = 150 \leq 165 = p/100\,x_Q$$

Sampling Weights

Often tables are created from data sets based on sample surveys. Indeed, NSIs collect a lot of their information this way. There are two reasons why disclosure risk assessment may also be necessary for a sample survey, since, especially in business surveys, the common approach is to sample with unequal probabilities. The first is that large companies are often sampled with probability one. Typically, it will be the case that, for some of the aggregates foreseen for a publication, all respondents have been sampled with probability one. In the absence of non-response, for such an aggregate, the survey data are actually the data of the full population. Secondly, even if sampling probabilities are smaller than one, if an aggregate relates to a strongly skewed population, and the sampling error is small, then the probability is high that the survey estimate for the aggregate total may also be a close estimate for the largest unit in that population.

Therefore, it makes sense to assess the disclosure risk for survey sample estimates, if the sampling errors are small and the variables are strongly skewed. The question is then how to determine technically if a sample survey aggregate should be considered sensitive or not.

For sample survey data, each record has a sample weight associated with it. These sample weights are constructed such that the tables produced from this data sets will resemble the population as much as possible, as if the table had been based on a full census.

Making a table one has to take into account these sample weights. If the data file has a sample weight, specified in the metadata, the table can be computed taking these weights into account. For making a cell total, this is a straightforward procedure; however, the sensitivity rules have to be adapted to the situation where we do not know the units of the population with the largest values. Essentially this means that, e.g., in case of a *p%* rule, we should replace each element in the inequality based on a full population, i.e.,

$$X - x_2 - x_1 < p/100\,x_1$$

by appropriate estimators. Similarly, we could replace each population parameter in any of the other sensitivity rules by (weighted) estimates.

A first option, as is implemented in τ-ARGUS (see Hundepool *et al.* 2011), is to approximate the full population of the cell in question using the sampling weights. That is, the sampling weights are considered to represent the population frequency of similar sampling units. A sampling weight of two is thus interpreted as that in the population two units exist with similar values of the variable in question. Example 4.2.8 illustrates this idea.

Example 4.2.8 *Consider a cell with two contributions* A *and* B, *with values 100 and 10 and weights 4 and 7, respectively. Then the (weighted) cell value will be equal to* $(4 \times 100) + (7 \times 10) = 470$. *Considering the sampling weights to represent the number of similar population units, the individual contributions of the population can be approximated by 100, 100, 100, 100, 10, 10, 10, 10, 10, 10 and 10. The largest two contributors are now 100 and 100. These are regarded as the largest two values for application of the safety rules. That is, in the formula of the p% rule, the estimates* $\hat{X} = 470$, $\hat{x}_1 = 100$ *and* $\hat{x}_2 = 100$ *are used.*

In this option, it is essential that the sampling weights can be interpreted as the 'number of similar population units' for the cell in question. In fact, this is the only pre-requisite for this option to make sense. On the other hand, this procedure cannot be applied in combination with the holding concept, because naturally for the approximated contributions it is unknown which possible holding they belong to.

If the weights are non-integers, a simple extension is applied in τ-ARGUS. Since we only need estimators for some of the largest population units, the extension is top–down: the largest population unit is estimated by a weighted sum of the largest sampling units such that the sum of the weights equals one. Then the second largest is estimated, again by a weighted sum of the (remaining) sampling units, such that the sum of the weights equals one, etc. The following example should explain this idea.

Example 4.2.9 *Consider a cell with three sampled contributions* A, B *and* C *with values 100, 50 and 20 and weights 1.6, 2.2 and 6, respectively. In the formula of the p% rule, we have to obtain estimates for* X, x_1 *and* x_2. *Obviously,* $\hat{X} = 1.6 \times 100 + 2.2 \times 50 + 6 \times 20 = 380$. *For the largest and second largest population units, we get* $\hat{x}_1 = 100$ *and* $\hat{x}_2 = 0.6 \times 100 + 0.4 \times 50 = 80$.

Another approach could be to use the largest values of the sample directly as estimates for the largest population units:

$$\hat{X} - x_2^s - x_1^s < p/100 \, x_1^s \tag{4.16}$$

where $\hat{X}$ denotes the estimated population total, and x_i^s, $(i = 1, 2)$ the largest two observations from the sample.

The difference between this and the above procedure is that with Equation (4.16), we consider as intruder the respondent with the second largest contribution observed in the sample, whereas according to the above procedure, whenever the sampling weight of the largest respondent is 2 or more, an 'artificial' intruder is assumed who contributes as much as the largest observation from the sample.

Considering formulation (4.4) of the dominance rule, it is straightforward to see that we can adapt Equation (4.16) to the case of dominance rules by

$$x_1^s + \cdots + x_n^s > k/100\,\hat{X}. \tag{4.17}$$

It is also important to note in this context that sampling weights should be kept confidential, because otherwise we must replace Equation (4.16) by

$$\hat{X} - w_2 x_2^s - w_1 x_1^s < p/100\,x_1^s, \tag{4.18}$$

where $w_i (i = 1, 2)$ denote the sampling weights. Obviously, according to Equation (4.18) more aggregates will be sensitive.

4.3   Disclosure risk assessment II: Secondary risk assessment

When some 'small', primary confidential cells within detailed tables have to be protected (by cell suppression or by some perturbative method), it may still be possible to publish sums for larger groups, i.e., the margins of those detailed tables. However, we must make sure that protection of the primary sensitive cells cannot be undone, by some differencing, for example. Referring to this type of risk as 'secondary disclosure risk', this section will introduce the main concepts of secondary risk assessment.

Modelling secondary disclosure risks requires that – apart from its 'inner' cells – margins and overall totals are considered to be part of a table. From this assumption, it follows that there is always a linear relationship between cells of a table. Consider, for instance, Example 4.1.3 of Section 4.1.3. The sector result (say: X_T) is the sum of the three sub-sector results (X_1 to X_3), i.e. the linear relationship between these four cells is given by the relation $X_T = X_1 + X_2 + X_3$. As we have seen in Example 4.1.3, if it has been established that a disclosure risk is connected to the release of certain cells of a table, then it is not enough to prevent publication of these cells. Other cells (so-called 'complementary' or 'secondary' cells) must be suppressed, or be otherwise manipulated in order to prevent the value of the protected sensitive cell being recalculated through, e.g. differencing.

4.3.1   Feasibility interval

When a table is protected by cell suppression, it is always possible for any particular suppressed cell of a table to derive upper and lower bounds for its true value, by making use of the linear relations between published and suppressed cell values. This holds for tables with non-negative values. In case of tables containing negative values as well, it holds when some (possibly tight) lower bound other than zero is available to data users in advance of publication for all cells in the table. The interval given by these bounds is called the *feasibility interval*.

Example 4.3.1 *This example[3] illustrates the computation of the feasibility interval in the case of a simple two-dimensional table where all cells are assumed to contain non-negative values, as shown in Table 4.9.*

Table 4.9   Simple table.

Example	A	B	Total
I	X_{11}	X_{12}	7
II	X_{21}	X_{22}	3
III	3	3	6
Total	9	7	16

For this table the following linear relations hold:

$$X_{11} + X_{12} = 7 \qquad \text{(R1)}$$
$$X_{21} + X_{22} = 3 \qquad \text{(R2)}$$
$$X_{11} + X_{21} = 6 \qquad \text{(C1)}$$
$$X_{12} + X_{22} = 4 \qquad \text{(C2)}$$
$$\text{with } X_{ij} \geq 0 \text{ for all } (i, j).$$

Using linear programming methodology, it is possible to derive systematically for any suppressed cell X_{ij} in a table an upper bound $X_{ij}^{\max}$ and a lower bound $X_{ij}^{\min}$ for the set of feasible values. A feasible value for a cell is a value that a cell can attain, maintaining all linear relations of the original table. In the example above, for cell (1,1) these bounds are $X_{11}^{\min} = 3$ and $X_{11}^{\max} = 6$.

However, in this simple instance, we do not need linear programming technology to derive this result. Because of the first column relation (C1), X_{11} must be less than or equal to 6, and because of the second row relation (R2), X_{21} must be less than or equal to 3. Therefore, and because of the first column relation (R1), X_{11} must be at least 3, i.e., $X_{11}^{\min} = 3$.

Of course, a feasibility interval for a cell depends strongly on the relations considered for its computation. If in the instance above, we compute a feasibility interval for X_{11} considering only (R1), the result will be [0, 7], instead of [3, 6], i.e., a much wider interval.

Note that, through the remainder of this chapter and for notational simplicity, we will more generally talk of a *table relation* rather than of 'row relations' or 'column relations'.

A general mathematical statement for the linear programming problem to compute upper and lower bounds for the suppressed entries of a table is given in Fischetti and Salazar (2000). Within a disclosure control process, computing the bounds of the feasibility intervals subject to a set of table relations is referred to as an ***audit*** of a protected table.

---

[3] Taken from Geurts (1992), Table 10, p. 20.

Table 4.10   Upper protection levels.

Sensitivity rule	Upper protection level
$(1, k)$ rule	$(100/k)\, x_1 - X$
(n, k) rule	$(100/k)(x_1 + x_2 + \ldots + x_n) - X$
$p\%$ rule	$(p/100)\, x_1 - (X - x_1 - x_2)$
(p, q) rule	$(p/q)\, x_1 - (X - x_1 - x_2)$

Note that feasibility intervals can in principle also be computed using this approach, when cell perturbation (as opposed to cell suppression) is used to protect the table. To be able to do this, the user should know something about the perturbation method. For example, in case of simple rounding, the rounding base should be known. In general, the user should be given information such that she can deduce meaningful (i.e. rather narrow) a priori intervals of each perturbed cell. Those intervals are referred to as *implicit intervals* in the following. The table relations can then be used to derive tighter a posteriori bounds of the perturbed cells.

4.3.2   Protection level

The protection provided by a set of suppressions (the *suppression pattern*), or by a perturbation technique that supplies users with implicit intervals, should only be considered *valid* if the bounds for the feasibility interval of any sensitive cell cannot be used to deduce bounds on an individual respondent's contribution to that cell that are too close according to the sensitivity rule employed. For a mathematical statement of that condition, we determine safety bounds for primary suppressions. We call the deviation between those safety bounds and the true cell value *upper and lower protection levels*. The formulas of Table 4.10 can be used to compute upper protection levels in case of concentration rules. Out of symmetry considerations, the lower protection level is often set identical to the upper protection level. The bounds of the 'protection interval' are computed by subtracting the lower protection from and adding the upper protection level to the cell value. As discussed in Section 4.2.2, minimum frequency rules should only be used instead of concentration rules if it is enough to prevent exact disclosure only. In such a case, the protection levels should also be chosen such that exact disclosure is prevented. That is, a minimal protection level is sufficient.

Example 4.3.2 proves the formula given in Table 4.10 for the case of the $p\%$ rule.

Example 4.3.2 *Let $X + U$ be the upper bound of the feasibility interval for a cell with cell value X. The second largest respondent can then deduce an upper bound x_1^U by subtracting its own contribution from that upper bound: $x_1^U = X + U - x_2$. According to the definition of the $p\%$ rule, proper protection means that $x_1^U \geq (1 + p/100)\, x_1$. So, if the feasibility interval provides upper protection, it must hold that $U \geq (1 + p/100)\, x_1 - X + x_2 = (p/100)\, x_1 - (X - x_1 - x_2)$.*

If the distance between the upper bound of the feasibility interval and the true value of a sensitive cell is below the upper protection level computed according to the formulas of Table 4.10, then this upper bound could be used to derive an estimate for individual contributions of the sensitive cell that is too close according to the safety rule employed, which can easily be proven along the lines of Cox (1981). We say then that this cell is subject to *upper inferential disclosure*. More generally, if the feasibility interval for a cell does not cover its protection interval, we say that the cell is *underprotected*, *not sufficiently protected* or *subject to inferential disclosure*. If the feasibility interval consists of only the true value of the sensitive cell, we say that the cell is subject to *exact disclosure*.

The distance between the upper bound of the feasibility interval and the true value of a sensitive cell must exceed the upper protection level; otherwise the sensitive cell is not properly protected. This safety criterion is a necessary, but not always sufficient criterion for proper protection. It is a sufficient criterion when the largest respondent makes the same contribution also within the combination of suppressed cells within the same aggregation (a row or column relation of the table, for instance), and when no individual contribution of any respondent (or coalition of respondents) to such a combination of suppressed cells is larger than the second largest respondent's (or coalition's) contribution.

4.3.3 Singleton and multi cell disclosure

Cases where the criterion is not sufficient arise typically when the only other suppressed cell within the same aggregation is a sensitive cell too (and not the marginal cell of the aggregation), or when the same respondent can contribute to more than one cell within the same aggregation. In these cases, it may turn out that the individual contribution of a respondent may still be disclosed, even though the upper bound of the feasibility interval is well away from the value of the sensitive cell to which this respondent contributes. The most prominent case is that of two cells with only one single contributor each (a.k.a. *singletons*) that are the only suppressed cells within the same table relation. No matter how large the cell values are, because both contributors obviously know their own contribution, they can use this additional knowledge to disclose each other's contribution. More generally, a singleton can undo the protection of 'its own' cell and can thus compute the feasibility interval for another sensitive cell using a special set of table relations, derived from a table where the 'own' cell is unprotected. For an analytical discussion of these issues, see Cox (2001). In principle, the protection of a table should not be considered as valid when certain respondents can use their insider knowledge (e.g. on their own contribution to a cell) to disclose individual contributions of other respondents (see Daalmans and de Waal 2011).

4.3.4 Risk models for hierarchical and linked tables

When using cell suppression methods (see Section 4.4.2) to protect a set of linked tables, or sub-tables of a hierarchical table, it is not enough to select secondary suppressions for each table (or sub-table) separately. Otherwise it might, for instance, happen that the same cell is suppressed in one table because it is used as secondary suppression, while within another table it remains unsuppressed. A user comparing the two tables would then be able to disclose confidential cells in the first table.

Even when, instead of cell suppression, cell perturbation (see Section 4.5) is used to protect the table, this kind of risk must eventually be taken into account considering the implicit intervals mentioned at the end of Section 4.3.1.

Regarding hierarchical and linked tables, the computation of feasibility intervals needs special attention. Intervals computed using the set of table relations of the complete hierarchical table (or the complete set of linked tables), tend to be tighter than those computed separately for each (sub-)table using their own (smaller) sets of table relations.

In practice, protection strategies implicitly relate to either of the following risk models (or a combination thereof).

Risk model I: Relation

The 'Relation' model is based on the assumption that it is most important to prevent a certain kind of spontaneous disclosure, requiring only one simple, straightforward computation: it can occur, if a feasibility interval for a protected sensitive cell does not cover the protection interval, even if this feasibility interval can be computed easily taking into account just one table relation. For instance, in the case of cell suppression, by assuming zero as lower bound and computing the upper bound by subtracting the known (unsuppressed) cell values from the relation (i.e. row or column) total.

Risk model II: Sub-table

This model assumes the following scenario: an interested data user might compute the feasibility intervals of a protected sensitive cell by taking into account the set of table relations of just one sub-table of an eventually much larger hierarchical table. For a skilled user with some linear programming knowledge, this would be a task that does not require much effort. However, we assume a skilled user here, so the risk is certainly lower than the one of the *Relation* model above.

Risk model III: Table

The assumption of this model is that a data user might take into account the full set of table relations of a protected hierarchical table, or even of a set of protected linked tables, when computing feasibility intervals for protected cells. This is certainly possible for a skilled user. However, in large hierarchical tables creating the required input information for a suitable analysis program is a major effort indeed, which makes this scenario less likely and hence this kind of risk lower than that of the *Sub-table* model above.

4.4   Non-perturbative protection methods

We begin this section presenting global recoding of categorical variables in the microdata as a non-perturbative pre-tabular method. It is a technique that is frequently used, though usually not just for confidentiality reasons.

The most prominent protection method for magnitude tabular data is cell suppression, which has a long tradition in the practice of official statistics. By its nature, it is a method that can only be applied *after* tabulation. Section 4.4.2 will introduce the basic concept and 4.4.3 the concepts of the most important algorithms. Section 4.4.4 outlines a class of heuristic algorithms allowing to handle cell suppression problems of large hierarchical tables and sets of linked tables efficiently.

4.4.1   Global recoding

In the context of tabular data protection, global recoding means that several categories of a categorical variable are collapsed into a single one. This will reduce the amount of detail of any table using this variable as spanning variable. Usually, as a result the number of unsafe cells in the tables will be reduced.

Using only global recoding to protect *all* sensitive cells in a table often results in huge loss of data utility. While it is certainly not an uncommon strategy in official statistics, it is hence usually applied in combination with other protection methods like cell suppression. Another reason to combine this method with other protection methods is that often publication requirements on the table design do not allow for global recoding. Moreover, it is usually in contrast with the aim of many statistical agencies to squeeze out the maximum amount of information of the data in the tables they release.

4.4.2   The concept of cell suppression

For business statistics, the most popular method for tabular data protection is *cell suppression*. In tables protected by cell suppression, all values of cells for which a disclosure risk has been established are eliminated from the publication. In practice, this means that those values will be replaced by a special symbol like, e.g., a cross ($\times$).

A cell suppression procedure involves two steps. Firstly, in the primary risk-assessment step (see Section 4.2), all primary sensitive cells are determined. In the second step, some non-sensitive cells are selected as secondary suppressions to protect the primary sensitive cells from disclosure by differencing (see Section 4.3). The joint set of primary and secondary suppressions is called a *suppression pattern*.

Cell suppression is a protection technique that requires a very careful evaluation of secondary disclosure risks (c.f. Section 4.3). Releasing multiple linked tables from the same survey is especially prone to secondary disclosure. When a protection concept building on cell suppression fails to represent existing links between some of the tables or sub-tables, this may cause a risk that users comparing the protected tables find that they can undo some of the protection. This might happen if, for example, the suppression status of cells that are logically identical and contained in two different tables is inconsistent. For example, a cell being (secondarily) suppressed in one table and published in the other.

Therefore, in advance of publishing tabular data from a survey, a table model needs to be established. That is, the secondary disclosure risks due to hierarchical and linked table relations should be closely regarded. Section 4.8 offers some guidelines on this. Often this is a straightforward task if tables are quite systematically structured. For example, when they are constructed on the basis of established hierarchical business or geography classifications. Otherwise, if tables are constructed using a complex variety of spanning variables, it can become tedious.

The problem of secondary disclosure can even become much worse if there is no complete dissemination plan, or if there are other reasons why some tabulations of a survey are protected (and published) sooner than others. Or if there are different parties disseminating tabulations using the same source data.

As said, the second step is the problem of finding an optimum set of suppressions. This is known as the *secondary cell suppression problem*. The goal is to find a feasible set of secondary suppressions with a minimum loss of information connected to it. For a mathematical statement of the secondary cell suppression problem, see, e.g., Fischetti and Salazar-González (2000).

In practice, a major problem is how to avoid singleton and multi cell disclosure risks described in Section 4.3.4. A simple approach would build on a worst-case scenario where all singletons collude to recompute another sensitive cell. However, this is not realistic and it would lead to suppression patterns with much too much information loss. Daalmans and de Waal (2011), on the other hand, suggest a theoretically sound model which the authors themselves declare to be computationally intractable in practice. For the algorithms explained in the following two sections, heuristic methods avoiding at least to some extent singleton and multicell risks have been proposed in the literature and are mentioned in the respective paragraphs.

4.4.3   Algorithms for secondary cell suppression

In this section, we briefly outline the concepts of some prominent algorithms for secondary cell suppression. For a more exhaustive survey of the respective literature, see, for example, Duncan et al. (2011) or Castro (2012).

Integer linear programming approach

Fischetti and Salazar-González (2000) present an integer linear programming (ILP) approach for an optimal solution of the secondary cell suppression problem stated above. It solves the problem of finding a set of secondary suppressions with a minimum sum of costs associated to their suppression, subject to constraints expressing that all primary sensitive cells are sufficiently protected. That is, that the feasibility intervals of the sensitive cells computed on the basis of this suppression pattern all cover the respective protection intervals. See Section 4.6 for considerations on how to assign cell costs in a suitable way. The algorithm is very efficient. For details, see

Salazar-González (2008). However, computing times grow rapidly when the table size increases. A heuristic for how to solve singleton/multicell problems is outlined at the end of the following paragraph relating to the linear programming approach.

Linear programming approach

A computationally much cheaper linear programming approach has been described in Sande (1984), Robertson (1993), Robertson (1994) and Frolova *et al.* (2009). Instead of the computation intensive ILP problem, only a linear programming relaxation is solved. Define a 'congruent table' as a table with the same set of table relations as the table that has to be protected. For each sensitive cell in the original table, the algorithm will determine a congruent table, where the value of this cell, as well as the values of some other cells, has been altered. Sensitive cells will change by at least their respective protection levels. The algorithm will minimise the sum of the deviations between true and altered cell values, weighted by some cell costs. These cell costs represent the amount of information that is associated with that cell.

So far, this would also describe an algorithm introduced in Fischetti and Salazar-González (2003) which they refer to as the *partial cell suppression method*, assuming that, instead of suppressing the set of cells that have been altered, one would publish the interval given by the upper and lower deviation of altered cell values. There is a perception that the majority of users of statistical publications prefer the statistical agencies to provide actual figures rather than intervals. Therefore, this method has not yet been implemented in any software package. Sande and Robertson on the other hand suggest making the algorithm suitable for cell suppression in the following way. They propose to carry out the algorithm twice. In the first run, cell costs should be assigned that increase slightly with the cell value, like, for instance, the logarithm of the cell value. In the second run, only cells can be altered that have been altered also in the first run. This time a decreasing cost function is used which will make the algorithm change fewer cells. The secondary suppressions are then those altered in the second run.

Versions of this algorithm are underlying the CONFID and CONFID2 packages of Statistics Canada and also the commercial package ACS (see Sande 1999).

Robertson (2000) proposes a heuristic which avoids singleton and multicell disclosure risks efficiently. However, sufficient protection is guaranteed only when one adopts the (weakest) risk model *Relation* as introduced in Section 4.3.4. That is, the heuristic ensures that two respondents contributing to different sensitive cells within the same table relation will not be able to disclose each other's contribution using their own contribution along with the table relation they both belong to. This heuristic was first implemented in CONFID and CONFID2. A similar approach is used in conjunction with the ILP approach of Fischetti and Salazar, as implemented in τ-ARGUS.

Network flow heuristics

Network flow heuristics for secondary cell suppression build on the fact that a suppressed cell in a two-dimensional table is safe from exact disclosure if, and only if, the cell is contained in a 'cycle', or 'alternating path' of suppressed cells, where a cycle is defined to be a sequence of non-zero cells with indices $\{(i_0, j_0), (i_1, j_0), (i_1, j_1), (i_2, j_1), \ldots, (i_n, j_n), (i_0, j_n)\}$, where all i_k and j_l for $k = 0, 1, \ldots, n$ and $l = 0, 1, \ldots, n$, respectively (n: length of the path), are different, i.e., $i_{k_1} \neq i_{k_2}$ unless $k_1 = k_2$ and $j_{l_1} \neq j_{l_2}$ unless $l_1 = l_2$.

The network flow cell suppression heuristic is based on the solution of a sequence of shortest path sub-problems that guarantee a feasible suppression pattern. This heuristic does not guarantee an optimal feasible suppression pattern, but will often be close to the optimal solution. See Castro (2003) and Castro (2006) for details about an implementation included in the τ-ARGUS software package. An earlier implementation of a network flow heuristic is included in a cell suppression package of the US Census Bureau (see Jewett 1993).

In the context of cell suppression, network flow heuristics have a major disadvantage. That kind of heuristics can only be used when the tables that have to be protected can be modelled as at most two-dimensional tables with at most one hierarchically sub-structured spanning variable.

The implementation of Jewett (1993) offers a heuristic to deal with singleton and multicell disclosure. That heuristic dynamically assigns for each target primary suppression so-called capacity constraints to potential secondary suppressions. This solves the problem and does not lead to oversuppression. However, in connection with Castro's algorithm, no solution to avoid this kind of risk is available so far.

Hypercube method

A Hypercube method for cell suppression builds on the fact that a suppressed cell in a simple n-dimensional table without sub-structure cannot be disclosed exactly, if that cell is contained in a pattern of suppressed, non-zero cells, forming the corner points of a hypercube. An implementation of such a method has been described in depth in Repsilber (1994) and Repsilber (2002). For briefer descriptions, see Giessing and Repsilber (2002) or Hundepool et al. (2011). A Hypercube method will successively construct for any primary suppression all possible hypercubes with this cell as one of the corner points.

If the disseminator requires protection against inferential disclosure, for each hypercube, a lower bound can be calculated for the width of the feasibility interval of the primary suppression resulting from the suppression of all corner points of the particular hypercube. To compute that bound, it is not necessary to implement the time-consuming solution to the linear programming problem and it is possible to consider bounds on cell values that are assumed to be known a priori to an intruder. If it turns out that the bound for the feasibility interval width is sufficiently large, that hypercube becomes a feasible hypercube. The algorithm then selects from the set of feasible hypercubes, the one with the smallest sum of cell costs associated to its corner points. These corner points will then be added to the suppression pattern. Repsilber (1994) proposes to assign cell costs dynamically as a logarithmic transformation of the cell value but also depending on the level of a cell, to avoid suppression of marginal cells.

Unlike the 'cycle criterion' of the network flow heuristics, the 'Hypercube criterion' is a sufficient but not a necessary criterion for a 'safe' suppression pattern. Thus, it may happen that the 'best' suppression pattern may not be a set of hypercubes. In that case, of course, the Hypercube method will miss the best solution and lead to some overprotection.

Regarding the problem of singleton disclosure, the implementation of Repsilber (1994) makes sure that a single respondent cell will never appear to be corner point of one hypercube only. It is ensured that it is a corner point of at least two (feasible) hypercubes. This implies protection not only under the weak risk model *Relation* but also under the more rigorous model *Sub-table* of Section 4.3.4.

4.4.4   Secondary cell suppression in hierarchical and linked tables

As explained in Section 4.1.2, hierarchical tables are specially linked tables: at least one of the spanning variables exhibits a hierarchical structure, i.e. contains (many) sub-totals. In the following, we explain how the Hypercube method and the ILP approach, as described in the previous paragraphs, can be applied to hierarchical and linked tables in a practical way.

Both algorithms have also been implemented in a way to deal with hierarchical and generally linked tables directly. However, applying these methods in practice to large hierarchical and linked tables becomes too computation extensive. Network flow heuristics on the other hand are usually fast enough, but are limited to tables with at most two dimensions of which only one can be hierarchical. If it would make sense to build a linear programming approach like the one of, e.g., Sande and Robertson (see Section 4.4.3) into the kind of heuristic described later is an open research issue. For large sub-tables with more than two or even three dimensions for which computation of the optimal ILP suppression pattern tends to be time consuming, it might eventually be an interesting alternative.

The idea of the heuristic is to protect sub-tables separately. Before protecting a particular sub-table, any complementary suppression belonging also to one of the other sub-tables is noted and suppressed in the table at hand as well. Then the cell suppression procedure for the current (sub-)table is applied. This approach is called a *backtracking procedure*. Within this backtracking process for a hierarchical table, the cell-suppression procedure will usually be repeated several times for each sub-table. However, the total computational effort for this process will be much smaller compared to the computational effort needed to protect the entire hierarchical table all at once. The backtracking approach will provide a sub-optimal solution that minimises the information loss per sub-table, but not necessarily the global information loss of the complete set of hierarchically linked tables.

It must also be stressed, that a backtracking procedure is not a global one according to the denotation in Cox (2001). See Cox (2001) for a discussion of residual disclosure risks related to non-global methods for secondary cell suppression. However, adopting the risk model *Sub-table* of Section 4.3.4, one can accept the protection provided by such a backtracking procedure.

Of course, for a backtracking procedure, the order sequence of the sub-tables has certainly an impact on the result. In the following, we give a short description of a typical approach. For a more detailed description of the algorithm known as the *Modular* method of the τ-ARGUS package that integrates the ILP approach of Fischetti and Salazar into a backtracking scheme, including some examples, see, e.g., de Wolf (2002).

Typically, backtracking procedures deal with hierarchical tables using a top–down approach. The first step is to determine the primary unsafe cells in the base table consisting of all the cells that appear when crossing the hierarchical spanning variables. This way all cells, whether representing a (sub-)total or not, are checked for primary suppression. Knowing all primary unsafe cells, the secondary cell suppressions have to be found in such a way that each (sub-)table of the base table is protected and that the different tables cannot be combined to undo the protection of any of the other (sub-)tables. The basic idea behind the top–down approach is to start with the highest levels of the variables and calculate the secondary suppressions for the resulting table. The suppressions in the interior of the protected table are then transported to the corresponding marginal cells of the tables that appear when crossing lower levels of the two variables. All marginal cells, both suppressed and not suppressed, are then 'fixed' in the calculation of the secondary suppressions of that lower level table, i.e. they are not allowed to be (secondarily) suppressed. This procedure is then repeated until the tables that are constructed by crossing the lowest levels of the spanning variables are dealt with.

A suppression pattern at a higher level only introduces restrictions on the marginal cells of lower level tables. Calculating secondary suppressions in the interior while keeping the marginal cells fixed, is then independent between the tables on that lower level. That is, all these (sub-)tables can be dealt with independently of each other using, for example, the Hypercube method or the ILP approach of Fischetti and Salazar. Moreover, primary suppressions in the interior of a lower level table are dealt with at that same level: secondary suppressions can only occur in the same interior, since the marginal cells are kept fixed.

However, when several empty cells are apparent in a low-level table, it might be the case that no feasible solution can be found if one is restricted to suppress interior cells only. Unfortunately, backtracking is then needed: protecting the sequence of sub-tables has to be repeated. The algorithm will then keep the secondary suppressions of the first step (or of the previous repetition) and add additional ones, if necessary.

Obviously, all possible (sub-)tables should be dealt with in a particular order, such that the marginal cells of the table under consideration have been protected as the interior of a previously considered table.

For generally linked tables, to determine a suitable order sequence to handle the (sub-)tables is less obvious. In de Wolf and Giessing (2008) several options are discussed. In the following, we briefly outline two of them, referred to as *extended modular* (in the cited paper called *adapted modular*) and *traditional*.

In order to discuss these options, it is convenient to introduce the term *cover table*. In general, this is the smallest table that has all linked tables that need to be protected as genuine sub-tables.

The extended modular approach is just the modular approach as described earlier, while skipping all sub-tables of the cover table that do not appear as a sub-table in any table of the set of linked tables that needs to be protected. That is, only those sub-tables of the cover table will be protected that appear as sub-table of at least one of the considered linked tables. For an application of this extended modular approach in combination with the ILP approach, see de Wolf and Hundepool (2010).

The traditional approach is to protect all sub-tables of one table and then proceed to the next table. For a set of two linked tables, this means we carry over the secondary suppressions of the first table to the second, protect the second table, carry over suppressions to the first table and protect the first table again. This is then repeated until no changes will be found. For sets of more than two linked tables a suitable looping scheme is not that obvious, see e.g., the discussion in Giessing (2009). For applications to linked tables, Repsilber's implementation of the Hypercube method builds on the scheme denoted as *simple linked tables sequence* in Giessing (2009).

Regarding linked tables, the recent version of τ-ARGUS offers the backtracking procedures outlined here, in the variant of de Wolf and Hundepool (2010) of the ILP method and of Repsilber's implementation of the Hypercube method, see Repsilber (1994).

It is important to observe that the methodology outlined here for linked tables can only work well when protecting a set of linked tables *at the same time*. The problem of assigning secondary cell suppressions in a set of tables that must be consistent with the cell suppressions of another set of linked tables *released earlier* and linked to the 'current' set may well turn out to be intractable due to infeasibility problems. Note that this is a major concern with cell suppression, raised, for example, in the context of modern infrastructure facilities to generate user requested tables. It is one of the main reasons why nowadays there is some discussion on replacing cell suppression methodology by perturbative protection methods.

4.5   Perturbative protection methods

In the case of perturbative protection methods, the distinction between pre-tabular methods manipulating the microdata before they are summed up for tabulation and post-tabular methods which are applied during or after tabulation is quite substantial.

What could be considered as important advantage of pre-tabular methods is the fact that, once the perturbed data set has been created, all tables derived from this data set will satisfy two paradigms, that of *table additivity* and that of *consistency of linked tables*. The table additivity paradigm means that all table relations that hold for tables computed for the original data will also hold for the perturbed data. Consistency of linked tables, on the other hand, means that any pair of (logically) identical cells of two linked tables will always have identically perturbed (or unperturbed) cell values.

For post-tabular perturbative methods, it can be difficult or even impossible to guarantee in general that neither of these paradigms is violated. In Section 4.5.2, we describe a method that guarantees table additivity, but it is very likely that feasibility problems occur when consistency of linked tables is requested, unless the set of tables is processed simultaneously. Section 5.4.2 on the other hand mentions a method referred to as 'ABS cell perturbation' that guarantees full consistency of linked tables, unless one requires that additivity is restored to the perturbed tables[4]. The advantage of post-tabular methods is that they can be designed to preserve data quality for each particular requested table in an optimal way according to some criteria. See Section 4.6.3 for a discussion of criteria to rate data quality.

It should be stressed that with strongly skewed business data, in order to provide sufficient protection, some cells with large cell values may have to change by amounts that are comparatively big relative to the cell values of other cells (in particular cells with relatively small cell values) in the table(s). Considering the perturbation as some kind of random noise (even if the perturbation is not in fact determined as outcome of a stochastic process) this means that the variance of the noise should definitely *not* be more or less constant for all cells. A typical (non-stochastic) method that yields more or less constant variance of the noise would be controlled rounding (see Section 5.4.3), where all cell values are rounded to 'adjacent' multiples of a user-defined, fixed rounding base like 3, or 10 or 1000, etc. Therefore, we do not discuss controlled rounding in this section.

With a perturbative method, the effect of data protection on data quality is not obvious to the users of the data. It may happen that for a non-negligible number of table cells the perturbations are so large that they affect the quality of the data severely. That is, more severe as compared to other quality affecting issues like sampling error, misclassification, non-response, data editing, etc. Therefore, it might be necessary to release quality indicators along with the perturbed data, e.g., flagging cells with especially large deviations. Otherwise data users might generally lose confidence in the quality of the data presented in the tables.

But one should also bear in mind the disclosure risk potential, if these quality indicators can be used to infer what we have denoted as 'implicit intervals' in Section 4.3.1. Typical cases of this kind of disclosure risk potential are methods that supply users of the data with an interval for the true value of each perturbed value, like, for instance, rounding methods. Methods like controlled rounding (see Section 5.4.3), or the partial cell suppression method Fischetti and Salazar-González (2003) mentioned in Section 4.4.3, take those risks into account explicitly. They guarantee that feasibility intervals computed considering the intervals provided by the algorithm will provide sufficient protection. That is, those feasibility intervals will cover the respective protection intervals. However, in the case of linked tables, it can happen that different intervals are produced for identical cells. The methods will then not guarantee that the intersection of these intervals which could of course easily be computed by a user still covers the protection interval. So, in the case of linked tables, there might be disclosure risks, which cannot be handled by the existing implementations of those algorithms. And unlike the case of cell suppression, for those methods no heuristic procedures have yet been developed to handle linked tables problems.

---

[4] See Giessing (2011) for suggestions on how to extend this method to make it suitable for skewed magnitude tabular data.

Generally speaking, up to now, in contrast to the case of frequency tables, literature on perturbative tabular data protection methods that are suited for skewed magnitude tabular data is comparatively scarce. In this chapter, we provide only brief presentations of one pre-tabular and one post-tabular method which have been under discussion or in use for some time now. We would like to point out that research and development of suitable methods is certainly a field that deserves more attention.

4.5.1 A pre-tabular method: Multiplicative noise

Evans *et al.* (1998) proposed pre-tabular multiplicative noise (see Section 3.7.2) for the protection of enterprise tabular data. Zayatz (2007) notes that such a methodology is used at the US Census Bureau for tabular magnitude data protection for several data products. A research report Nayak *et al.* (2010) examines statistical properties of two variants. According to the simple variant, multiplicative noise with a mean of one and constant variance is assigned to the microdata. In this case, the conditional (unit level) coefficient of variance of the noisy (micro)data is a constant, defined by the noise variance. For the second variant, a balanced noise method (see Massell and Funk 2007), the research report proves that for any set of units, the perturbed total is an unbiased estimate of the original total. Moreover, it states that the balancing mechanism works indeed, i.e., it reduces the noise variance of the cell totals in a reference table.

4.5.2 A post-tabular method: Controlled tabular adjustment

Controlled tabular adjustment or CTA is a relatively new protection method for magnitude tabular data, suggested for instance in Cox and Dandekar (2002), Castro (2006), and Castro and Giessing (2006a, 2006b).

CTA methodology aims at finding the closest additive table to the original table ensuring that adjusted values of all confidential cells are safely away from their original value (considering the protection intervals) and that the adjusted values are within a certain range of the real values.

Several variants of CTA have been discussed in the literature (see, e.g., Cox and Dandekar 2002, Cox *et al.* 2004, Castro 2006). They suggest to obtain CTA by using (mixed integer) linear programming methodology. The main differences between those alternatives are on one hand the way in which the deviation sense of the primary suppressions is determined (heuristically vs. optimal). On the other hand, the definition of constraints matters (forcing the adjusted values to be within a 'certain range' of the real values). And finally, there is the issue of information loss or cell costs, i.e., the distance metric used to determine what is 'close' to the original table. Typically, weighted metrics are used. Implementations usually offer a choice of cost functions.

As an alternative to the above linear programming based approaches, Cox *et al.* (2006) and Ichim and Franconi (2006) suggest methodology to achieve CTA using statistical methods like iterative proportional fitting (IPF) or calibration techniques.

4.6   Information loss measures for tabular data

There are two different purposes for information loss measures in magnitude tabular data protection. The first is to guide and control a post-tabular protection process. The second is to inform on the outcome of a protection process. In both cases, there are differences in the information loss assessment used in connection with perturbative and non-perturbative data.

The challenge of tabular data protection is to preserve as much information in the table as possible, while creating the required uncertainty about the true values of the sensitive cells, as explained in the previous sections. It is necessary to somehow rate the information content of data, in order to be able to express the task of selecting an 'optimal set' of secondary suppressions as a mathematical programming problem. In case of perturbative methods, it is again needed to express the task of adjusting a table in an optimal way as a mathematical programming problem. Information loss is expressed in these mathematical models as the sum of costs associated to either the secondary suppressions, or to the non-sensitive cells subject to adjustment.

4.6.1   Cell costs for cell suppression

For cell suppression, the idea of equating a minimum loss of information with the smallest number of suppressions is probably the most natural concept. This would be implemented technically by assigning identical costs to each cell. Yet experience has shown that this concept often yields a suppression pattern in which many large cells are suppressed, which is often undesirable. In practice, other cost functions, based on cell values, power transformations thereof or cell frequencies yield better results. Note that several criteria, other than the numeric value, may also have an impact on a users perception of a particular cells importance. Think, e.g., on its place within the table (totals and sub-totals are often rated as highly important), or its category (certain categories of variables are often considered to be of secondary importance). The Hypercube method of Repsilber (1994) uses an elaborate internal weighting scheme in order to possibly avoid the suppression of totals and sub-totals which is essential for the performance of the method.

Sometimes even dynamic cost schemes are used to make an algorithm prefer which cells to select as secondary suppressions. For example, a cell of a size class that is adjacent to a suppressed cell might be preferred, because then often the two cells (the two size classes) can be collapsed 'locally'. That is, the (hopefully) non-sensitive total value of the two can be recalculated by some differencing and provides meaningful information.

Very common functions that determine cell costs (called 'cost functions') are the following:

- Unity, i.e., all cells have equal weight (minimising the number of suppressions).

- The number of contributors (minimising the number of contributors to be hidden).

- The cell value itself (preserving as much as possible the largest and often most important cells).

- The cell value of another variable in the data set (that variable will be the indication of the importance of the cells).

Especially in the latter two cases the cell value could give too much preference to the larger cells when the data is highly skewed. But is a 10 times bigger cell really 10 times as important? Sometimes there is a need to be a bit more moderate. In that case, a transformation of the cost function is desired. Popular transformations are taking the square-root or a log-transformation. Box and Cox (1964) have proposed a system of power transformations:

$$y = \frac{x^\lambda - 1}{\lambda}.$$

In this formula $\lambda = 0$ yields a log-transformation and the -1 is needed for making this formula continuous in λ. To avoid negative values, a simplified version of the transformation has been proposed in Hundepool *et al.* (2011):

$$y = \begin{cases} x^\lambda & \text{for } \lambda \neq 0 \\ \log(x) & \text{for } \lambda = 0. \end{cases}$$

For example, choosing $\lambda = 1/2$ will give the square-root transformation.

Note that the effects of the cost functions as mentioned above would be expected to appear for unstructured (non-hierarchical) tables handled by the ILP algorithm (see Section 4.4.3). For larger hierarchical tables, some explorative 'playing' with cell cost transformations can be useful.

4.6.2 Cell costs for CTA

The linear programming-based CTA algorithms minimise the sum of distances between true and perturbed cell values, weighted by cell costs. That is, they do not consider the sum of the cell costs directly. This has to be taken into account when developing cost functions for CTA.

Cox *et al.* (2004) discuss several cost functions to achieve 'quality preserving' CTA solutions. Castro and Giessing (2006b) propose a cost function to reflect the notion that cells on higher hierarchical levels or in the margins of tables are 'more important'.

4.6.3 Information loss measures to evaluate the outcome of table protection

For cell suppression, usually the loss of information due to secondary suppression is captured in terms of the number and/or the sum of the cell values of secondary suppressions. To provide more detailed information, this kind of statistic is sometimes presented not just globally on the table level, but in a detailed breakdown according to the hierarchical structures of the protected table(s).

For perturbative methods, measures evaluating the outcome are vital to maintain the user's trust in the data. A first, global measure would be a statistic based on the distribution of the cells as a function of the (relative) deviation between true and perturbed cell values. However, when such a global statistic is released, indicating that for some cells there is a rather strong effect, users might be tempted to question the quality of all cells. Such behaviour could be avoided with local information loss measures, providing information on the individual cell level.

In contrast to a global measure, a local measure will inform the data user on the reliability of each particular (adjusted) cell value. In addition, Castro and Giessing (2006a) discuss criteria that could be used by data providers to decide whether an adjustment is so small that there is actually no need for the cell to be flagged as 'adjusted'.

In order to be able to publish a single value in the tables while at the same time clarifying to the users that these values are not the original values, Giessing *et al.* (2007) suggest to round the values resulting from CTA. The rounding base would be chosen in such a way that the rounding interval includes both the true and the adjusted value. This strategy requires however some occasional post-processing to ensure that the protection of the sensitive cells is sufficient. In this way, one can take into account also disclosure risks connected to the release of the rounding intervals.

4.7   Software for tabular data protection

We begin this section with an overview of software for the methods and algorithms presented in the current chapter for risk assessment and protection. With respect to the latter, we concentrate on implementations of post-tabular methods presented in Table 4.4.

There are two software packages for magnitude tabular data protection that have been developed to serve not only the needs of the agency which has implemented the tool, but to be used by a wider community. One is the τ-ARGUS package (see Hundepool *et al.* 2011), the other one is the R-package *sdcTable* (see Meindl 2009 and Meindl 2011). According to the author, the R-package is basically an open source re-implementation of τ-ARGUS. Up to now, neither a detailed methodological comparison of the two packages is available, nor has a rigorous empirical comparison of the two packages taken place. In the following, we focus on the established τ-ARGUS package for which more documentation and documented test results are available. Apart from those two packages, we would also like to mention Statistics Canada's tabular data confidentiality software based on SAS, CONFID2 (see Frolova *et al.* 2009)[5], a package of the US Census Bureau[6]

---

[5] An older package, implementing basically the same secondary cell suppression algorithm, is the commercial package ACS (see Sande 1999).

[6] With respect to the USBC package, in the following we only refer to the implementation of the network algorithm (see Jewett 1993), although there might be extensions that we are not aware of.

and the package CIF[7] distributed by Eurostat, available to registered users of the Eurostat Circa site at http://circa.europa.eu/Members/irc/dsis/structbus/library?l=/ confidentiality/confidentiality_1\&vm=detailed\&sb=Title\&cookie=1. In addition to those, there probably exist numerous home-brewed software tools which have never been presented internationally.

The software packages cover one or more of the process steps:

1. Table definition/table computation.

2. Disclosure risk assessment I: determine primary sensitive cells, protection levels, etc.

3. Table protection.

4. Disclosure risk assessment II: audit of protected tables.

5. Output of protected tables and information loss information.

Steps 1, 2 and 5 involve no really complex tasks. In principle, they could easily be integrated into the usual table production process. Except that for older tabulation tools, it might be a problem to obtain the n largest contributions for each cell, which is a pre-requisite for assessment of primary sensitivity according to a concentration rule. In other words, older tabulation tools might not be able to provide the metadata necessary to be able to deal with Step 2.

Step 4 is a little more demanding, requiring some linear programming. However, algorithms for the audit are well known and not too difficult to implement with standard LP solver packages.

Therefore, when discussing tabular data disclosure control software, the emphasis is on Step 3: the algorithms implemented for table protection.

Regarding the post-tabular perturbative method, we consider in this chapter (i.e., Controlled Tabular Adjustment), Castro and Gonzalez (2009) describe developed software, implementing an algorithm for optimal CTA. Another implementation based on a simple heuristic to fix the direction for the adjustment of sensitive cells is available in the R-package *sdcTable* (see Meindl 2009).

4.7.1 Empirical comparison of cell suppression algorithms

We will compare the implementations of secondary cell suppression algorithms offered by the previously mentioned packages, with respect to some practical issues, disclosure risk management and performance regarding information loss.

The empirical results we will use are based on the conclusions of two evaluation studies comparing the τ-ARGUS algorithms Modular ILP, Hypercube and Hyper0. We will refer to these studies as study A and study B.[8]

---

[7] The package CIF is basically a graphical user interface to Repsilber's Hypercube method mentioned in Section 4.4.3. The Hypercube method is also offered by the τ-ARGUS package.

[8] For further details on these studies, see Giessing (2004) (study A) and Giessing *et al.* (2006) (study B).

However, these studies did not involve the American packages[9]. Both studies are based on tabulations of a strongly skewed magnitude variable. For study A, we used two- and three-dimensional hierarchical tables, with a total number of cells varying between 460 and 150 000. For study B, results have been derived for two-dimensional tables with a seven-level-deep hierarchical structure (given by the NACE economy classification) and a four-level-deep structure of the other dimension given by the variable Region. We also do not report results for the network algorithm offered by the package because it does not provide any protection against singleton disclosure, and is therefore not usable in production[10].

Practical issues of software implementations

The most important practical issues for a software package are probably license costs, interface issues, technical requirements (like platform availability) and speed.

License costs
Regarding license costs, the two packages τ-ARGUS and *sdcTable* are available for free. However, unlike the hypercube and network algorithm, the τ-ARGUS algorithms ILP and Modular ILP require a relatively expensive licence for a commercial LP solver (either XPRESS or CPLEX). The R-package *sdcTable* on the other hand offers implementations of those algorithms in connection with free open-source LP solvers. As for the packages of Statistics Canada and the US Census Bureau, we have no information on the policy of these agencies regarding sharing of the software. Anyway, the CONFID2 package also relies on a commercial LP solver: the SAS/OR® LP solver.

Interface and platforms
τ-ARGUS comes along with a graphical user interface which is a great advantage when introducing the software to new and inexperienced users. Once use of a package has been established, for expert users it will be more important that the tool is easy to integrate into the production environment. In that respect, all packages offer tools for batch mode or similar convenience. While τ-ARGUS is available only for the Windows platform, *sdcTable* can also be used on other platforms. We do not have platform information regarding the other packages.

Speed
Speed differs widely for the secondary cell suppression algorithms offered by the packages. Network and Hypercube methods are generally comparatively fast. The ILP method offered by τ-ARGUS is in most cases too slow for typical hierarchical

---

[9] We are not aware of any recent comparative studies involving either of the packages.

[10] Experience is that the Modular ILP is usually fast enough to handle the two-dimensional applications that the network method can deal with. Hence, no capacity has been devoted on a respective extension of the network algorithm, since it will anyway be unable to deal with three-dimensional tables or two-dimensional tables with hierarchical structure in both dimensions.

table applications in statistical agencies, which is also why it was not considered in the empirical studies. For hierarchical tables, the Modular ILP algorithm is much faster. However, for very large two- and three-dimensional tables, execution times can take several hours (choice of the LP solver also matters here). The Hypercube method normally deals with such applications within a few minutes, which is its main strength. As we do not have results regarding the LP algorithm of CONFID2 on the same or similar tables, we cannot exactly compare it, but results reported in Frolova *et al.* (2009) indicate that speed may be in a similar range as that of the Modular ILP algorithm of τ-ARGUS.

Risk concepts and empirical disclosure risks

Next we will compare properties of the implementations of secondary cell suppression algorithms regarding disclosure risk limitation. Recollecting what we denoted as singleton disclosure risk and the risk models for hierarchical and linked tables of Section 4.3.4, Table 4.11 shows under which risk model the respective algorithm implementations grant proper protection in general. Moreover, it shows additional information regarding the protection of singletons against disclosure by another singleton.

At the end of the paragraph on protection levels in Section 4.3.2, we have distinguished between the more rigorous protection against risks of *inferential* disclosure (when the feasibility interval of a sensitive cells does not completely cover the protection interval) and weaker protection only against *exact* disclosure (when the feasibility interval consists of only the true cell value). The behaviour of the methods related to this 'type of risk' is given in the last column of Table 4.11. Recall that the weakest

Table 4.11   Risk model and type of risk taken into account by secondary cell suppression algorithm implementations.

Package[a]	Algorithm	Risk model		Type of risk
		In general	Regarding singletons	
τ- ARGUS	ILP	Table	Relation	Inferential
	Modular ILP	Sub-table	Relation	Inferential
	Hypercube	Sub-table	Sub-table	Inferential
	Hyper0[b]	Sub-table	Sub-table	Exact
	Network	Table	No	Inferential
CONFID2	LP	Table	Relation	Inferential
USBC	Network	Table	Table	Inferential

[a]The table has no entries on the CIF package because it is based on the same implementation of the Hypercube method as implemented in τ-ARGUS.

[b]Hyper0 is indeed not a separate algorithm. It can be selected from the τ-ARGUS suite of algorithms by modifying a parameter of the Hypercube algorithm. The only difference between Hyper0 and Hypercube is that the relaxed variant Hyper0 only considers risks of exact disclosure but not of inferential disclosure.

risk model *Relation* means that only risks of disclosure by differencing between cells occurring in the same table relation are considered by the implementation. The stronger *Sub-table* and *Table* models also request protection against the type of disclosure risk that occurs when sets involving more than one relation of a sub-table (or table) are taken into account.

For use in the production of statistics at statistical offices, we advise *not* to use any method that does not provide any protection against singleton disclosure (entry 'No' in the 'Regarding singletons' column of Table 4.11).

It should be stressed here that home-brewed software often supports the weakest risk model *Relation* in the general case and not only in case of regarding singleton risks. Often, there is no protection at all against singleton disclosure, and the packages usually only address protection against exact disclosure.

In theory the risk model *Sub-table* is inferior to the risk model *Table*. However, how much does it matter in practice? Especially, considering that this kind of risk is rather a risk *potential* and not comparable to the disclosure risk that would result from a publication of cells found 'at risk' in tables protected according to the *Sub-table* model. After all, the effort connected to the computation of feasibility intervals based on sets of table equations beyond those relating to the same sub-table would be quite substantial for an intruder. Moreover, in a part of the disclosure risk cases only insiders (other respondents) would actually be able to disclose the individual contribution.

Empirical disclosure risks

The studies A and B we mentioned, provide some empirical results on the disclosure risk. Study B found between about 4% (protection by Modular ILP) and about 6% (protection by Hyper0) of sensitive cells at risk of inferential disclosure in an audit based on the *Table* risk model. Results from study A are similar.

Under the relaxed risk model *Sub-table*, i.e. when the audit is carried out separately for each sub-table for the Modular ILP no cells at risk are found. For Hyper0, study B found about 0.4% cells at risk of inferential disclosure, when computing feasibility intervals for several sub-tables of a detailed tabulation.

For those sub-tables, when protected by the Modular ILP, for about 0.08% of the singletons, it turned out that another singleton (after undoing the suppression of 'his own' cell) would be able to disclose its contribution when doing the audit under the *Sub-table* model. Hyper0 on the other hand provides singleton protection not only under the relaxed *Relation* model, hence no such case was found for tables protected by Hyper0.

Performance regarding information loss

Using the number and the sum of the cell values of secondary suppressions to measure information loss, studies A and B report the following results:

- Best results are achieved by the Modular ILM method, while using either variant of the Hypercube method leads to an often quite considerable increase in the amount of secondary suppressions compared to the result of the modular method.

- The size of this increase varies between hierarchical levels of the aggregation. On the higher levels of a table, the increase tends to be even larger than on the lower levels. In study B, on the national level we observed an increase of 75% for Hypercube (41% for Hyper0) on the state level. On the district level, the increase was about 28% on the higher levels of the economy classification for Hypercube (9% for Hyper0), and 20% for Hypercube (14% for Hyper0) on the lower levels. In study A, we observed increases mostly around 30% for Hypercube. For the smallest two-dimensional and the largest three-dimensional instance even an increase of 50% and 60%, respectively.

- When variable Region was taken down to the community level, we found that this additional detail in the table caused an increase in suppression at the district level of about 70–80% with Hyper0 and about 10–30% with Modular.

The main conclusion of the studies is that, while both the Modular ILP method and the Hypercube method satisfy the same standard regarding disclosure risk management, the Modular ILP gives much better results regarding information loss. Even compared to a variant of Hypercube with relaxed protection standard (Hyper0), it performs clearly better. This holds true especially for the stronger aggregated parts of large and detailed hierarchical tables. Although longer computation times for this method (compared to the fast Hypercube method) can be a nuisance for the modular ILP method, the results clearly justify this additional effort. The (computational) costs for protecting large hierarchical tabulations should be set relative to the total costs involved in processing a survey. Doing this, it would not be justified to use the Hypercube (or Hyper0) method only in order to avoid the costs for the commercial license which is necessary to run the optimisation tools of Modular.

4.8   Guidelines: Setting up an efficient table model systematically

In Section 4.3, we have introduced the concept of secondary disclosure risk assessment. We have explained how the protection of primary sensitive cells can be undone through some clever differencing between published cells when, according to their definition, some cells are logically the sum of other cells. More generally, we have pointed out that when this kind of additive relation between cells exits, users of the data can compute (possibly narrow) feasibility intervals. This can be done using a priori knowledge in conjunction with straightforward linear programming based on the table relations. The a priori knowledge on the table cells could exist of generally known bounds to suppressed or perturbed cells. Another form of a priori knowledge would be published values of logically identical cells in linked tables. Hence, in order to manage secondary disclosure risks of this kind, all the linear relations and identities between the cells of the set of published tables have to be modelled.

In the absence of confidentiality concerns, a statistician creates a table in order to show certain properties of a data set. Or she publishes a table to enhance comparison between different variables. A table might literally mix apples and oranges. Moreover, statisticians may wish to present a number of different 'properties', publishing multiple tables from a particular data set.

In this section, we provide some guidance on how to model in a systematic way, a set of tables planned to be released, as a starting point for a disclosure control discussion. We provide suggestions how in some typical constellations the table model can be efficiently relaxed. The outcome of this kind of analysis would be a single set (or eventually multiple sets) of linked tables, together with a mapping protocol. Tables in this set are called 'relaxed' tables. Disclosure control should be carried out only for the tables of the relaxed model. Afterwards, the protection of cells of the relaxed tables would be mapped to the cells of the original tables according to the mapping protocol.

A trivial, but fundamental consideration in this respect is that a linked tables setting in the sense of Section 4.3.4 is constituted only, if each of the tables involved presents sums of the same quantitative response variable. This means that when systematically modelling the tables, the first consideration has to be, which tables present the same (set of) response variables. In this way, the tables can be grouped. Each group of tables relates to a specific set of response variables. Therefore, we can assume without loss of generality that each table of a group of (eventually linked) tables is given by the definition of (categorical) spanning variables only. The group itself is determined by a set of quantitative response variables presented by all tables in the group.

We will discuss how to define those tables efficiently and in particular how to describe the relation structure of the spanning variables. In Section 4.8.2, we explain ways of dealing with multiple response variables. As an effect of those techniques, an initial group of tables may further separate in sub-groups that again are determined by a set of quantitative response variables. This time it would not be the set of response variables *presented* by all tables in the group, but the set of response variables *used for disclosure control* of all tables in the group.

Those sets of eventually linked, hierarchical tables can then be protected using, e.g. procedures mentioned in Section 4.4.4.

4.8.1 Defining spanning variables

Spanning variables of a table are categorical classifications that either define the breakdown of a population of statistical units into categories, or define a categorical breakdown of quantitative items collected for those units. A population of statistical units can, for example, be the population of enterprises of a given sector of the economy. Another example would be the local units of enterprises located in a geographical area. A typical example for a categorical breakdown of a quantitative item is a breakdown by gender of workforce information collected from enterprises. In the following, we refer in both cases to the *domain* of a spanning variable as its 'total' category in a specific table.

In our context, each spanning variable is defined by its domain, and by the relations between its categories, in other words by the structure of the variable. In the introduction (Section 4.1.2), we have presented well-established models to provide meta-information on the structure of hierarchical variables by, e.g., using a specific coding scheme, or by providing an indented list of categories.

It should be stressed that when using a protection algorithm that adopts the *Sub-table* risk model (c.f. Table 4.11), introducing a hierarchy of sub-totals into a classification with many categories helps to speed up a protection process significantly. It can even lead to a more favourable outcome regarding information loss.

Hierarchical Standard Classifications

In case of typical standard classifications like for the economic activity of a company, or for a geographical variable, often the same standard classification is used to define several tables. Usually, not all tables are displayed at the same amount of detail. Typically, a state-level table, for example, will offer more detail regarding economic activity than a district level table. In such a case, it is advisable to think of a *base hierarchy* for this classification that involves all the levels that are used in the definition of any of the tables in the set. It is not necessary to define all tables in the set using this base hierarchy down to the level of its most detail. But starting from the desired lowest level in a given table, all the levels above that level should also belong to the hierarchy of this variable as defined for this table. Otherwise, the protection methods will not ensure that a sub-total on level m protected in one of the tables is not disclosed when adding up the perhaps unprotected corresponding cells of another table where the definition of the hierarchy of this variable involves level $m + 1$, but not level m.

Once a base hierarchy has been defined for a classification, it will usually be enough to define the respective spanning variable for a table through this base hierarchy, its lowest level in the table, and the variable domain in the table.

A typical problem of table models based on a detailed standard classification can be that they are too detailed, requiring a lot of protection and thus causing a lot of information loss even at higher levels. It can be a good strategy to redefine such a large multidimensional table in a more specific way, as set of several smaller linked tables. Assume, for example, a three-dimensional table, where one of the spanning variables is a standard classification for geography. Assume the table is defined at its lowest level of detail, say the district level. It may then turn out that at this level a very large number of cells is either empty or primary sensitive. In such a case, it may make sense not to publish the full table. Instead, one could publish the three-dimensional marginal tables down to the district level of geography, a three-dimensional table at state level and perhaps some three-dimensional tables at district level for a few selected states ('domains' or 'sub-domains' in our notation) with particularly dense district populations of the statistical units.

4.8.2   Response variables and mapping rules

As explained earlier, it makes sense for disclosure control to analyse those groups of tables, where each table displays the same set of quantitative response variables. There are some issues connected to the response variable that matter for disclosure analysis. Has the variable been collected in a sample survey? Which (primary) sensitivity rule should be used? Which are the statistical units the response variable relates to? Do the response variables (or a subset thereof) define a categorical breakdown of another one, thus defining a spanning variable?

Response variables not defining a spanning variable

For the response variables that do not define a spanning variable, there are two popular techniques:

1. Apply primary and secondary disclosure control for the response variable in question.

2. Select another (usually well correlated) *lead variable*. Apply primary and secondary disclosure control for the lead variable. Map the resulting protection of the lead variable to the current response variable. For cell suppression, this means to suppress cells presenting the current response variable, if and only if the corresponding cell for the lead variable was suppressed during disclosure control.

Use of lead variables

As pointed out in Section 4.2.2, concentration rules like the dominance and $p\%$ rule make sense only if it is assumed that intruders are able to identify the largest contributors to a cell. While this is perhaps true for typical lead variables like 'turnover', it may not hold for other variables. For example, if the cell values are investments in some commodity, not always the largest companies have the largest values. Simply applying the sensitivity rules could result in a situation where a rather small company dominates the cell which will enforce protection for this cell. If we assume that this smaller company is not very well known and visible, there is no reason to protect this cell.

In addition to this, one can imagine several alternative techniques, also based on the selection of a lead variable and useful to coordinate protection of different tables with the same spanning variables. The basic idea would be to transfer the protection of a table presenting the lead variable to the same table (i.e. defined by the same spanning variables) presenting the response variable in question. This might include only primary protection or also secondary protection. Then add some primary protection based on the response variable in question and protect the current table.

This model is sometimes applied to *periodical data sets* as well. In that case, it is applied to avoid 'flipping' protection. For example, consider tables that are produced on a monthly basis. A cell that is not protected in month m could then be used to estimate a (nearly) identical, secondarily protected cell in month $m + 1$. Which in turn could lead to disclosure of a primary sensitive cell in that table for month $m + 1$.

Response variables defining a spanning variable

So far, we have discussed techniques to protect response variables which do not provide a categorical breakdown of another one, thus defining a spanning variable. In the following, we do now consider this case. That is, we consider the case where one response variable defines a domain, and other response variables are sub-domains of that domain. For example, the total investment of companies (domain), and their investment into several commodities (sub-domains). Two frequently practiced approaches for this scenario are the following:

1. Consider this spanning variable as a regular spanning variable of one joint table presenting all the thus related response variables. Protect according to the usual procedure.

2. Omit this spanning variable, i.e. protect a set of separate tables, each presenting another of the related response variables.

Omitting a spanning variable
We distinguish two typical cases where the technique of omitting a spanning variable can be useful:

1. The domain table results are considered less relevant, or have to be protected when sub-domain data are already published[11].

2. Only one of the sub-domain tables is published.

In the first case, protect the sub-domain tables first. In the second case, usually the domain table should be protected first. In both cases, when finally processing the last table, apply suitable additional rules for primary protection taking into account the outcome of the protection of the respective previous tables to achieve that:

- It must not be possible to disclose protected sub-domain table cells by differencing between unprotected domain table cells and unprotected corresponding cells of other sub-domains. Consider especially the case of two singleton sub-domain cells, when specifying those rules.

- It must not be possible to disclose protected domain table cells by adding published sub-domain cells.

The second goal can imply a requirement for the domain table to specify certain cells ineligible for secondary protection.

The technique of omitting a spanning variable is not only useful for spanning variables where domains and sub-domains relate to different response variables. It can also be applied when the domain relates to the full population of statistical units, and the sub-domains are respective sub-populations. For instance, EU-level data as

---

[11] Note that this technique can be quite useful when dealing with non-nested hierarchies. For example, in the case when the non-nested hierarchy can be divided naturally into one nested main hierarchy and a number of relations that actually group selected categories across different branches of the main hierarchy into new domains. Omitting these relations as spanning variables of tables relating to those new domains, the new-domain results would be considered as less relevant or to be published later. This can avoid or limit additional information loss caused by protecting the new-domain tables in the 'old' tables based on the main hierarchy.

domain and member-state-level data as sub-domains. An instance for a case with only one published sub-domain table could be farms as domain and farms organised in bio-organic associations as sub-domain.

4.9   Case studies

The subject of the case study consists of two sets of tables. The first set is of tables that the German Federal Statistical Office (FSO) releases in its own publications. The second set is of tables delivered to Eurostat for use and publication in the series of European Structural Business Survey (SBS) publications for two sectors of the economy, i.e., mining and production (NACE sectors B and C). Effectively, this set thus consists of two tables. The German FSO protects the tables in both sets using cell suppression. Of course, cell suppression in the tables of both sets must be coordinated.

For the sake of simplicity, let us assume that the German FSO's own publication involves the same response variables as those requested for the European tables.[12] The European tables present up to 24 variables, like, for instance, turnover, number of enterprises, salaries, working hours, etc.

4.9.1   Response variables and mapping rules of the case study

Ignoring a pre-release of preliminary results, the set of tables can be divided into two groups. Eleven response variables are presented in three tables (one National, two European ones), the other 13 variables are presented in two tables (one National, one European).

The statistical units are enterprises. All response variables are observations from a sample survey. The $p\%$ rule in the variant expressed by (4.16) in the paragraph on survey weights in 4.2.2 is applied for primary sensitivity assessment.

There is at least one case, where a subset of the response variables defines a categorical breakdown of another one, thus defining a spanning variable. This is the case where net salary + social insurance = total salary costs.

However, the approach chosen is to use turnover as lead variable for all the response variables. For the salaries case, this means in principle to omit the relation presented above as spanning variable and use the protection derived for turnover for all three variables. Of course, three identical patterns for the three variables will not cause any problems regarding the disclosure risk types mentioned in the paragraph on omitting a spanning variable in Section 4.8.2.

---

[12] As usual, in real life the situation is much more complex. Some variables appear only in the EU-delivery tables, others only in the national tables. Some appear in both EU-delivery tables and in the German national publication, some in either of the two EU-delivery tables and in the German national table.

Table 4.12   Coding scheme for NACE.

Level	Name of level	Digits in code	Example
0	Sector	–	C
1	2-Digit	$1-2$	10
2	3-Digit	$1-3$	101
3	4-Digit	$1-4$	1011

4.9.2   Spanning variables of the case study

The spanning variables of the tables are the European economic activity classification NACE and several variants of a size class grouping of the number of employees.

Meta-information on the structure of the NACE classification is provided by using a specific coding scheme, as described in Table 4.12. Four spanning variables are derived from this standard classification and used in the sets of tables of this case study. See Table 4.13, for a definition of these spanning variables. Table 4.14 shows meta-information on the structure of four size-class groupings for the variable 'number of employees' used to define the tables of our case study. For each of the four groupings, a list of the respective ranges is provided. The fourth grouping (SCL2) has a hierarchical sub-structure. In that case, the list of ranges is indented using '@' as indentation character.

4.9.3   Analysing the tables of the case study

Now we have defined the spanning variables that are used in our case study, we can discuss the structure of the tables.

The first group of tables
For each of the NACE sectors B and C, for the first group of tables (presenting the first set of 11 response variables) three tables (one National, two European delivery tables) have to be protected. The two European delivery tables are a two-dimensional table defined by cross combination of NACE-3 (B or C, respectively) and SCL_EU and a one-dimensional table defined by NACE-4 (B or C, respectively). The National

Table 4.13   Spanning variables derived from NACE.

Name of spanning variable	Level of most detail	Domain	Code for domain total
NACE3-B	2	Mining	B
NACE4-B	3	Mining	B
NACE3-C	2	Production	C
NACE4-C	3	Production	C

Table 4.14   Size-class spanning variables, defined using variable 'Number of Employees'.

Name of spanning variable	SCL_EU	SCL_N	SCL1	SCL2
Domain	All	≥ 20	All	All
Range[a,b]			$0-19$	$0-19$
	$0-9$			$@0-9$
	$10-19$			$@10-19$
				≥ 20
	$20-49$	$20-49$	$20-49$	$@20-49$
	$50-249$		$50-249$	$@50-249$
		$50-99$		$@@50-99$
		$100-249$		$@@100-249$
	≥ 250			$@\geq 250$
		$250-499$	$250-499$	$@@250-499$
		$500-999$	$500-999$	$@@500-999$
		≥ 1000	≥ 1000	$@@\geq 1000$

[a]Categories are intervals on number of employees. For example, $20-49$ means $20 \leq$ number of employees ≤ 49.
[b]Only SCL2 has a hierarchical sub-structure. This structure is given using an indented code list with '@' as character for indentation.

table is a two-dimensional table given by cross combination of NACE-4 (B or C, respectively) and SCL_N.

Comparing SCL_EU to SCL_N in Table 4.14, we find that SCL_N, the national structure, is more detailed for larger enterprises (with 50 and more employees), and is defined only for a sub-domain, i.e. it does not cover the small enterprises (less than 20 employees).

On the other hand, as NACE-4 of course completely covers NACE-3, results for NACE-3 are published according to both size class groupings. This means that we must make sure that a protected 50–249 result, for example, cannot be computed by adding corresponding unprotected 50–99 and 100–249 results. Therefore, when using a cell suppression software, we have to provide meta-information to make the secondary cell suppression algorithm properly consider the relation '50–249 = 50–99 + 100–249'. This can be achieved by creating a joint hierarchy, e.g. SCL2 of Table 4.14 which contains all categories of both, SCL_EU and SCL_N, and defines (by indentation) the hierarchical relations.

Using this joint hierarchy for all NACE categories down to NACE-3 is adequate. But on the NACE-4 level, it might cause oversuppression. On that level, only the respective total domain results for all enterprises are published (i.e. in the one-dimensional EU delivery table) and (in the national publication) results are released according to SCL_N (covering only the sub-domain of enterprises with 20 and more employees). This again means we have to create meta-information to make the secondary cell suppression algorithm ensure that a sensitive size class 0–19 cell on the 4-digit NACE level (which is not contained neither in the national, nor in the European delivery table) cannot be disclosed by subtracting the eventually unsuppressed '20 and more' sub-domain result of the national table from the corresponding German 4-digit NACE result of the EU table.[13] So, for the NACE-4 level, we need another joint classification, e.g., SCL1. SCL1 contains all categories of SCL_N. Its domain is the total domain of all enterprises, and it explicitly involves a category for the small enterprises '0–19', thus defining a relation to be taken into account explicitly by a cell suppression algorithm.

We can now define two new two-dimensional tables as cross-combinations of NACE-3 with SCL2 and of NACE-4 with SCL1. These two tables (or four tables, respectively, considering that there is always a version for sector B and for sector C) together contain all cells of the national table and of the two EU-delivery tables.

Looking at these tables, we find we have two sets of linked tables, one set containing the two tables for B, the other one the two tables for C. Both sets are structured identically, except that the hierarchy of NACE sector B of course is different from the hierarchy of NACE sector C.

The second group of tables

For the second group of the 13 remaining response variables, only two tables (one National, one European delivery table) have to be protected. The European delivery table is the one-dimensional table defined by NACE-4 (for B or C, respectively). The National table is the same as the one for the first group of tables above. With the arguments above, we find that for dealing with this second group of tables, we only have to protect the table given by cross-combination of NACE-4 with SCL1 (for, e.g., the lead response variable), considering again of course that there is a version for sector B and for sector C of the table.

4.9.4   Software issues of the case study

Each pair of linked tables (in the case of the first group) and of course the single table (in the case of the second group) can be handled by the algorithms for hierarchical and linked tables explained in Section 4.4.4. For example by the '*extended modular*' method of de Wolf and Giessing (2008) implemented now in τ-ARGUS in combination with the ILP method de Wolf and Hundepool (2010), by the '*traditional*' method as implemented in a wrapper package for the τ-ARGUS modular ILP algorithm written in SAS (Schmidt and Giessing (2011)), and by Repsilber's implementation of the Hypercube method also offered by τ-ARGUS.

---

[13] Even if the EU publication does not explicitly contain National results, if one of those is not explicitly flagged confidential, the protection methods applied to the EU-level data will not ensure that it cannot be disclosed by differencing between a published EU-total and corresponding published results of other Member States.

Using the results of Sections 4.9.2 and 4.9.3, it is now straightforward work to protect the data using either of the packages. All that has to be done is, for each of the two sectors, to provide the two linked tables (or the single ones for the two sectors, in case of the second group) in the data format required by the respective packages, along with the meta-information on the structure of the size class variables SCL1 and SCL2 (and of course also on NACE-3 and NACE-4) in the respective τ-ARGUS metadata format. After selecting a rule for primary sensitivity, the cell suppression procedure can be started.

Once this is finished, we have protected tables for the response variable 'turnover'. What remains is to map the suppressions assigned in the two B and C tables for the first group to the cells relating to the remaining other 10 response variables, and similarly to map the B and C table results for the second group to the cells relating to the 13 response variables of the second group.

5

Frequency tables

5.1 Introduction

This chapter discusses disclosure control for frequency tables, that are tables of counts (or percentages) where each cell value represents the number of respondents in that cell.

Traditionally, frequency tables have been the main method of dissemination for census and social data by National Statistical Institutes (NSIs). These tables contain counts of people or households with certain social characteristics. Frequency tables are also used for business data where characteristics are counted, such as the number of businesses. Because of their longer history, there has been relatively more research on protecting frequency tables, as compared with newer output methods such as microdata.

Section 5.2 of the chapter outlines the common types of disclosure risk and how the consideration of these risks leads to the definition of unsafe cells in frequency tables. The process of aggregating individual records into groups to display in tables reduces the risk of disclosure compared with microdata, but usually some additional Statistical Disclosure Control (SDC) protection is needed for unsafe cells in tables. Disclosure control methods are used to reduce the disclosure risk by disguising these unsafe cells. A range of different SDC methods is described in Section 5.3. A well-established method of SDC for frequency tables is rounding and Section 5.4.3 discusses alternative techniques such as conventional and random rounding, small cell adjustment and controlled rounding. Controlled rounding is currently the recommended method to protect frequency tables. Section 5.5 describes different information loss measures that can be used to evaluate the impact that different disclosure control methods have on the utility of frequency tables. Section 5.6 provides information on the software package τ-ARGUS and how it can be used to protect frequency

Statistical Disclosure Control, First Edition. Anco Hundepool, Josep Domingo-Ferrer, Luisa Franconi, Sarah Giessing, Eric Schulte Nordholt, Keith Spicer and Peter-Paul de Wolf.
© 2012 John Wiley & Sons, Ltd. Published 2012 by John Wiley & Sons, Ltd.

tables. Finally, Section 5.7 introduces some case studies from different countries on the different approaches to protecting tables constructed from census data.

5.2   Disclosure risks

Disclosure risks for frequency tables primarily relate to 'unsafe cells'; that is, cells in a table which could lead to a statistical disclosure. There are several types of disclosure risk and the associated unsafe cells can vary in terms of their impact. A risk assessment should be undertaken to evaluate the expected outcomes of a disclosure. In order to be explicit about the disclosure risks to be managed one should also consider a range of potentially disclosive situations and use these to develop appropriate confidentiality rules to protect any unsafe cells.

The disclosure risk situations described in this section primarily apply to tables produced from registration processes, administrative sources or Censuses, e.g. data sources with a complete coverage of the population or sub-population. Where frequency tables are derived from sample surveys, and the counts in the table are weighted, some protection is provided by the sampling process. The sample a priori introduces uncertainty into the zero counts and other counts through sample error.

It should be noted that when determining unsafe cells, one should take into account the variables that define the population within the table, as well as the variables defining the table. For example, a frequency table may display income by region for males. Although sex does not define a row or column, it defines the eligible population for the table and therefore must be considered as an identifying variable when thinking about these disclosive situations.

Disclosure risks are categorised based on how information is revealed. The most common types of disclosure risk in frequency tables are described later.

Identification as a disclosure risk involves finding yourself or another individual or group within a table. Many NSIs will not consider that self-identification alone poses a disclosure risk. An individual who can recall their circumstances at the time of data collection will likely be able to deduce to which cell in a published table their information contributes. In other words, they will be able to identify themselves but only because they know what attributes were provided in the data collection, along with any other information about themselves which may assist in this detection.

However, identification or self-identification can lead to the discovery of rareness, or even uniqueness, in the population of the statistic, which is something individuals might not have known about themselves before. This is most likely to occur where a cell has a small value, e.g. a 1, or where it becomes in effect a population of 1 through subtraction or deduction using other available information. For certain types of information, rareness or uniqueness may encourage others to seek out the individual. The threat or reality of such a situation could cause harm or distress to the individual, or may lead them to claim that the statistics offer inadequate disclosure protection for them, and therefore others.

Identification or self-identification may occur from any cells with a count of 1, i.e. representing one statistical unit. Table 5.1 presents an example of a low-dimensional

Table 5.1   Marital status by sex.

Marital status	Male	Female	Total
Married	38	17	55
Divorced	7	4	11
Single	3	**1**	4
Total	48	22	70

table in a particular area where identification may occur. The existence of a 1 in the highlighted cell indicates that the female who is single is at risk from being identified from the table. Often, with a small cell, it may not be possible to find an attribute disclosure (learning something *new* about an individual) but the individual who has self-identified may perceive a risk that someone else might be able to find out something about them.

Identification itself poses a relatively low disclosure risk, but its tendency to lead to other types of disclosure, together with the perception issues it raises means several NSIs choose to protect against identification disclosure. There are two types of perception that are considered as threats: (i) the perception that because there are small counts that no protection (or at least insufficient protection) has been applied and (ii) the perception of the individual who has identified themselves, that others can also identify them – and find out something new about them. Section 5.3 discusses protection methods which focus on reducing the number of small cells in tables.

5.2.1   Individual attribute disclosure

Attribute disclosure involves the uncovering of new information about a person through the use of published data. An individual attribute disclosure occurs when someone who has some information about an individual could, with the help of data from the table (or from a different table with a common attribute, perhaps even from a different data source), discover details that were not previously known to them. This is most likely to occur where there is a cell containing a 1 in the margin of the table and the corresponding row or column is hence dominated by zeros. The individual is identified on the basis of some of the variables spanning the table and a new attribute is then revealed about the individual from other variables. Note that identification is a necessary pre-condition for individual attribute disclosure to occur, and should therefore be avoided.

This type of disclosure is a particular problem when many tables are released from one data set. If an intruder can identify an individual, then additional tables provide more detail about that person. In continuation of the example shown in Table 5.1, the cell disclosing the single female as unique will ultimately turn into a marginal cell in a higher dimensional table such as Table 5.2 mentioned later and her number of hours worked is revealed. The table shows how attribute disclosure arises due to the

Table 5.2   Marital status and sex by hours worked.

Marital status/ Hours worked	Male			Female			Total
	More than 30	16–30	15 or less	More than 30	16–30	15 or less	
Married	30	6	2	14	3	0	55
Divorced	3	4	0	2	2	0	11
Single	2	0	1	0	0	1	4
Total	35	10	3	16	5	1	70

zeros dominating the column of the single female, and it is learned that she is in the lowest working hours band.

The occurrence of a 2 in the table could also lead to identification. This would be the case if an individual contributed to the cell and he/she could easily identify the other individual in the cell.

An example of potential attribute disclosure from the 2001 UK Census data involves 184 persons living in a particular area in the UK. Uniques (frequency counts of 1) were found for males aged 50–59, males aged 85+, and females aged 60–64. An additional table showed these individuals further disseminated by health variables, and it was learned that the single male aged 50–59 and the single female aged 60–64 had good or fairly good health and no limiting long-term illness, while the single male aged 85+ had poor health and a limiting long-term illness. Without disclosure control, anyone living in this particular area had the potential to learn these health attributes about the unique individuals. Full coverage sources – like the Census – are a particular concern for disclosure control because of their compulsory nature there is an expectation to find all individuals in the output. Although there may be some missing data and coding errors etc, NSIs work to minimise these, and in terms of Census the data issues are unlikely to be randomly distributed in the output. Certain SDC techniques can be adjusted to target particular variables (or tables) with more or less inherent data error, for example, providing more cell suppression for variables which are known to be of better quality and have fewer data issues.

5.2.2   Group attribute disclosure

Another disclosure risk involves learning a new attribute about an identifiable group or learning that a group does not have a particular attribute. This is termed group attribute disclosure, and it can occur when all respondents fall into a subset of categories for a particular variable, i.e. where a row or column contains mostly zeros and a small number of cells that are non-zero. This type of disclosure is a much neglected thread to the disclosure protection of frequency tables, and in contrast to individual attribute disclosure, it does not require individual identification. In order

Table 5.3   Martial status by hours worked.

Marital status	Full time	Part time	Total
	Hours worked		
Married	**6**	**0**	**6**
Divorced	5	1	6
Single	2	2	4
Total	13	3	16

to protect against group attribute disclosure, it is essential to introduce ambiguity in the zeros and ensure that all respondents do not fall into just one or a few categories.

Table 5.3 shows respondents in a particular area broken down by hours worked and income. From the table, we can see that all married individuals work full time, therefore any individual in that area who is married will have their hours worked disclosed.

The table also highlights another type of group attribute disclosure referred to as 'within group disclosure'. This occurs for the divorced group and results from all respondents falling into two response categories for a particular variable, where one of these response categories has a cell value of 1. In this case, the divorced person who works part time knows that all other divorced individuals work full time.

5.2.3   Disclosure by differencing

Differencing involves an intruder using two or more overlapping tables and subtraction to gather additional information about the differences between them. A disclosure by differencing occurs when this comparison of two or more tables enables a small cell (0, 1 or 2) to be calculated. Disclosures by differencing can result from three different scenarios which will be explained in turn:

1. *Disclosure by geographical differencing.* It may result when there are several published tables from the same data set and they relate to similar geographical areas. If these tables are compared, they can reveal a new, previously unpublished table for the differenced area. For instance, 2001 Output Areas (OAs) are similar in geographical size to 1991 Enumeration Districts (EDs), and a new differenced table may be created for the remaining area. A fictitious example of this is presented in Tables 5.4–5.6.

The above example demonstrates how simple subtraction of the geographical data in Table 5.4 from Table 5.5 can produce disclosive information for the area in Table 5.6. Similar issues arose in the planning for 2011 UK Census, where OAs were mostly kept consistent with those in 2001, that lay within administrative wards. Where ward boundaries had altered, there would have been possibilities to difference the sum of the constituent OAs to discover information about a small sliver. In the

Table 5.4   Single-person households and hours worked in Area A
(2001 OA definition).

	Single-person household male	Single-person household female
More than 30	50	54
16–30	128	140
15 or less	39	49

end, to avoid this, ward statistics were best fitted to OAs, rather than to the exact ward boundaries.

2. *Disclosure by linking.* It can occur when published tables relating to the same base population are linked by common variables. These new linked tables were not published by the NSI and therefore may reveal the statistical disclosure control methods applied and/or unsafe cell counts.

A fictitious example of disclosure by linking is provided in Tables 5.7–5.10, which are linked by employment status and whether or not the respondents live in the area. Note that Table 5.8 is not strictly necessary to obtain this disclosure.

Table 5.10 shows the new data which can be derived by combining and differencing the totals from the existing tables. The linked table discloses the female living and working in Area A as a unique.

Importantly, when linked tables are produced from the same data set, it is not sufficient to consider the protection for each table separately. If a cell requires protection in one table, then it will require protection in all tables, otherwise the protection in the first table could be undone.

3. The last type of disclosure by differencing involves differencing of *sub-population tables.* Sub-populations are specific groups into which data may be subset before a table is produced (e.g. a table of fertility may use a sub-population of females). Differencing can occur when a published table definition corresponds to a sub-population of another published table, resulting in the production of a new, previously unpublished table. If the total population is known and the sub-population of females is gathered from another table, the number of males can be deduced.

Table 5.5   Single-person households and hours worked in area A
(1991 ED definition).

	Single-person household male	Single-person household female
More than 30	52	55
16–30	130	140
15 or less	39	49

Table 5.6   New differenced table (via geographical differencing).

	Single-person household male	Single-person household female
More than 30	2	1
16–30	2	0
15 or less	0	0

Table 5.7   Area of residence or workplace.

	Working in area	Living in area	Living and working in area
Area A	49	102	22

Table 5.8   Employment status in Area A.

	Number of persons
Employed	85
Not employed	17
Total	102

Table 5.9   Males working and living in Area A.

	Living and working in Area A	Living in Area A and working elsewhere	Working in Area A and living elsewhere
Males	21	58	23

Table 5.10   New differenced table (via linking).

	Living and working in Area A	Living in Area A and working elsewhere	Working in Area A and living elsewhere
Males	21	58	23
Females	**1**	5	4
Total	22	63	27

Table 5.11   Hours worked by sex in Area A.

	< 20	20–39	40–59	60–69	Over 70
Male	6	9	5	8	4
Female	10	38	51	42	32

Tables based on categorical variables which have been recoded in different ways may also result in this kind of differencing. To reduce the disclosure risk resulting from having many different versions of variables, most NSIs have a set of standard classifications which they use to release data. An example using the number of hours worked is shown in Tables 5.11–5.13.

The example indicates how a new table can be differenced from the original tables, in particular a new hours worked group (for 20–24 hours) which reveals that the male falling into this derived hours worked group is unique.

More information on disclosure by differencing can be obtained from Brown (2004) and Duke-Williams and Rees (1998).

5.2.4   Perception of disclosure risk

In addition to providing actual disclosure control protection for sensitive information, NSIs need to be seen to be providing this protection. The public may have a different understanding of disclosure control risks and their perception is likely to be influenced by what they see in tables. If many small cells appear in frequency tables, users may perceive that either no SDC, or insufficient SDC methods, have been applied to the data. Section 5.3 discusses SDC methods, but generally some methods are more obvious in the output tables, than others. To protect against negative perceptions, NSIs should be transparent about the SDC methods applied. Managing perceptions is important to maintain credibility and responsibility towards respondents. Negative perceptions may impact response rates for censuses and surveys if respondents perceive that there is little concern about protecting their confidentiality. More emphasis has been placed on this type of disclosure risk in recent years due to declining response rates and decreasing data quality. It is important to provide clear explanations to the public about the protection afforded by the SDC method, as well as guidance on the impact of the SDC methods on the quality and utility of the outputs. Explanations should provide details of the methods used but avoid stating the exact parameters as this may allow intruders to partly or wholly unpick the protection.

Table 5.12   Hours worked by sex in Area A.

	< 25	25–39	40–59	60–69	Over 70
Male	7	8	5	8	4
Female	10	38	51	42	32

Table 5.13   New differenced table (via sub-populations).

	< 20	20–24	25–39	40–59	60–69	Over 70
Male	6	**1**	8	5	8	4
Female	10	0	38	51	42	32

5.3   Methods

There are a variety of disclosure control methods which can be applied to tabular data to provide confidentiality protection. The choice of which method to use needs to balance how the data are used, the operational feasibility of the method and the disclosure control protection it offers. SDC methods can be divided into three categories which will be discussed in turn further: (1) those that adjust the data before tables are designed (pre-tabular), (2) those that determine the design of the table (table redesign) and (3) those that modify the values in the table (post-tabular). Further information on SDC methods for frequency tables can also be found in Willenborg and de Waal (2001) and Doyle *et al.* (2001).

5.3.1   Pre-tabular

Pre-tabular disclosure control methods are applied to microdata before they are aggregated and output in frequency tables. These methods include: record swapping, overimputation, data switching, PRAM and sampling (see Chapter 3 for details of the methods). A key advantage of pre-tabular methods is that the output tables are consistent and additive since all outputs are created from protected microdata. Pre-tabular methods by definition only need to be applied once to the microdata and after they are implemented for a microdata set (often in conjunction with threshold or sparsity rules) they can be used to allow flexible table generation. This is because pre-tabular methods provide some protection against disclosure by differencing and any uncovered slivers will have already had SDC protection applied.

Table 5.14   Summary of disclosure risks associated with frequency tables.

Disclosure risk	Description
Identification	Identifying an individual in a table (working out specifically who the individual is)
Attribute disclosure (individual and group)	Finding out previously unknown information about an individual (or group) from a table
Disclosure by differencing	Uncovering new information by comparing more than one table.
Perception of disclosure	The public's feeling of risk based on what is seen in released tables

Table 5.15   Summary of tabular disclosure control methods.

	Pre-tabular	Table re-design	Post-tabular
Tables and totals will be additive and consistent	Yes	Yes	No
Methods are visible to users and can be accounted for in analysis	No	Yes	Yes
Methods need to be applied to each table individually	No	Yes	Yes
Flexible table generation is possible	Yes	No	No

Disadvantages of pre-tabular techniques are that one must have access to the original microdata. Also, a high level of perturbation may be required in order to disguise all unsafe cells. Pre-tabular methods have the potential to distort distributions in the data, but the actual impact of this will depend on which method is used and how it is applied. It may be possible to target pre-tabular methods towards particular areas or sensitive variables. Generally pre-tabular methods are not as transparent to users of the frequency tables and there is no clear guidance that can be given in order to make adjustments in their statistical analysis for this type of perturbation.

5.3.2   Table re-design

Table redesign is recommended as a simple method that can minimise the number of unsafe cells in a table and preserve original counts. It can be applied alongside post-tabular or pre-tabular disclosure control methods, as well as being applied on its own. As an additional method of protection, it has been used in many NSIs including the UK and New Zealand. As table redesign alone provides relatively less disclosure control protection than other methods, it is often used to protect sample data, which already contains some protection from the sampling process.

Table redesign methods used to reduce the risk of disclosure include:

- aggregating to a higher level geography or to a larger population sub-group;
- applying table thresholds;
- collapsing or grouping categories of variables (reducing the level of detail);
- applying a minimum average cell size to released tables.

The advantages of table redesign methods are that original counts in the data are not damaged and the tables are additive with consistent totals. In addition, the method is simple to implement and easy to explain to users. However, the detail in the table will be greatly reduced, and if many tables do not pass the release criteria it may lead to user discontent.

5.3.3   Post-tabular

Statistical disclosure control methods that modify cell values within tabular outputs are referred to as post-tabular methods. Such methods are generally clear and transparent to users, and are easier to understand and account for in analyses, than pre-tabular methods. However, post-tabular methods suffer the problem that each table must be individually protected, and it is necessary to ensure that the new protected table cannot be compared against any other existing outputs in such a way which may undo the protection that has been applied. In addition, post-tabular methods can be cumbersome to apply to large tables. The main post-tabular methods include cell suppression, cell perturbation and rounding.

5.4   Post-tabular methods

5.4.1   Cell suppression

Cell suppression is a non-perturbative method of disclosure control (described in detail in Chapter 4), but the method essentially removes sensitive values and denotes them as missing. Protecting the unsafe cells is called primary suppression, and to ensure these cannot be derived by subtractions from published marginal totals, additional cells are selected for secondary suppression.

Cell suppression cannot be unpicked provided secondary cell suppression is adequate and the same cells in any linked tables are also suppressed. Other advantages are that the method is easy to implement on unlinked tables and it is highly visible to users. The original counts in the data that are not selected for suppression are left unadjusted.

However, cell suppression has several disadvantages as a protection method for frequency tables, in particular, information loss can be high if more than a few suppressions are required. Secondary suppression removes cell values which are not necessarily a disclosure risk, in order to protect other cells which are a risk. Disclosive zeros need to be suppressed and this method does not protect against disclosure by differencing. This can be a serious problem if more than one table is produced from the same data source (e.g. flexible table generation). When disseminating a large number of tables, it is much harder to ensure the consistency of suppressed cells, and care must be taken to ensure that the same cells in linked tables are always suppressed.

5.4.2   ABS cell perturbation

ABS cell Perturbation is a new cell perturbation algorithm which was developed by the Australian Bureau of Statistics to protect the outputs from their 2006 Census (it is also planned to protect their 2011 Census outputs). The method is designed to protect tables by altering all cells by small amounts. The cells are adjusted in such a way that the same cell is perturbed in the same way even when it appears across different tables. This method adds sufficient 'noise' to each cell so if an intruder tried to gather information by differencing, they would not be able to obtain the real data.

The method consists of a two stage process:

1. Stage one adds the perturbations to the cell values, and results in a consistently perturbed but non-additive table. All microdata records are assigned a record key and when creating a table the record keys for all records contributing to each internal cell are summed. A function is applied to this sum to produce the cell key. Lookup tables (which are determined by the organisation) are then used, by means of the true cell value and the cell key, to determine the amount each cell should be perturbed. This means that the same cell is always perturbed in the same way. The perturbation can be set to zero for a pre-determined set of key outputs (e.g. the total age by sex population counts). Table margins are perturbed independently using the same method.

2. Stage two adds another perturbation to each cell (excluding the grand total) to restore table additivity. The stage two perturbations are generated using an iterative fitting algorithm which attempts to balance and minimise absolute distances to the stage one table (although not necessarily producing an optimal solution).

The ABS cell perturbation method is a more informed post-tabular method of disclosure control since it utilises pre-tabular microdata information during the perturbation stage. The method is very dependent upon the lookup table used, but it is flexible in that look-up tables can be specifically designed to suit needs, and different look-up tables could potentially be used for different tables. Furthermore, the lookup table can be designed to model other post-tabular methods (e.g. small cell adjustments or random rounding). The method provides protection for flexible tables and can be used to produce perturbations for large high dimensional hierarchical tables.

However, while consistency is maintained during the first stage of the perturbation process, it is lost when the additivity module is applied. The method must be applied to each table separately, and is less transparent than other methods. The method may also require careful specification of look-up tables for different types of data or output, particularly sparse tables.

5.4.3   Rounding

Rounding involves adjusting the values in all cells in a table to a specified base so as to create uncertainty about the real value for any cell. It adds a small, but usually acceptable, amount of distortion to the original data. Rounding is considered to be an effective method for protecting frequency tables, especially when there are many tables produced from one data set. It provides protection to small frequencies and zero values (e.g. empty cells). The method is simple to implement, and for the user it is easy to understand as the data is visibly perturbed.

Care must be taken when combining rounded tables to create user-defined areas. Cells can be significantly altered by the rounding process and aggregation compounds these rounding differences. Furthermore, the level of association between variables is affected by rounding, and the variance of the cell counts is increased.

Table 5.16   Population counts by sex.

	Male	Female	Total
Area A	1	0	1
Area B	3	3	6
Area C	12	20	32
Total	16	23	39

There are several alternative rounding methods including: conventional rounding, random rounding, controlled rounding and semi-controlled rounding, which are outlined later. Each method is flexible in terms of the choice of the base for rounding, although common choices are 3 and 5. All rounded values (other than zeros) will then be integer multiples of 3 or 5, respectively.

Conventional rounding
When using conventional rounding, each cell is rounded to the nearest multiple of the base. The marginal totals and table totals are rounded independently from the internal cells. An example of conventional rounding is provided below; Table 5.16 shows counts of males and females in different areas, while Table 5.17 shows the same information rounded to a base of 5.

The example shows the male's unsafe cell in Area A in Table 5.16 is protected by the rounding process in Table 5.17.

The advantages of this method are that the table totals are rounded independently from the internal cells, and therefore consistent table totals will exist within the rounding base. Cells in different tables which represent the same records will always be the same. While this method does provide some confidentiality protection, it is considered less effective than controlled or random rounding. Tables are not additive (e.g. row 3 of Table 5.17 does not sum to 35) and the level of information is poor if there are many values of 1 and 2. The method is not suitable for flexible table generation as it can be easily 'unpicked' when differencing and linking tables. For these reasons, conventional rounding is not recommended as a disclosure control method for frequency tables. Conventional rounding is sometimes used by NSIs for quality reasons (e.g. rounding data from small sample surveys to emphasise

Table 5.17   Population counts by sex (conventional rounding).

	Male	Female	Total
Area A	0	0	0
Area B	5	5	5
Area C	10	20	35
Total	15	25	40

Table 5.18   Random rounding.

Original value	Rounded value (probability)
0	0 (1)
1	0 (2/3) or 3 (1/3)
2	3 (2/3) or 0 (1/3)
3	3 (1)
4	3 (2/3) or 6 (1/3)
5	6 (2/3) or 3 (1/3)
6	6 (1)

the uncertain nature of the data). The distinction between rounding performed for disclosure control reasons and rounding performed for quality reasons should always be made clear to users.

Random rounding

Random rounding shifts each cell to one of the two nearest base values in a random manner. Each cell value is rounded independently of other cells, and has a greater probability of being rounded to the nearest multiple of the rounding base. For example, with a base of 5, cell values of 6, 7, 8 or 9 could be rounded to either 5 or 10. Marginal totals are typically rounded separately from the internal cells of the table (i.e. they are not created by adding rounding cell counts) and this means tables are not necessarily additive. Various probability schemes are possible, but an important characteristic is that they should be unbiased. This means there should be no net tendency to round up or down and the average difference from the original counts should be zero.

If we are rounding to base 3, the residual of the cell value after dividing by 3 can be either 0, 1 or 2. See Table 5.18.

- If the residual is zero no change is made to the original cell value.

- If the residual is 1, then with a probability of 2/3 the cell value is rounded down to the lower multiple of 3 and with a probability of 1/3 the cell value is rounded up to the higher multiple of 3.

- If the residual is 2, the probabilities are 2/3 to round up and 1/3 to round down.

As an example, Table 5.19 shows a possible solution for Table 5.16 using random rounding to base 5.

The main advantages of random rounding are that it is relatively easy to implement, it is unbiased and it is clear and transparent to users. Table totals are consistent within the rounding base because the totals are rounded independently from the internal cells. All values of 1 and 2 are removed from the table by rounding, which prevents cases of perceived disclosure as well as actual disclosure. The method may also provide some protection against disclosure by differencing as rounding should obscure most of the exact differences between tables.

Table 5.19   Population counts by sex (with random rounding).

	Male	Female	Total
Area A	0	0	0
Area B	5	0	5
Area C	10	20	35
Total	15	20	40

Controlled rounding
However, random rounding has disadvantages including the increased information loss which results from the fact that all cells (even safe cells) are rounded. In some instances, the protection can be 'unpicked' and in order to ensure adequate protection, the resulting rounded tables need to be audited. Although the method is unbiased, after applying random rounding there may be inconsistencies in data within tables (e.g. rows or columns which do not add up like row 3 of Table 5.19 does not sum to 35) and between tables (e.g. the same cell may be rounded to a different number in different tables).

Unlike other rounding methods, controlled rounding yields additive rounded tables. It is the statistical disclosure control method that is generally most effective for frequency tables. The method uses linear programming techniques to round cell values up or down by small amounts, and its strength over other methods is that additivity is maintained in the rounded table, (i.e. it ensures that the rounded values add up to the rounded totals and sub-totals shown in the table). This property not only permits the release of realistic tables which are as close as possible to the original table, but it also makes it impossible to reduce the protection by 'unpicking' the original values by exploiting the differences in the sums of the rounded values. Another useful feature is that controlled rounding can achieve specified levels of protection. In other words, the user can specify the degree of ambiguity added to the cells, for example, they may not want a rounded value within 10% of the true value. Controlled rounding can be used to protect flexible tables although the time taken to implement the method may make it unsuitable for this purpose.

Table 5.20 shows a possible rounding solution for Table 5.16, using controlled rounding to base 5. The disadvantages of controlled rounding are that it is a complicated method to implement, and it has difficulty coping with the size, scope and

Table 5.20   Population counts by sex (controlled rounding).

	Male	Female	Total
Area A	5	0	5
Area B	0	5	5
Area C	10	20	30
Total	15	25	40

magnitude of the very large census tabular outputs. It is hard to find control-rounded solutions for sets of linked tables, and in order to find a solution cells may be rounded beyond the nearest rounding base. In this case, users will know less about exactly how the table was rounded and it is also likely to result in differing values for the same internal cells across different tables.

Semi-controlled rounding

Semi-controlled rounding also uses linear programming to round table entries up or down, but in this case, it controls for the overall total in the table, or it controls for each separate output area total. Other marginal and sub-totals will not necessarily be additive. This ensures that either the overall total of the table is preserved (or the output area totals are all preserved), and the utility of this method is increased compared with conventional and random rounding. Consistent totals are provided across linked tables, and therefore the method can be used to protect flexible tables, although the time it takes to implement may make it unsuitable. Disadvantages of semi-controlled rounding relate to the fact that tables are not fully additive, and it is difficult to find an optimal solution. There are some more specialised rounding methods which have been used at various times by NSIs to protect census data, two of these methods are described as follows:

1. *Small cell adjustment.* It was used (in addition to random record swapping (a pre-tabular method)) to protect 2001 Census tabular outputs for England, Wales and Northern Ireland. This method was also used by the ABS to protect their tabular outputs from the 2001 Census.

    Applying small cell adjustments involves randomly adjusting small cells within tables upwards or downwards to a base using an unbiased prescribed probability scheme. During the process:

- small counts appearing in a table are adjusted;

- totals and sub-totals are calculated as the sum of the adjusted counts. This means all tables are internally additive;

- tables are independently adjusted so counts of the same population which appear in two different tables, may not necessarily have the same value;

- tables for higher geographical levels are independently adjusted, and therefore will not necessarily be the sum of the lower component geographical units;

- output is produced from one database which has been adjusted for estimated undercount so the tables produced from this one database provide a consistent picture of this one population.

Advantages of this method are that tables are additive, and the elimination of small cells in the table removes cases of perceived as well as actual identity disclosure. In addition, loss of information is lower for standard tables as all other cells remain the same; however, information loss will be high for sparse tables. Other disadvantages include inconsistency of margins between linked tables since margins are calculated

Table 5.21 Summary of SDC rounding methods.

	Conventional rounding	(Semi-) controlled rounding	Random rounding
Internal cells add to table totals (additvity)	No	Yes	No
Method provides enough SDC protection (and cannot be unpicked)	No	Yes	In some situations this method can be unpicked
Method is quick and easy to implement	Yes	It can take time for this method to find a solution	Yes

using perturbed internal cells, and this increases the risk of tables being unpicked. Furthermore, this method provides little protection against disclosure by differencing, and is not suitable for flexible table generation. Where tables were sparse, the utility of the output suffered greatly. In the 2001 UK Census, these were most often origin-destination tables with small counts.

2. *Barnardisation*. It is a form of cell perturbation which modifies each internal cell of every table by $+1, 0$ or -1, according to probabilities. Zeros are not adjusted. The method offers some protection against disclosure by differencing, however table totals are added up from the perturbed internal cells, resulting in inconsistent totals between tables. Typically, the probability p is quite small, and therefore, a high proportion of risky cells are not modified. The exact proportion of cells modified is not revealed to the user. This is generally a difficult method to implement for flexible output. The SDC-aspects of the different rounding methods are summarised in Table 5.21.

5.5 Information loss

As described in Sections 5.3 and 5.4.3, there are a number of different disclosure control methods used to protect frequency tables. Each of these methods modifies the original data in the table in order to reduce the disclosure risk from small cells (0's, 1's and 2's). However, the process of reducing disclosure risk results in information loss. Some quantitative information loss measures have been developed by Shlomo and Young (2006a and 2006b) to determining the impact various SDC methods have on the original tables.

Information loss measures can be split into two classes: (1) measures for data suppliers used to make informed decisions about optimal SDC methods depending on the characteristics of the tables and (2) measures for users in order to facilitate adjustments to be made when carrying out statistical analysis on protected tables. Here, we focus on measures for data suppliers. Measuring utility and quality for SDC methods is subjective. It depends on the users; the purpose of the data and the required statistical analysis; and the type and format of the statistical data. Therefore, it is useful to have a range of information loss measures for assessing the impact of the SDC methods.

The focus here is information loss measures for tables containing frequency counts; however, some of these measures can easily be adapted to microdata. Magnitude or weighted sample tables will have the additional element of the number of contributors to each cell of the table.

When evaluating information loss measures for tables protected using cell suppression, one needs to decide on an imputation method for replacing the suppressed cells similar to what one would expect a user to do prior to analysing the data, (i.e. we need to measure the difference between the observed and actual values, and for suppressed cells the observed values will be based on user inference about the possible cell values). A naive user might use zeros in place of the suppressed cells whereas a more sophisticated user might replace suppressed cells by some form of averaging of the total information that was suppressed, or by calculating feasibility intervals.

A number of different information loss measures are described as follows, and more technical details can be found in Shlomo and Young (2006a) and Shlomo and Young (2006b):

- An exact Binomial Hypothesis Test can be used to check if the realisation of a random stochastic perturbation scheme, such as random rounding, follows the expected probabilities (i.e. the parameters of the method). For other SDC methods, a non-parametric signed rank test can be used to check whether the location of the empirical distribution after the application of the SDC method, has changed.

- Information loss measures that measure distortion to distributions are based on distance metrics between the original and perturbed cells. Some useful metrics are also presented in Gomatam and Karr (2003). A distance metric can be calculated for internal cells of a table. When combining several tables, one may want to calculate an overall average across the tables as the information loss measure. These distance metrics can also be calculated for totals or subtotals of the tables.

- SDC methods will have an impact on the variance of the average cell size for the rows, columns or the entire table. The variance of the average cell size is examined before and after the SDC method has been applied. Another important variance to examine is the 'between' variance when carrying out a one-way analysis of variance (ANOVA) test based on the table. In ANOVA, we examine the means of a specific target variable within groupings defined by independent categorical variables. The goodness of fit statistic R^2 for testing the null hypothesis that the means are equal across the groupings is based on the variance of the means between the groupings divided by the total variance. The information loss measure therefore examines the impact of the 'between' variance and whether the means of the groupings have become more homogenised or spread apart as a result of the SDC method.

- Another statistical analysis tool that is frequently carried out on tabular data is a test for independence between categorical variables that span the table. The test for independence for a two-way table is based on a Pearson Chi-Squared Statistic and the measure of association is the Cramer's V statistic. For multi-way tables, one can examine conditional dependencies and calculate expected cell frequencies based on the theory of log-linear models. The test statistic for the fit of the model is also based on a Pearson Chi-Squared Statistic. SDC methods applied to tables may change the results of statistical inferences. Therefore, we examine the impact to the test statistics before and after the application of the SDC method.

- Another statistical tool for inference is the Spearman's Rank Correlation. This is a measure of the direction and strength of the relationship between two variables. The statistic is based on ranking both sets of data from the highest to the lowest. Therefore, one important assessment of the impact of the SDC method on statistical data is whether we are distorting the rankings of the cell counts.

In order to allow data suppliers to make informed decisions about optimal disclosure control methods, ONS have developed a user-friendly software application that calculates both disclosure risk measures based on small counts in tables and a wide range of information loss measures (as described above) for disclosure controlled statistical data as described in (Shlomo and Young 2006b). The software application also outputs R-U (Risk-Utility) maps.

5.6   Software

5.6.1   Introduction

τ-ARGUS (see Hundepool *et al.* 2011) is a software package which provides tools to protect tables against the risk of statistical disclosure (τ-ARGUS is also discussed in Chapter 4). Controlled rounding is easy to use in τ-ARGUS and the controlled rounding procedure (CRP) was developed by JJ Salazar (see Salazar-González *et al.* 2006). This procedure is based on optimisation techniques similar to the procedure developed for cell suppression. The CRP yields additive rounded tables, where the rounded values add up to the rounded totals and sub-totals shown in the table. This means realistic tables are produced and it makes it impossible to reduce the protection by 'unpicking' the original values by exploiting the differences in the sums of the rounded values. The CRP implemented in τ-ARGUS also allows the specification of hierarchical structures within the table variables.

Controlled rounding gives sufficient protection to small frequencies and creates uncertainty about the zero values (i.e. empty cells). In general, cell suppression leaves empty cells unmodified.

Zero-restricted controlled rounding

In zero-restricted controlled rounding cell counts are left unchanged if they are multiples of the rounding base or shifted to one of the adjacent multiples of the rounding base. The modified values are chosen so that the sum of the absolute differences between the original values and the rounded ones are minimised (under an additivity constraint). Therefore, some values will be rounded up or down to the most distant multiple of the base in order to help satisfy these constraints. In most cases a solution can be found, but in some cases it cannot and the zero-restriction constraint in CRP can be relaxed to allow the cell values to be rounded to a non-adjacent multiple of the base. This relaxation is controlled by allowing the procedure to take a maximum number of *steps*.

For example, consider rounding a cell value of 7 when the rounding base equals 5. In zero-restricted rounding, the solution can be either 5 or 10. If 1 step is allowed, the solution can be 0, 5, 10 or 15. In general, let z be the integer to be rounded in base b, then this number can be written as

$$z = ub + r,$$

where ub is the lower adjacent multiple of b (hence u is the floor value of z/b) and r is the remainder. In the zero-restricted solution the rounded value, a, can take values:

$$\begin{cases} a = ub & \text{if } r = 0; \\ a = \begin{cases} ub \\ (u+1)b \end{cases} & \text{if } r \neq 0. \end{cases}$$

If K steps are allowed, then a, can take values:

$$\begin{cases} a = \max\{0, (u+j)\}b, & j = -K, \ldots, K & \text{if } r = 0; \\ a = \max\{0, (u+j)\}b, & j = -K, \ldots, (K+1) & \text{if } r \neq 0. \end{cases}$$

5.6.2   Optimal, first feasible and RAPID solutions

For a given table, there can exist more than one controlled rounded solution, and any of these solutions is a *feasible* solution. The Controlled Rounding Program embedded in τ-ARGUS determines the *optimal* solution by minimising the sum of the absolute distances of the rounded values, from the original ones. Denoting the cell values, including the totals and sub-totals, with z_i and the corresponding rounded values with a_i, the function that is minimised is

$$\sum_{i=1}^{N} |z_i - a_i|,$$

where N is the number of cells in a table (including the marginal ones). The optimisation procedure for controlled rounding is a rather complex one (*NP*-complete problem), so finding the optimal solution may take a long time for very large tables. In fact, the algorithm iteratively builds different rounded tables until it finds the optimal solution. In order to limit the time required to obtain a solution, the algorithm can be stopped when the first feasible solution is found. In many cases, this solution is quite close to the optimal one and it can be found in significantly less time.

The RAPID solution is produced by CRP as an approximated solution when a feasible one cannot be found. This solution is obtained by rounding the internal cells to the closest multiple of the base and then computing the marginal cells by addition. This means that the computed marginal values can be many jumps away from the original value. However, a RAPID solution is produced at each iteration of the search for an optimal solution, and it will improve (in terms of the loss function) over time. τ-ARGUS allows the user to stop CRP after the first RAPID solution is produced, but this is likely to be very far away from the optimal one.

5.6.3 Protection provided by controlled rounding

The protection provided by controlled rounding can be assessed by considering the uncertainty (about the true values achieved) when releasing rounded values; that is the existence interval that an intruder can compute for a rounded value. We assume that the values of the rounding base, b, and the number of steps allowed, K, are known by the user together with the output rounded table. Furthermore, we assume that it is known that the original values are positive frequencies (hence non-negative integers).

Zero-restricted rounding

Given a rounded value, a, an intruder can compute the following existence intervals for the true value, z:

$$z \in [0, b-1] \qquad \text{if } a = 0;$$
$$z \in [a-b+1, a+b-1] \quad \text{if } a \neq 0.$$

For example, if the rounding base is $b = 5$ and the rounded value is $a = 0$, a user can determine that the original value is between 0 and 4. If the rounded value is not 0, then users can determine that the true value is between ± 4 units from the published value.

K-step rounding

As mentioned above, it is assumed that the number of steps allowed is released together with the rounded table. Let K be the number of steps allowed, then an intruder can compute the following existence intervals for the true value z:

$$z \in [0, (K+1)b-1] \qquad \text{if } a < (K+1)b;$$
$$z \in [a-(K+1)b+1, a+(K+1)b-1] \quad \text{if } a \geq (K+1)b.$$

For example, assume that for controlled rounding with $b = 5$, $K = 1$ and $a = 15$, then a user can determine that $z \in [6, 24]$.

Very large tables

The procedure implemented in τ-ARGUS is capable of rounding tables up to 150 000 cells on an average computer. However, for larger tables a partitioning procedure is available, which allows much larger tables to be rounded. Tables with over six million cells have been successfully rounded this way.

5.7   Case studies

5.7.1   UK Census

In the 2001 UK Census, the ONS used record swapping and small cell adjustment as the main methods of disclosure control for tabular outputs. Households were selected at random and swapped with other households that matched on a set of key characteristics. The swap rate was kept low in order to maintain data utility and there was a subsequent decision to also use small cell adjustment to apply more protection for the small counts in tables. There was hence a mixture of pre-tabular and post-tabular methods used, though the post-tabular adjustment was not applied in Scotland. Hence, there was an inconsistency between Scottish outputs and those from England, Wales and Northern Ireland. ONS has come under considerable criticism for the approach from users, both for the inconsistency between the different countries' approaches (and the consequent difficulty in forming UK outputs) and for the effect on tables which were not consistent with each other (different totals for a cell or marginal appearing in different tables).

In November 2006, the Registrars General (RsG) of the UK Census Offices issued an agreement relating to the conduct of the 2011 Census (see ONS 2011). ONS are bound by legal requirements under the Statistics and Registration Service Act (SRSA) 2007 which came into force in April 2008 http://www.legislation.gov.uk/ukpga/2007/18.

The aim was to design a UK SDC strategy that was consistent with both the RsG statement and the SRSA, while taking into account user feedback from the 2001 approach. ONS reviewed a wide range of methods during a thorough evaluation. This concluded that record swapping would again be the primary method for tabular outputs. However, unlike in 2001, the 2011 method would target 'risky' records so as to provide the adequate level of protection without the need for the post-tabular method.

Other strengths of the method were that:

- tables would be additive and consistent;
- the method would be easy to explain;
- the method had been used before (albeit in a different form) in other censuses.

Households were assessed for riskiness. Firstly, individuals were assessed for uniqueness or rarity on a number of variables and given a score. Based on the individual scores, each household was assigned a household risk score. The variables selected for scoring (targeting) were based on those thought to be visible or in the public domain and those that would be considered sensitive. For the latter, the Data Protection Act 1998 http://www.legislation.gov.uk/ukpga/1998/29 listed those that should be considered sensitive.

The required level of swapping was assessed by considering the level of uncertainty introduced into attribute disclosures by swapping. Moreover, there is often natural uncertainty in outputs from data collection and processing: respondent error, respondents lying, data capture errors, even before imputation and swapping. In the case of the UK Census, it was possible to assess uncertainty introduced by non-response and imputed records. Where records are imputed, the cases are not real and so there is no risk associated with them. When aligned with record swapping, the uncertainty was assessed through two success measures:

1. *Success measure 1*. The percentage of attribute disclosures in the raw data that are removed through imputation and/or swapping.

2. *Success measure 2*. The percentage of (apparent) attribute disclosures in the output tables that are not real, i.e., they have been introduced by the processes of imputation and/or swapping.

The RsG had a requirement of uncertainty in the outputs, but had not specified what constituted 'sufficient' uncertainty. That is a judgement call for the responsible statistician. A sample of standard tables was formed and the swapping rates varied until the required levels of uncertainty appeared to be reached.

A similar methodology was used for protecting residents in communal establishments, except that individuals were swapped between communal establishments. Much more detail on this is available in Frend *et al.* (2011).

5.7.2   Australian and New Zealand Censuses

In common with most other NSIs, including UK, Australia and New Zealand are both bound by legislation on confidentiality. The Australian Census and Statistics Act (1905) states that the Australian Bureau of Statistics (ABS) shall not release statistics *'in a manner that is likely to enable the identification of a particular person or organisation'*. Similarly, in New Zealand, the Statistics Act (1975) states *'All statistical information published.......shall be arranged in such a manner as to prevent any particulars published from being identifiable by any person'*. Each NSI aims to maximise data while preserving respondent confidentiality. While the aims may be similar; however, different histories and attitudes within the different countries lead to different approaches. The demands of census users are different too. While in UK, the primary user requirement is for additivity and consistency between tables, and hence the preference for pre-tabular methods, both ABS and Statistics New Zealand (SNZ) have pre-dominantly made use of post-tabular methods to protect confidentiality. Whereas UK had to react to users' fundamental concerns related to the small cell adjustment employed in 2001, ABS and SNZ both selected methods that built on the methods used in their last censuses. All three countries went through detailed evaluations of potential methods to reach that point.

For the 2006 Australian Census of Population in Housing, ABS developed and implemented a new confidentiality method. The method involved creating a record key for each unit record, a random number specific to the record. When a table was created, the record keys for each unit record in a cell were combined to give a key relating to the cell. The cell key was then used to determine the perturbation that would be applied to the cell via a fixed look-up table. Hence, wherever the same cell appeared in different tables, the perturbation applied would be consistent. The method allowed much more flexibility for users in that they could define their own tables in such a way that each table would be protected consistently. For more detailed information about the method, see Fraser and Wooton (2006).

The SNZ approach for the New Zealand 2006 Census involved suppressing tables that did not conform to a set of rules. Initially, tables would be suppressed where:

- the geographic classification was the most detailed (i.e. 'meshblocks', averaging 100 people);

- tables were detailed and included the income classification;

- the mean cell size was low (this was defined as the population for a geography divided by the number of cells in the table).

In addition, all published counts were rounded to base 3.

Subsequent to the 2006 Census, SNZ responded to user needs for information that allowed time series with previous censuses. In 2007, SNZ began releasing more detailed tables with lower mean cell sizes, with small cell counts suppressed. The very detailed travel to work tables were also released to local government users under a licence agreement.

Both ABS and SNZ moved towards their respective 2011 Censuses by evaluating the previous approaches. Neither encountered the level of user discontent experienced in UK and each adopted a similar approach to 2011. Consistent SDC methods between censuses (and indeed in different waves of other data collections) are beneficial because users develop expectations about which data will be available and how they are presented. Of course, consistency needs to be balanced against potential improvements that can be made from updating or changing the SDC methods used. ABS and SNZ conduct national censuses every 5 years (for the UK this is 10 years) and this interval provides greater potential for improvements both in technology and in best practice.

Users of census data are key players in driving the type of methods that are used to protect confidentiality. Following their 2006 Censuses, both ABS and SNZ consulted with users. A particular outcome for ABS was the apparent desire for users to create bespoke tables with bespoke geographies. In response, ABS developed two web-based products, CDATA Online (freely available, allowing users to create their own tables from several underlying topic-based data cubes) and Census TableBuilder (allowing users, registered with ABS, to create more detailed tables from most of the census variables). The use of the cell perturbation method described above allowed these outputs to be protected automatically. SNZ found that users accepted their use of random rounding, despite the non-additivity that results from the method. They were also extremely keen for SNZ to recognise 'professional users' who could be licensed to use more detailed tables (that is a move that appears to be gathering momentum in many countries).

The ABS has extended the 2006 Census methodology to protect survey information. For 2011, ABS retained the cell perturbation approach, and have also considered the scope to combine pre-tabular and post-tabular methods by treating outliers in the microdata before the tables are created and confidentialised.

The SNZ approach of random rounding to base 3 again formed the main protection for census tables for the 2011 round, with other rules providing additional protection on occasions where random rounding was not sufficient (e.g. where tables were sparse with a large proportion of small counts). The mean cell size rules stated that the total unrounded subject population for a geographic area divided by the number of cells in a table must be greater than two. In the case of tables where the mean cell size is less than or equal to 2, all cells with an unrounded count of 5 or fewer are suppressed. Secondary suppression is not required since the rounding will provide sufficient protection so that a suppressed count cannot be unpicked.

Despite the ABS and SNZ having similar legal obligations to ONS, the three approaches are quite different. The decision process towards each approach is similar in the three countries. NSIs can choose what they consider to be the most appropriate methods to balance disclosure risk against data utility, and it is clear that there is no prescribed approach to doing this. The factors which influence the choice of SDC methods used in each country include some that are country specific:

- Lessons learned from previous censuses.

- The attitude and risk appetite of the NSI.

- The needs of users, the impact of the methods on these and data utility (and how users define 'good' data utility).

- The feasibility of approach in conjunction with technology available.

- The applicability in relation to geographic classifications (see Forbes *et al.* 2009).

6

Data access issues

6.1 Introduction

Although very sophisticated methods have been developed to make safe microdata files, the needs of serious researchers for more detailed data cannot be met by these methods. It is simply impossible to release these very detailed microdata sets to users outside the control of the National Statistical Institutes (NSIs) without breaching the necessary confidentiality protection.This is especially true for business microdata. Nevertheless, the NSIs recognise the serious and respectable requests by the research community for access to the very rich and valuable microdata sets of the NSIs. Therefore, different initiatives have been taken by the NSIs to meet these needs.

The first step was the creation of so-called safe research data centres, a special room in the NSI, where researchers can analyse the data sets, without the option to export any information without the consent of the NSI. In parallel to this initiative, there are options for remote execution. Remote execution facilities are various kinds of systems where researchers can submit scripts for $SAS^{®}$, $SPSS^{®}$, etc. to the NSI. The most recent initiative is remote access, where users can 'log in' to a research data centre from a remote desktop.

As all these options allow the researcher to access unprotected sensitive data in some way, all possible precautions have to be taken. These options are certainly not available to the general public, but only to selected research institutes like universities and similar research institutes. In addition, strict contracts have to be signed between the NSI, the researcher and preferably the research institute itself. This enables the NSI to take action against the institute itself as well as against the researcher, if there were to be a breach of the access conditions. A common repercussion for the institute could be a ban for the whole institute to access these facilities. In some cases, there may even be legal sanctions if the breach is willful, negligent or malicious. So it will be in the interest of the institute to ensure a correct behaviour of the researcher.

Statistical Disclosure Control, First Edition. Anco Hundepool, Josep Domingo-Ferrer,
Luisa Franconi, Sarah Giessing, Eric Schulte Nordholt, Keith Spicer and Peter-Paul de Wolf.
© 2012 John Wiley & Sons, Ltd. Published 2012 by John Wiley & Sons, Ltd.

6.2   Research data centres

In order to meet the needs of the researcher community to analyse the rich data sets compiled by the NSIs, while safeguarding the confidentiality constraints, the first solution was to create special rooms in the premises of the NSIs (research centres). The NSI makes available special computers for the researchers. On this computer, the necessary software for the research will be installed by the NSI together with the necessary data sets. Ideally, these computers have no connection whatsoever to the internet and there is no email. Also drives for removable discs (floppy or CD) are not available and the use of memory sticks has to be blocked. The access to the internal production network of the NSI has to be blocked as well, preventing the possibilities of the researchers to access other sensitive information. Installing a printer is a risk as well as is the use of phones, though if there is sufficient supervision, these two options could be made available. Nevertheless, some kind of supervision on the research centre is always needed.

The data sets to be used in the research centre have to be anonymised (i.e. at least the names, address, etc. have been removed). It is also advisable to restrict the variables available to the set that is needed for the specific research.

On these computers, the researchers should nevertheless be able to fully analyse the data files and complete their analysis. However, the outputs cannot be released to the researchers unless they have been cleared by NSI staff. Unfortunately this is not a trivial task and there is normally an agreement to send on outputs when cleared within 5 working days or some other agreed period of time. This will be discussed in Section 6.6.

The concept of research centres is meeting many research needs and several NSIs have adopted this idea. Without being complete, this initiative has been implemented in the USA, Canada, The Netherlands, Italy, Germany, Denmark, Eurostat and several other countries.

The concept of research centres has proved to be very successful. Many good research papers and theses have been completed for which these centres were indispensable. However, there are some drawbacks. The most important one is that the researchers have to come physically to the premises of the NSIs. Even in a small country like the Netherlands, this is seen as a serious problem. Also, the researcher cannot just try some further analysis when he is back at his normal working place, because he has to travel to the NSI first. And the fact that he cannot work in his normal working environment is considered a drawback.

6.3   Remote execution

As modern communication techniques have become available, the NSIs have investigated the possibilities to use these techniques. The first initiative is remote execution. In this concept, the researchers will get a full description of all the metadata of the data sets available for research. However, the data set available will remain on the computers of the NSIs. The researchers will prepare scripts for analysing the data sets (with SPSS$^{\circledR}$, SAS$^{\circledR}$, etc.) and send them to the NSI (by email or via some internet page). The NSI will then check the script (e.g. for commands like List Cases, but also other unwanted actions like printing the residuals of a regression) before running it and after a second check send back the results to the researcher.

For the researcher, this system has the advantage that he no longer has to travel to the NSI. He can send a script whenever he wants, though he cannot run the script directly since this is done by the NSI. Correcting errors in a script can thus take much more time, depending on the turn around time of the NSI. This process could be accelerated if the NSI will make available a fake data set which corresponds to the original file in terms of structure but not in content. The main objective of this data set is to avoid all unsuccessful submissions due to syntax errors, etc.

For the researcher, remote execution has several advantages (no need for travel) but also some drawbacks (slow turn around time). For the NSIs, it is very time consuming, as they have to check so many scripts and results. It is not uncommon in statistical analysis that several scripts are submitted and executed. But then the outcome proves to be not the optimal model and a new script is submitted. However, the NSI does not know in advance which script is successful and has to check everything. This is very time consuming if the NSI takes this seriously.

Current examples of this kind of systems are the Luxembourg Income Study (LISSY, see http://www.lisdatacentre.org/data-access/) and the Australian Remote Access Data Laboratory (RADL, see http://www.abs.gov.au/).

6.4 Remote access

Recent new developments are systems for remote access. The aim is to combine the flexibility for researchers to do all their analysis in a research centre while removing the constraints of travelling to the NSIs. Modern developments in the internet make it possible to set up a safe controlled connection, a virtual private network (VPN). A VPN is a technique to set up a secure connection between the server at the NSI and a computer of the researcher. It uses firewalls and encryption techniques. Also, additional procedures to control the login procedure like software tokens or biometrics can be used to secure the connection. The most well-known product behind this is Citrix$^{\circledR}$, but other systems exist as well. Citrix has been developed to set up safe access to business networks over the internet without giving access to unauthorised persons. This will safeguard the confidentiality of the information on this network. Some NSIs are now using Citrix to set up a safe connection between the PC of the researcher and a protected server of the NSI. This approach is followed now by Denmark, Sweden, the Netherlands and Slovenia.

The main idea of a remote facility is that it should resemble the 'traditional' on-site research data centres as much as possible, concerning confidentiality aspects. The following aspects have to be taken into account:

- only authorised users should be able to make use of this facility;
- microdata should remain at the NSI;

- desired output of analyses should be checked for confidentiality;

- legal measures have to be taken when allowing access.

The key issue is that the microdata set remains in the controlled environment of the NSI, while the researcher can do the analysis in his institute. In fact, it is an equivalent of the research centre. The Citrix connection will enable the researcher to run SPSS, SAS, etc. on the server of the NSI. The researcher will only see the session on his screen. This allows him to see the results of his analysis but also the microdata themselves. This is completely equivalent to what he can see, if he would be at the research centre. Citrix will only send the pictures of the screens to the PC of the researcher, but no actual microdata are sent to him. Even copying the data from the screen to the hard disk is not possible. If the researcher is satisfied with some analysis and wants to use the results in his report, he should make a request to the NSI to release these results to him. The NSI has to check the output for disclosure risks and if this is approved, the NSI will send the results to the researcher.

As the researchers will work with very sensitive data, all measures should be taken to ensure the confidentiality of the data. Therefore, also legal measures have to be taken, binding not only the researcher himself but also the institute.

6.5 Licensing

Another access option for microdata releases available to NSIs is to release data under licence or access agreements. A spectrum of different data-access arrangements can be provided. A variety of factors should be taken into account when granting approval for access – including the purpose of the access, the status of the user, the legal framework, the status of the data, the availability of facilities and the history of access. The levels of control over use and user applied within the licence should be balanced by the level of detail and/or perturbation in the microdata.

6.6 Guidelines on output checking[1]

6.6.1 Introduction

When NSI grant access to confidential data in one of the options described in the previous sections, the confidential data remain under the control of the NSI. The researchers can perform the analysis they want to do, they can view the results, but the results cannot be downloaded. Everything in still under control and in the possession of the NSI. When researchers want to publish the results, they must ask for permission to release the results. This can only be done after the checking of the results. This is a very difficult task as the researchers can perform very different kinds of complex analysis. However, it is not to be avoided and NSIs should see it as a part of their task.

---

[1] These guidelines are based on the results of the ESSnet on SDC. The UK, Italian, German and Dutch Statistical Office contributed to these guidelines.

NSIs must then find the right balance between enabling these researchers to undertake useful research while always ensuring the confidentiality of the data. One common method of protecting the confidentiality of the data is to check all results that researchers want to take out of the controlled environment of the Research Data Centre (RDC). Throughout this section, this process will be referred to as output checking. This provides the reader with guidelines for, and best practices of, this output checking process. The aim of the guidelines is threefold:

1. to allow experienced NSIs to learn from each other by sharing best practice;

2. to provide NSIs that have little or no experience in RDCs with practical advice on how to set up an efficient and safe output checking process;

3. to facilitate harmonisation of output checking methods across NSIs.

The third aim is particularly crucial in light of the current focus on the development of European infrastructures for microdata access, as it is crucial that NSIs agree on a common method for checking output. Cross-border data access will be most efficient if NSI X can check output that was generated with data sets from country Y as well. However, only with common guidelines will NSIs delegate the task of output checking to each other.

The remainder of this section deals with a general approach to output checking. In Section 6.6.3, possible types of output will be categorised. For each category of output both a simple rule and guidance for more experienced researchers will be outlined. Section 6.6.4 then deals with a number of practical issues (procedural, organisational), which are necessary for an efficient high-quality output checking process. Section 6.6.5 focuses on the training for researchers that use the RDC and will be producing the output.

6.6.2   General approach

We first need to discuss the general approach to output checking.

The Safe Data zoo
An RDC, remote access and remote execution are safe environments where accredited researchers can access the most detailed microdata to undertake any research that they desire (as long as it serves the public good). This makes output checking in an RDC totally different from disclosure control of official NSI publications. The official publications are of a well-defined form (usually a table) and the intruder scenarios are limited in number, whereas the output of an RDC can be anything. Researchers twist, transform and link the original data in different and complex new ways. This makes it very difficult to come up with a set of rules for that cover every possible output, as one expert vividly states it: designing rules for output checking is like designing cages for a zoo – that will keep the animals both contained and

alive – without knowing in advance which animals will be kept in the cages (see Ritchie 2007b).

Safe and unsafe classes of output

To bring some order to output checking in an RDC, all output can be classified into a limited number of categories (for instance, tables, regression, etc.) (see Ritchie 2007a). Each class of output is then labelled 'safe' or 'unsafe'. This classification is done solely on the functional form of the output, not on the data specifically used.

- 'Safe' outputs are those which the researcher should expect to have cleared with no or minimal further changes; for example, the coefficients estimated from a survival analysis. Analytical outputs and estimated parameters are usually 'safe'. The exceptions where a 'safe' output is not released should be well defined and limited in number.

- 'Unsafe' outputs will not be cleared unless the researcher can demonstrate, to the output checker's satisfaction, that the particular context and content of the output makes it non-disclosive. For example, a table will not be released unless it can be demonstrated that there are enough observations, or the data have been transformed sufficiently, so that the publication of that table would not lead to identification of outputs. Linear aggregations of data are almost always 'unsafe'.

Note that the burden of proof differs: for safe outputs, the output checker must provide reasons why the output cannot be released against normal expectations; for unsafe outputs, the researcher has to make a case for release. In both situations, the ultimate decision to release an output remains with the output checker. Output checking is always context specific. It is not possible to say, *ex ante*, that something will or will not be released. The purpose of the safe/unsafe classification is to give guidelines on the likelihood of an output being released and, if it is unsafe, to suggest ways that it can be made safe. Not all outputs have yet been classified. The default classification is 'unsafe'.

Two types of error

With this in mind, one needs to realise that the optimal way to check output is the one that maximises use of the data sets while minimizing the risk for disclosure. So, phrased differently, a less than perfect output check can lead to two types of errors:

1. Confidentiality errors: releasing disclosive output

2. Inefficiency errors: not releasing safe output

Rules and guidelines for output checking can prevent both errors. But the trick is to find the right rule.

Consider, for instance, a rule that sets a minimum for the number of units in each cell of a table (threshold rule). If the minimum cell count is set too high, this will lead to inefficiency errors: the output will be safe, but will contain less information than could be obtained from the data file. The researcher might have had to group classes that he was interested in, to reach the threshold. On the other hand, if the minimum is set too low, this will lead to unsafe outputs being released.

Finding the correct disclosure control rules for use in an RDC is especially difficult. One has to realise that the exact shape of the output is not known beforehand. The idea of an RDC is to give researchers the maximum amount of freedom in analysing the data files, so they will produce output of all sizes and shapes (tables, models, etc.).

To deal with this problem, a two-model approach has been developed. The first model is called the principles-based model. This model minimises both confidentiality and inefficiency errors. The other is called the rule-of-thumb model. For this model, the focus is on preventing confidentiality errors and inefficiency errors are accepted. Both models will be discussed in more detail in the following paragraphs.

Principles-based model

The principles-based model centres on a good collaboration between researchers and RDC staff. Because this model also aims to prevent inefficiency errors, simple rules for checking output are not appropriate. The reason is that simple rules can never take into account the full complexity of research output. To give maximum flexibility to the researcher, no output is ruled in or out in advance. All output needs to be considered in its entire context before deciding on its safety. For instance, a table that contains very small cell counts (maybe even some cell counts of 1) is not necessarily unsafe. If, for instance, the original data had been transformed beforehand, the information that the 'risky' cells disclose might not be traceable to the individual. What is needed is a clear understanding of the governing principles behind disclosure control, therefore, both the researcher and the RDC staff need training in disclosure control.

The principles-based model has the obvious advantage that it leaves a maximum amount of flexibility to the researcher. Therefore, data files will be used to their fullest extent. However, the model also has some possible drawbacks:

- The model relies on serious training of NSI staff and researchers. Researchers have to be willing to invest their time and effort on a topic, which is not naturally within their field of interest.

- The model spreads the responsibility for clearing an output. In a rules-based model, the responsibility lies with the people that design the rules. In this principles-based model, the responsibility lies with each individual checker. There are no strict rules to follow and each checker has to make his own decision on clearing the output, based on his experience and understanding of the underlying principles.

To circumvent these drawbacks, an alternative model is presented: the rule-of-thumb model.

Rule-of-thumb model

In this model, the main focus is on preventing confidentiality errors, and some inefficiency errors are taken for granted. This model typically leads to very strict rules. The chance that an output, that passes these rules, is non-disclosive is very high. The advantage is that the rules can be applied more or less automatically by both researchers and staff members with only limited knowledge of disclosure control.

It is important to stress the fact that, although the rules of thumb are very strict, this is not a 100% guarantee that all output that passes these rules is indeed non-disclosive. There is a very small chance that a disclosive output slips through. This is because the rules are rigid and do not take the full context of the output into consideration. The rule-of-thumb model is useful for a number of situations:

- Naïve researchers whose output is usually far from the cutting edge of disclosure control (for instance policy makers who just want tabular output with limited detail).

- Inexperienced NSIs starting up an RDC. In this case, both users and RDC staff could have too little experience to be able to work with the principles-based model. The rule-of-thumb model provides them with a starting point that ensures maximum safety. In using the rule-of-thumb model, they build up experience and subsequent case law along the way. At some point in time, they might feel confident enough to set up the principles-based model and open up the way to clearing more complex output.

- Automatic disclosure control for RDCs. This will mainly be useful for more controlled types of data access like remote execution. In remote execution, researchers write their scripts without having access to the real data file (sometimes dummy data sets are provided for this purpose). They then send the finished script to the RDC, where a staff member (or an automated system) runs it on the full data sets. The results are then returned to the researcher.

Even for RDCs using the principles-based model, the rules of thumb are usually the starting point when checking any particular output. Using the rules of thumb, attention is quickly focused on the parts of the output that breach these rules. These parts can then be considered more carefully using the full principles-based model to decide whether they can be released or not.

6.6.3   Rules for output checking

Classification of output

As described earlier, the Safe Data Zoo can be somewhat structured by classifying all output into a limited number of classes. The undermentioned table lists the different classes of output. Each class is marked as either safe or unsafe (see Section 6.6.2 for an explanation of the safe-unsafe classification).

Type of statistics	Type of output	Class
	Frequency tables	Unsafe
	Magnitude tables	Unsafe
	Maxima, minima and percentiles (incl. median)	Unsafe
	Mode	Safe
Descriptive statistics	Means, indices, ratios, indicators	Unsafe
	Concentration ratios	Safe
	Higher moments of distributions (incl. variance, covariance, kurtosis, skewness)	Safe
	Graphs: pictorial representations of actual data	Unsafe
	Linear regression coefficients	Safe
	Non-linear regression coefficients	Safe
Correlation and regression analysis	Estimation residuals	Safe
	Summary and test statistics from estimates (R^2, χ^2, etc.)	Safe
	Correlation coefficients	Safe

The overall rule of thumb

As discussed earlier, the rule-of-thumb model is based on clear and simple (and strict) rules. Because these rules differ only slightly for different classes of output, an overall rule of thumb can be established. This overall rule of thumb is presented first; after that, when describing the rules for each class of output, the interpretation of this overall rule of thumb, for the specific class of output is given. The overall rule of thumb has four parts:

1. *10 units.* All tabular and similar output should have at least 10 units (unweighted) underlying any cell or data point presented. A common term for such a rule is a threshold rule (the cell count must exceed a specified threshold).

2. *10 degrees of freedom.* All modelled output should have at least 10 degrees of freedom and at least 10 units have been used to produce the model. Degrees of freedom = (number of observations) − (number of parameters) − (other restrictions of the model).

3. *Group disclosure.* In all tabular and similar output, no cell can contain more than 90% of the total number of units in its row or column to prevent group disclosure. Group disclosure is the situation where some variables in a table (usually spanning variables) define a group of units and other variables in the table divulge information that is valid for each member of the group. Even though no individual unit can be recognised, confidentiality is breached because the information is valid for each member of the group and the group as such is recognizable.

4. *Dominance.* In all tabular and similar data the largest contributor of a cell cannot exceed 50% of the cell total.

A practical problem: The dominance rule
As simple as it looks, even the overall rule of thumb has one major difficulty. This concerns element no. 4: the dominance rule.

In order to check this rule, the researcher must provide additional information on the value of the largest contributor for each cell. This often burdens the researcher with a lot of extra work. In addition, the extra information obviously has to be removed from the output before release, as releasing the value of the largest contributor of a cell would be very disclosive.

So, although the dominance rule is included in the rule of thumb, the current practice in many countries is that it is not actively checked. Even so, researchers are told that they should take it into consideration when creating their output.

Usually, an NSI decides to actively check it only in certain circumstances, for instance:

- magnitude tables on business data;

- output based on very sensitive variables;

- variables with a very skewed distribution.

Nevertheless, for those wishing to follow only the rules of thumb rather than the principles-based model, checking for dominance should be considered best practice.

The remainder of this section will examine each output classified above and discuss the interpretation of the overall rules of thumb and give some more detailed information for the principles-based model.

Frequency tables

Rule of thumb
In the case of frequency tables only parts 1 and 3 of the overall rule of thumb apply:

- Each cell of the table must contain at least 10 units (unweighted). Researchers should include the unweighted cell count in their tables to enable the checking of this rule.

- The distribution of units over all cells in any row or column is such than no cell contains more than 90% of the total number of units in that particular row or column (no concentration of units in only one cell of the row or column). This rule prevents group disclosure.

Detailed information for principles-based mode
Every cell in a table is potentially unsafe. There are no general rules for making a table safe. However, as noted above, there are various options for making a table safer. In such circumstances, a number of issues should be taken into consideration:

- whether the data set is itself disclosive (whether it has been transformed into a new variable; level of detail, etc.);

- whether units making up the data or subsets could be identified;

- closeness of the data to elements 1 (threshold), 3 (group disclosure) and 4 (dominance) of the rule of thumb;

- whether the rank ordering of contributors is known (in other words, is the largest/smallest/tallest, etc. known or guessable?);

- choice of the cell units; are these people, households, regions, etc.;

- sample choice;

- weighting.

The context in which analysis is undertaken is important. Factors that should be taken into consideration include:

- level of geographic disaggregation;

- detail of industrial/occupational classification;

- global context: domestic vs. international operations.

The most important criteria are establishing that no respondent can reasonably be identified, and that no previously unavailable confidential information about groups can be inferred.

The level of what is considered 'reasonably' is not defined precisely here—it is a function of the risk appetite of the responsible statistician or data custodian. Moreover, the aim of principles-based output checking is that all definitions are subject to the particular context. However, some factors which would be taken into consideration include identification are as follows:

- taking a significant amount of time and effort;

- requiring additional knowledge that most individuals would not be expected to have or be able to acquire easily;

- needing some technical ability;

- sensitivity and impact of any potential identification.

Magnitude tables

Rule of thumb

In the case of magnitude tables, elements 1, 3 and 4 of the overall rule of thumb apply:

- Each cell of the table contains at least 10 units (unweighted). Researchers should include the unweighted cell count in their tables to enable the checking of this rule.

- The distribution of units over all cells in any row or column is such than no cell contains more than 90% of the total number of units in that particular row or column (no concentration of units in only one cell of the row or column). This rule prevents group disclosure.

- In every cell, the largest contributor cannot exceed 50% of the cell total.

Detailed information for principles-based mode

There may be circumstances in which the cell count rule of ten is inappropriate. For example, a researcher may believe that an output is non-disclosive even though cell counts do not meet the required threshold of ten units. The onus is then on the researcher to explain why this is the case, and the individual responsible for checking output will be required to make a decision as to whether this output can be released.

The following factors are important in making this decision:

- the context of the output;

- accompanying information (this may be dependent on previous outputs);

- can the person checking output identify individuals or companies from the output, particularly if the output includes a dominant observation?

The guidelines in the section on frequency tables should be borne in mind when making this judgment.

Maxima, minima and percentiles

Rule of thumb

Maxima and minima are not released since they usually refer to only one unit.

Percentiles are treated as special cases of magnitude tables. Each percentile band should be treated as a table cell with membership of the cell determined by position in the rank. If a unit's position is known, then information about that unit can be gleaned. How useful/disclosive this is depends upon the size of the cell and the range of the band. The rules of thumb for magnitude tables apply.

Detailed information for principles-based model

For percentiles, the same principles as for magnitude tables apply.

In the principles-based model, for maxima and minima the rules for magnitude tables also apply. This means that if the minimum/maximum cannot be associated with an individual data point, they can be released.

Modes

Rule of thumb

The mode is safe when element 3 of the rule of thumb (group disclosure) is met. This prevents the case that all observations have the same value.

Detailed information for principles-based model

The rule of thumb applies.

Means, indices, ratios, indicators

Rule of thumb

Common SDC rules can be applied to means, indices, ratios and indicators, as each represents the synthesis of n observed values in a single value. The same considerations as for magnitude tables cells apply; in particular, elements 1 and 3 of the overall rule of thumb apply:

- Each single value should derive from the synthesis of at least 10 units (un-weighted). Researchers should include information in their output to enable the checking of this rule.

- For each single value to be released, the largest contributor included in the synthesis cannot exceed 50% of the total. Researchers should include information in their output to enable the checking of this rule.

Detailed information for principles-based model

When evaluating means, indices, ratios and indicators, the following considerations should be taken into account. First of all, index formula should be considered.

In general, a simple index summarises the individual variable values for the statistical units in a given population:

$$I = f(X.n). \tag{6.1}$$

For the output evaluation, the index formula f and the population size n must be specified.

Sometimes the index formula is very complex, involving a lot of attributes for each unit and combining values in such a way that reverse calculation of single values is unrealistic. As an example, consider the complexity of the Fisher price Index:

$$P_F = \sqrt{P_L.P_P} = \sqrt{\frac{\sum_{j=1}^{m} p_{1,j}.q_{0,j}}{\sum_{j=1}^{m} p_{0,j}.q_{0,j}} . \frac{\sum_{j=1}^{m} p_{1,j}.q_{1,j}}{\sum_{j=1}^{m} p_{0,j}.q_{1,j}}}. \tag{6.2}$$

Therefore, assuming that the index is not calculated upon very few units, complexity of data transformation is itself a reasonable protection against the disclosure of individual information from the value of the index. In the same way, complex indices (i.e. a combination of several simple indices, including ratios) are in general less disclosive than simple ones.

Nevertheless, index formula can also be very simple, e.g.:

$$I = \frac{\sum_{i=1}^{n} X_i}{n} \tag{6.3}$$

or

$$I = \frac{\sum_{i=1}^{n} X_i}{\sum_{i=1}^{n} Y_i} \tag{6.4}$$

or

$$Range = X_{MAX} - X_{MIN}. \tag{6.5}$$

Furthermore, the value of one or more arguments of the index formula could be easily publicly available. Note that 6.3 corresponds to the arithmetic mean.

For instance, in case 6.3 and 6.4 the denominator could be known: in such cases, the problem is reduced to the evaluation of the numerator ($\sum X_i$). This matches the issue of checking a cell of a magnitude table (if X is quantitative) or a frequency table (if X is dichotomous).

Only in this last particular case, where $X = 0, 1$, coherently with the case of frequency tables, the applied rule should be: $\sum X_i \geq 10$, and, since the frequency count of 0 values can also be derived, also: $n - \sum X_i \geq 10$. Example: with X dichotomous, a mean value with a population size of $n = 20$ corresponds to the frequency table cells:

$X = 0$	$X = 1$	Total
6	14	20

In this example, even if the mean of X is calculated upon 20 units (more than 10) and the corresponding frequency of values '1' is 14 (≥ 10, the rule for frequency cell counts), one can derive that there are only 6 (≤ 10) units having $X = 0$. However, it has to be considered whether, in this kind of situation, there is an effective risk of disclosing information upon individuals or not (see the section on frequency tables on page 217).

About case 6.5, being range a linear function of maximum and minimum, it should be released only if its components (max and min) could be released (see section on minima and maxima on page 219). In the same way, as stated earlier, when evaluating other comparison or dispersion indices, the formula (whether it is linear or not) and its arguments should be considered, along with details of the constituent variables.

The 'Indices' category comprises a wide range of statistical results (including means, totals, etc.). From the output checking point of view indices pose an Statistical Disclosure Control (SDC) problem practically quite easy to be dealt with, but theoretically analogous to that of tabular data. Therefore, the same items have to be taken into account (see section on frequency tables on page 217 and magnitude tables on page 218), particularly the type of variables involved (if they are publicly available as often is the case for social indices) and the characteristics of the sub-population of n units involved in the index computation (if they are selected according to sensitive or identifying spanning variables).

Concentration ratios

Rule of thumb
Concentration ratios are safe as long as they meet elements 1 (threshold), 3 (group disclosure) and 4 (dominance) of the rule of thumb.

Detailed information for principles-based model
Concentration ratios below CR10 (largest 10 units) are allowable provided the results can be shown to be non-disclosive, which would be the case if:

- the sample on which the output is based contains at least 10 units;

- rule 4: the dominance rule is met;

- percentages are displayed with no decimal places.

Higher moments of distributions

Rule of thumb
Higher moments (variance, skewness and kurtosis) are safe as long as part 2 of the rule of thumb is met:

- All modelled output should have at least 10 degrees of freedom and at least 10 units have been used to produce the model.
     Degrees of freedom = (number of observations) − (number of parameters) − (other restrictions of the model)

Detailed information for principles-based model
The simple rule applies. In discussions with researchers, it may be noted that with fewer than ten degrees of freedom the statistical value of the output is doubtful anyway.

Graphs (descriptive statistics or fitted values)

Rule of thumb
Graphs in themselves are not permitted as output. The underlying data may be submitted as output, which when cleared, may be used to reconstruct graphs in the researcher's own environment.

Detailed information for principles-based model
Graphical output is only allowed in cases where it is impossible to reproduce the graph without access to the full microdata; in other words, to build the graph from (tabular) output that would be cleared. In this case, the graph should meet the following conditions:

- Data points cannot be identified with units. When the graph consists of transformed or fitted data, this is usually not a problem.

- There are no significant outliers.

- The graph is submitted as a 'fixed' picture, with no data attached. This means that graphs should be submitted as files with one of the following formats: jpeg (.jpg or .jpeg), bitmap (.bmp) or windows metafile (.wmf).

Linear regression coefficients

Rule of thumb
Linear regression coefficients are safe provided at least one estimated coefficient is withheld (e.g. intercept).

Detailed information for principles-based model
Complete regressions can be released as long as:

- they have at least 10 degrees of freedom;
- they are not based solely on categorical variables;
- they are not on one unit (e.g. time series on one company).

Non-linear regression coefficients

Rule of thumb
As with linear regressions, non-linear regression coefficients are safe provided at least one estimated coefficient is withheld (e.g. the intercept).

Detailed information for principles-based model
Non-linear regressions differ from linear regressions because of the nature of the dependent variables, which are discrete. If a regression is estimated, and the same regression is repeated with one additional observation, then by using the variable's means in conjunction with the regression results, it may be possible to infer information about that one particular observation. However, in practice, this is unlikely. Changes in the number of observations included in a model are the result of:

- changing the sample explicitly, or
- changing the specification and so finding observations with inadmissible values (e.g. missing) dropping out.

The first is unlikely to lead to one observation more or less, as the information gain is negligible (unless the researcher is deliberately aiming to circumvent SDC rules). In the second case, different numbers of observations are not relevant because the specification has been changed.

Estimation residuals

Rule of thumb
No residuals and no plots of residuals should be released.

Detailed information for principles-based model
As with the rule of thumb, residuals should not be released. A reasonable request for a plot of residuals may be made by a researcher, for example, in order to demonstrate the robustness of the model. However, plots of residuals should be discouraged. Instead, a research should analyse plots within the safe setting, and a written description of the shape of the plot may be released to demonstrate robustness. If there is a need for a plot of residuals to be released, this should be assessed as for graphs, described on page 222.

Summary and test statistics

Rule of thumb

The following summary statistics can be released provided there are at least 10 degrees of freedom in the model:

- R^2 and variations;

- estimated variance;

- information criteria (like AIC and BIC);

- individual and group tests and statistics (like t, F, chi-square, Wald and Hausman).

Detailed information for principles-based model

No principles in addition to the rule-of-thumb guidelines.

Correlation coefficients

Rule of thumb

Correlation is a measure of linear relationships between variables. The checkers have to ensure that the released outputs cover first element of the rule-of-thumb, which means a minimum of 10 unweighted units underlying each correlation coefficient.

Detailed information for principles-based model

A very small number of cases exist where problems could arise (e.g. the correlation matrix includes 0 or 1). Even in these cases, the actual problem is the publication of a correlation coefficient connected with summary statistics. Correlations of binary variables are treated like linear regressions with binary explanatory variables or full saturated linear regressions. Therefore, the same items have to be taken into account as for linear regression coefficients (see section on linear regression on page 222). Nevertheless, in the majority of analyses, correlation coefficients could be classified as safe.

6.6.4   Organisational/procedural aspects of output checking

General remarks

The previous section dealt with rules for output checking. An efficient, high-quality output checking process is only possible when effective organisational and procedural practices are implemented in addition to these rules. In this chapter, the most important of these will be discussed. Practical guidelines are given for each, split into a minimum requirement and a best practice. The difference between the minimum requirement and the best practice is often just a different balance between cost and benefit. The minimum can be seen as the best value-for-money measure, while the best practice is the operational excellence every RDC should aim for.

Legal basis

The aim of this section is to outline the legal framework for output checking. Output (and therefore of output checking) is essential to an RDC. Because an RDC provides a researcher access to highly confidential data, some of the NSIs responsibility for confidentiality needs to be transferred to the researcher. This responsibility can be underpinned by drawing up contracts that include the dos and don'ts for researchers using an RDC.

Minimum standard

A legally enforceable agreement with the researcher needs to be drawn up. As a minimum standard, this contract should include the statement that data on individual persons/households/companies/organisations, etc. can NEVER leave the safe environment of the RDC, and that the researcher has a responsibility to ensure this.

Best practice

As a best practice a two-tier approach is suggested. The first level is a contract with the research organisation that the user is affiliated with. This contract makes the organisation as a whole responsible for the research project and for keeping the data safe. The agreement contains confidentiality statements, rules for an appropriate use of the data (what can/cannot be done with the data, including rules relating to NSI policy, e.g., not reproducing official statistics, not duplicating other research, etc.), and details of penalties for misuse of the data. This agreement ensures that if employee A of this organisation breaches confidentiality, it can have repercussions not only for employee A, but for the organisation as a whole. Given the seriousness of the breach, the NSI can then decide to suspend or ban the whole organisation and not just employee A.

The second level is a confidentiality statement signed by each individual researcher, binding him to safeguard data confidentiality. To link these statements to the institutional contracts, the confidentiality statement should also be signed by the organisation. This means it is signed by someone with institutional authority and the ability to discipline employers that misbehave, and not just a senior researcher or professor.

Of course, these agreements and statements have to be signed by the NSI as well. The person who signs on behalf of the NSI strongly depends on the national organisation. Four different situations have been identified:

1. the RDC director: the person in charge of the RDC has the power to give permission;

2. the NSI/Institute: President/Director/Legal unit/Board of Directors;

3. the Director of the department responsible for the data;

4. a Statistical/Confidentiality Committee: this can be internal or external to the NSI/institution.

The first two situations are the most frequently observed.

Access requests

An access request is a key component of the organisational and procedural aspect of running an RDC. At the access request stage the following points should be made clear:

- who is applying to access the RDC (is he/she or his/her institute eligible?);
- which results are going to be produced (are they feasible?);
- how results will be disseminated (is this consistent with the NSIs policy?).

Even where these issues do not directly affect output checking, an accurate preliminarily evaluation of the access request can make the checker's job easier (see, for instance, Section 6.6.4).

Minimum standard

Access is granted to 'researchers' for 'research purposes': RDCs must deal with qualified and eligible users. Eligibility is defined depending on national/European legislation or on admissibility criteria defined by the NSI. For instance, the requesting institute or the researcher him/herself have to be on a certified list, or the researcher must belong to an admissible institute. It is assumed that a bona fide user is not interested in breaching data confidentiality. Moreover, as technical and/or methodological support is not usually provided, it is important that a user is qualified to undertake their research. Practically, an access request should contain at least:

- the researchers' name and position and the institution to which they are affiliated;
- the data to be used in the research;
- a description of the research project.

Best practice

National legislation/organisation is, of course, crucial in determining RDC access procedures. The following summarises best practices currently in use in many countries:

- the research project description should provide:
  - an outline of the scientific research purposes;
  - a preliminary description of the expected results: outputs must be within the scope of the projects;
  - information on how project results will be disseminated, and particularly if they will be made publicly available;
  - motivation for the use of microdata: an explanation of why the research objective cannot be realised other than by analysing confidential microdata.

The following issues may be taken into account when developing access procedures, but are strictly dependent on national legislation/organisation and do not reflect minimum standard or best practice, but should be considered by anyone setting up an RDC:

- The application form: this may represent the agreement between RDC and the user or it may be a preliminary step towards a formal agreement that addresses all legal considerations. As an application may be returned for revisions and/or clarifications. RDC administrative/technical staff often assist applicants in this preliminary phase.

- Entities entitled to access data: usually, access to RDCs is granted to universities, research institutes or similar entities undertaking research for the public good. In some countries, a register of eligible institutions exists. Others use a quite formal procedure to verify eligibility of institutions applying for data access, for instance, the Netherlands. Depending on national circumstances, some countries also allow government departments to access confidential statistical data.

- Research aims: the aims of a research project should be scrutinised before access is granted. This is to identify projects that are not viable or can be undertaken with less sensitive data, and those that are explicitly intended to embarrass the NSI or damage its ability to make statistics. Any project that disregards the NSI's policies will be rejected (for example, some national legislation does not allow RDC output that is similar to official statistics).

Access to sampling frames and sensitive variables

RDCs provide access to a range of data, which are likely to have been generated using different sampling frames: for example, business registers may be used for company surveys, while post/zip codes are often used to generate samples for surveys on households/individuals. By their nature, sampling frames contain identifiable information, such as names, addresses or even tax reference numbers. A general principle to minimise the risk of disclosure is that such identifiable sampling information, and any other variables that can be used to directly identify individuals, should not be made available to researchers. This leaves the characteristics of variables as the only possible source of disclosure, which is why disclosure control procedures are implemented on all output. By removing sampling frame data, the risk of accidental disclosure is minimised.

Minimum standard

No direct identifiers should be included in data files that are made available to researchers. Names and addresses of individuals and companies, and other identifiers including administrative codes such as health insurance numbers should be excluded. Where a unique observation reference number is necessary (for example, when constructing a panel), the identifier should be replaced by a unique but meaningless reference number. Depending on national legislation and organisation, some RDCs may choose to provide access to identifying information in special circumstances, for example, the provision of post/zip codes for spatial research. In all cases, irrespective of the variables made available the usual disclosure control rules should apply to output.

Best practice
The minimum standard applies.

Responsibility for quality of outputs

The aim of this section is to leave no doubt as to who is responsible for the content of outputs and the conclusions that are based on these outputs. In an RDC environment the NSI cannot take this responsibility. The aim of an RDC is to allow researchers to undertake analyses for which NSIs do not have the remit and/or resources. For an NSI to take responsibility for the output of these analyses, it would imply an involvement in the research project and a responsibility for the conclusions and the quality of output. It should always be made very clear that outputs from an RDC are NOT official statistics.

Minimum standard
For this issue there is no minimum standard. All RDCs should comply with the best practice.

Best practice
Researchers are required to include a disclaimer in all publications, papers, presentations, etc., that are (in part) based on research performed at the RDC. The exact wording of the disclaimer can be country specific but the disclaimer should contain the following elements:

- data from the NSI have been used in the research, with reference to the exact data sets used;

- the presented results are the sole responsibility of the author;

- the presented results do not represent official views of the NSI nor constitute official statistics.

Examples of the exact wording in a few countries are included in Section 6.7.1.

Checking for quality

Almost all RDCs check outputs only for statistical disclosure since quality of output is generally considered the responsibility of the researcher not the RDC. Having said that, many RDCs will give researchers guidance if they believe an output has some quality issues. Also, some RDCs might be tempted to judge an output on its possible negative impact on the NSI itself (for instance, outputs that show some quality problems in official statistics).

Minimum standard
An RDC should clearly distinguish between comments relating to confidentiality and to quality. Ideally, comments relating to confidentiality should be delivered formally, while comments on quality are by a more informal channel (for example, orally in person or by phone).

Best practice
Only check an output for confidentiality, not for quality or possible negative impact on the NSI itself. If there are concerns over output being used to embarrass an NSI, this should ideally be dealt with during the application phase, when a project can be rejected entirely for this reason (as is outlined above). For added security, an NSI could include a clause in their access agreement which requires users to approach the NSI first when they suspect errors or quality issues in the data used.

Output documentation

Researchers often produce numerous outputs; to understand the data, for reasons of statistical data editing and to specify models. Therefore, over time the amount of output which needs to be checked increases. As the majority of the research projects extend over several years and syntaxes may be sent irregularly, checkers need to frequently reacquaint themselves with the details of individual research projects. This section aims to outline the information that researchers should provide to ensure that the checkers are able to understand and assess complex output quickly and efficiently. The requirements for the documentation of output should be communicated with researchers as part of the access agreement.

Minimum standard
In order to maximise the efficiency of output checking, as a minimum standard all output should include:

- the researcher's name and institute;
- the name (and number where relevant) of the research project;
- the date on which the syntax is submitted;
- a brief description of the purpose of the output;
- the data files used;
- a description of original and self-constructed variables;
- in the case of sub samples the selection criteria and the size of the sub-sample;
- weighted results and unweighted frequencies.

If the output does not correspond to the minimum standard the checkers are advised to reject the output.

Best practice

For best practice an output should include all the information above, and

- a full record (log-file) of the analysis;

- a self-explanatory documentation/annotation of the steps of analysis;

- full labelling of all variables and all value labels.

Checking every output or not?

Creating an efficient output checking process means balancing the time spent with the risk avoided. The extremes are obvious for everyone: no checking leads to large risks and zero personnel costs for checking, while checking everything reduces risk, but generates very high personnel costs for checking. There may be some situations where the time saved by checking only a sample of all outputs outweighs the added risk. Obviously, there is a benefit in this for the researchers as well. It means that they receive most of their outputs immediately, without the time delay that would be necessary for checking them.

Minimum standard

As a minimum standard all outputs are checked in order to maximise data security.

Best practice

As a best practice an NSI should define an output checking strategy. As part of this strategy the relative risk and utility of sub-sampling output should be assessed. Some considerations to take into account when developing this strategy are:

- New researchers should have all their outputs checked. After a set number of non-disclosive outputs have been submitted, a researcher is labelled 'experienced'.

- It may be possible for experienced researchers to be checked randomly and by sub-sample. If disclosive output is submitted, the researcher loses their 'experienced' status and falls back to the situation where all their output is checked.

- 'Experienced' researchers will have a greater responsibility for the confidentiality protection.

- Outputs based on sensitive variables may not be suitable for sub-sampling.

Number or size of outputs

The aim of this section is to provide guidelines to prevent the RDC being buried under a pile of outputs. An RDC will typically be used by numerous researchers at the same time who all want to receive their cleared outputs as quickly as possible. However, output checking capacity is limited, so if researchers submit very many or very large outputs this increases clearing times for them and other RDC users. So, an incentive should be built in to ensure that outputs are focused and valuable and that no checking capacity is wasted. Output checking capacity is a resource that is shared by all users and should, therefore, be claimed in all fairness.

Minimum standard
The RDC should have (and make known) a policy that allows them to reject any output on the grounds of volume or quantity only, irrespective of its content. This policy should be explained to researchers from the start.

Best practice
A cost barrier (in time or money) is created to prevent many or large outputs. Possible solutions include:

- Allowing each research project only a limited number of free outputs (in the extreme: zero). Researchers would then pay a fixed fee (based on the average time it takes to clear an output) for each additional output.

- Charging an hourly rate for the time spent clearing an output that takes an inordinate amount of time (for instance, more that five times the average time it takes to clear an output).

- Placing large output at the back of the queue. This means that smaller outputs will be checked first even when they will be submitted at a later point in time. Without such a measure, the 'cooperative researchers' will be punished because they have to wait longer for their output, since the large (and therefore time consuming) output needs to be cleared first.

It should be realised that if a financial barrier is implemented, researchers could start bundling outputs together to form one large output, so they will only pay a fee once. Setting a sensible pricing level (i.e. not too high), could go a long way in preventing this.

Output clearance times

Most organisations give guidelines but do not make a strict commitment to RDC users on the time taken to check and return their output. Experience seems to suggest that the majority of output can be cleared in 5 working days. However, this depends both on the type and the size of the results submitted, as well as RDC staff resources. Defining a fixed time frame in which researchers may expect to receive cleared output may be seen as a pressure on checkers, but the aim has to be to avoid uncertainty among users over how long they can expect to wait for output.

Minimum standard
Provide users with an indication of the average time needed to check the output with reference to the type and the volume of the submitted output, without making a commitment when this specific output will be checked.

Best practice
Response times should be monitored, and exceptions should be documented. A commitment should be made on the maximum time a user can expect to wait for a response, either in the form of a released output or a request for further information. It is worth noting that response time can be influenced by the way the RDC is funded. If a fee is charged for output checking, it limits the amount of output submitted and makes the users more aware of what they really need as output but, on the other side, it places an obligation on the NSI to reply in a specified time frame.

Number of checkers

The aim of this guideline is to protect the RDC against mistakes and to ensure confidentiality of outputs. The guideline also ensures that the output checks are consistent across all checkers.

Minimum standard
As a minimum standard the output should be checked by one employee. The RDC has to ensure that the output corresponds to the legal rules for confidential data.

Best practice
The best practice is the 'four eyes principle'. Two options exist for implementing the four eyes principle. First, the two checkers are both staff members of the RDC and second the first check is accomplished by one RDC employee (expert in statistical analyses) and the second check is done by the subject-matter department (expert in data). This last procedure lowers the risk of disclosive output being released, because the two checkers each bring their own expertise and in that way complement each other.

Skill of checkers

Researchers often use complex statistical methods for their analyses, which generate a wide range of statistical and econometric outputs. Because output checking is time consuming and costly it is vital for efficiency that checkers fulfil a minimum standard of skill. The aim of this guideline is to ensure that checking is accurate, consistent and that checkers do not waste time understanding data and statistical methods.

Minimum standard
The RDC has to ensure that the employees fulfil the following requirements:

- the technical expertise to understand most of the output;
- a working knowledge of the data held in the RDC;
- periodic training in new statistical methods combined with regular reviews.

Best practice
In addition to the requirements above, for best practice:

- At least one checker should have active research experience.

- output checking should form the main part, but not all of checker's job. This enables RDCs to employ qualified staff who might otherwise be uninterested in a job that consisted only of checking other people's work.

- To facilitate the training and review process mentioned above, a copy of all outputs and discussions relating to those outputs should be archived.

6.6.5   Researcher training

Training for RDC users (use of the facility, legal aspects, disclosure control, etc.) is widely considered a good practice. Usually, a face-to-face contact is used to train the users, and this helps in building trust and increase security.

Minimum standard
For the minimum standard face-to-face training is not necessary; however, documentation should be provided covering:

- the researcher's legal responsibilities;

- instructions on how to use the facility;

- disclosure control principles to ensure researcher understand the issues.

Best practice
For best practice a standard face-to-face training module should be developed covering the issues outlined above. It may be designed for group or individual presentation, but the criteria for such a module would be that:

- It should give researchers basic tools to use the facility without assuming that they read the documentation.

- It can be delivered by more than one person without significant loss of consistency.

Training topics

Since the organisational and procedural aspects of running an RDC are country specific, these guidelines outline the topics that should be included in best practice researcher training. The exact format of a training course can be tailored to a country's needs. The topics are grouped together in two parts.

Part 1: Process
- Legal and ethical background

  - Laws and regulations governing RDC

  - Code of ethics

- The RDC: role and purpose
  - NSI has a duty to support research
  - concerns about confidentiality risks
- RDC security model: principles
  - valid statistical purpose   safe projects
  - trusted researchers   + safe people
  - anonymisation of data   + safe data
  - technical controls around data   + safe settings
  - disclosure control of results   + safe outputs
                                     = safe use
- Who can access the RDC
- RDC use:
  - diagram of interactions (user/data/staff/output), for example:

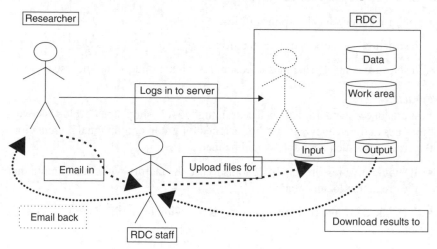

- What users cannot do:
  - attempt to remove data (including writing down data from the screen);
  - attempt to identify individuals, households, or firms;
  - use data which they are not allowed to;
  - use data for anything other than the proposed project.
- Penalties and disciplinary actions:
  - In violation of laws;
  - In violation of RDC rules/agreement.

- RDC House rules
  - booking;
  - charges and payment;
  - user's behaviour (keep workspace clean, do not misuse resources, etc.);
  - output description;
  - output releasing.
- Applying for access:
  - research project description (including preliminary output description);
  - details of people applying for access;
  - legal agreement with the institution and the researcher;
  - how to submit applications/time frame.
- Role of the RDC team

Part 2: Disclosure control

- General principle of disclosure control:
  - finding out confidential information;
  - associating that information with its source.
- Key concepts:
  - Two approaches: rules of thumb and principles based
  - Safe/unsafe classes of output
  - Primary disclosure
  - Secondary disclosure
- Rules
  - Rules of thumb for each class of output (with examples)
  - Extra attention on tabular output
    * Primary/secondary disclosure
  - Recommendations and hints for safer output.
- Practicalities of disclosure control at RDC:
  - Expectations on clearing times
  - Output description
  - Number and size of outputs.

6.7   Additional issues concerning data access

6.7.1   Examples of disclaimers

Researchers are required to include a disclaimer in all publications, papers, presentations, etc. that are (partly) based on research performed at the RDC. The exact wording of the disclaimer can be country specific but the disclaimer should contain the following elements:

- Data from the NSI have been used in the research

- The presented results are the sole responsibility of the author.

- The presented results do not represent official views of the NSI nor constitute official statistics.

Examples of the exact wording in a few countries are as follows:

United Kingdom
This work contains statistical data from ONS which is Crown copyright and reproduced with the permission of the controller of HMSO and Queen's Printer for Scotland. The use of the ONS statistical data in this work does not imply the endorsement of the ONS in relation to the interpretation or analysis of the statistical data. This work uses research data sets which may not exactly reproduce National Statistics aggregates.
Italy
The data used in the present work stem from the Italian National Statistical Institute (Istat) – survey xxxx. Data has been processed at Istat Research Data Centre (Laboratorio ADELE). Results shown are and remain entirely the responsibility of the author; they neither represent Istat views nor constitute official statistics.

6.7.2   Output description

Each output should be described by the researcher. This description should contain at least the following items:

- Researcher's name

- The date on which the output is submitted

- The research project that the output belongs to

- Brief description of the purpose of the output

- The data files used.

The following items could be added if need be:

- Any relation to earlier outputs (for instance, if it is a small adaptation of a previous output);
- Name and location of the output file;
- Email address to send the output after clearing it;
- The company that the researcher works for.

Demands on the output itself:

- Full labelling of all variables (including ones derived by the researcher);
- A description of self-constructed variables (specifying the analytical transformation or recoding applied);
- Use of sub-samples/sub-population: specify the selection criteria and size of the sub-sample/sub-population;
- Use of weights: show weighted and unweighted results;
- For magnitude tables, report the corresponding frequency table;
- For graphs, report the underlying data (or better still, just the underlying data without the graph);
- For indices, report the formula and each single factor composing the index;
- No hidden items (for instance, hidden columns in excel, hidden data behind an excel graph, hidden tables in access, etc.);
- Analytical results separated from descriptive tables.

6.8 Case studies

6.8.1 The US Census Bureau Microdata Analysis System

The Microdata Analysis System (MAS) is a system under development by the US Bureau of the Census at the time of this writing (Lucero *et al.* 2011). It will allow users to receive certain statistical analyses of Census Bureau data, such as cross tabulations and regressions, without ever having access to the data themselves.

The beta prototype of the MAS implements a Java interface within DataFERRETT, which submits requested analyses to an **R** environment. The system is being tested using the publicly available data from the Current Population Survey March 2008 Demographic Supplement.

User analyses must satisfy several statistical confidentiality rules; those that fail to meet these rules will not be output to the user. Confidentiality rules enforced by MAS include the following:

- *Confidentiality rules for universe formation.* The universe of a user analysis is the set of records the user wants to analyse. The MAS rules related to universe formation include the following:

  - *Minimum universe size requirements.* All universes formed on the MAS must pass two confidentiality rules: (1) the *No Marginal 1 or 2 Rule* and (2) the *Universe Gamma Rule.* The *No Marginal 1 or 2 Rule* requires that for a universe defined using m variables, there may not be an $m - 1$ dimensional marginal total equal to 1 or 2 in the m-way contingency table induced by the chosen variables. The *Universe Gamma Rule* requires that a universe must contain at least Γ observations, where Γ is a value kept confidential in the US Bureau of the Census.

- *Confidentiality by random record removal.* The *Drop q Rule* removes q observations from the chosen universe, where q is enough to thwart a differencing attack. A differencing attack is one in which an intruder attempts to reconstruct a confidential microdata record by subtracting the statistical analysis results obtained through two queries on similar universes.

- *Cutpoint methods.* Numerical variables are binned for presentation as categorical recodes. The cut points marking the boundaries of bins should ensure a pre-specified minimum number of observations between any two cut-point values. To guarantee this, MAS implements several cut-point determination strategies.

- *Confidentiality rules for regression methods.* The MAS implements a series of confidentiality rules for regression models, in addition to the above-mentioned universe restrictions. For example:

  - *Limited number of independent variables.* Users may only select up to 20 independent variables for any single regression equation.

  - *Limited data transformations.* Users are allowed to transform numerical variables only, and they must select their transformations from a pre-approved list. This prevents the user from performing transformations that deliberately overemphasise individual observations such as outliers. Currently, the allowable transformations are square, square root and natural logarithm.

  - *Synthetic residual plots.* To determine whether the regression adequately describes the data, diagnostics such as residual plots are necessary. Actual residual values pose a potential disclosure risk, since a data intruder can obtain the values of the dependent variable by simply adding the residuals to the fitted values obtained from the regression model. Therefore, the MAS does not pass the actual residual values back to the user. To help data users assess the fit of their ordinary least squares regression models, diagnostic plots are based on synthetic residuals and synthetic real values. These plots are designed to mimic the actual patterns seen in the scatter plots of the real residuals vs. the real-fitted values, or of the real residuals vs. the values of the individual variables.

6.8.2   Remote access at Statistics Netherlands

In this section, a description will be given of the remote access facility that is used at Statistics Netherlands. Both the technical and the functional aspects of the facility will be discussed.

Functional aspects

The main idea is that the remote access facility should resemble the 'traditional' RDC situation as much as possible, concerning confidentiality aspects. Moreover, it should resemble the look and feel of the RDC facility without the aspect of having to travel to the premises of Statistics Netherlands. i.e., the following aspects have to be taken into account:

- only authorised users should be able to make use of this facility
- microdata should remain at Statistics Netherlands
- desired output of analyses should be checked on confidentiality
- legal measures have to be taken when allowing access.

Only authorised users are allowed

At the RDC facility access of authorised users only is ensured because researchers cannot enter nor leave the premises of Statistics Netherlands unaccompanied. Moreover, only a selected group of researchers working at universities and similar institutes is allowed to make use of this facility.

The remote access facility is making use of biometric identification, to ensure that the researcher who is trying to connect to the facility is indeed the intended person. Whenever the researcher wants to access the facility, he will be identified by his fingerprint. This is checked at random times during the session as well. For more information about the process of logging on to the facility, see page 241. Obviously, again only a selected group of researchers will be given permission to make use of this facility.

Microdata remain at Statistics Netherlands

The network that is used by the remote access facility is separate from the production network (for the technical implementation, see Section 6.8.2). Moreover, the connection that a researcher makes with this facility is a terminal connection: he can only see on his screen what is running on a computer at Statistics Netherlands. He is not able to print or download any of the results of his analyses to his own computer at the institute where he is working. This ensures that the microdata and the intermediate results remain at Statistics Netherlands.

The network that is used by the 'traditional' RDC facility is not connected to the production network as well. Moreover, the computers that the researcher can use are such that no removable media can be used (no floppy drive, no USB ports) and no internet connection is possible (no email, no surfing, no ftp, etc.). This means that the microdata used by the researcher can only be accessed using a computer at the premises of Statistics Netherlands and that the researcher cannot take a copy of the data to his institute. He is able to view the (intermediate) results of his analyses on the screen, but he is not able to send those results to his institute by email or otherwise. Moreover, he is not allowed to take a printout of the results to his institute either, without having it checked by a member of Statistics Netherlands staff.

Checking output

Whenever a researcher wants to take output from a session at the 'traditional' RDC facility to his own institute, the output first needs to be checked by Statistics Netherlands' staff for confidentiality. Only then he will either be allowed to take a printout with him or the results will be sent to him by email.

A researcher that makes use of the remote access facility is not able to download or print his results either. Whenever he wants to have the results on his own computer at the institute, the results need to be checked by Statistics Netherlands staff for confidentiality. The results will then be sent to him by email.

Legal measures

Both in case of the 'traditional' RDC and the remote access facility, legal measures are taken to prevent misuse of the microdata. To that end, a contract will be signed by the institute where the researcher is working. Moreover, a statement of secrecy is signed by the researcher as well as the institute he works for.

Technical aspects

In Figure 6.1, the network representation of the remote access facility is given. An important fact is that there are three hardware firewalls involved denoted by FW1–FW3, controlling the connectivity between the 'outside' world, the remote access facility and the production network of Statistics Netherlands. Each horizontal line extending from a firewall in Figure 6.1 is a separate virtual local area network (VLAN), i.e., communication between the different VLANs is directed through at least one firewall. The three firewalls effectively guard three parts of the complete network of Statistics Netherlands. The first firewall (FW1) controls the access of the 'outside world' to the demilitarised zone (DMZ). The second firewall (FW2) is in between the DMZ and the backend, where several 'intermediate services' are situated, like the Citrix part of the remote access facility and the e-mail servers for Statistics Netherlands (not displayed in Figure 6.1). The third and final firewall (FW3) separates the backend with the actual production network of Statistics Netherlands. This way it is virtually impossible to directly connect from an external computer to the production network.

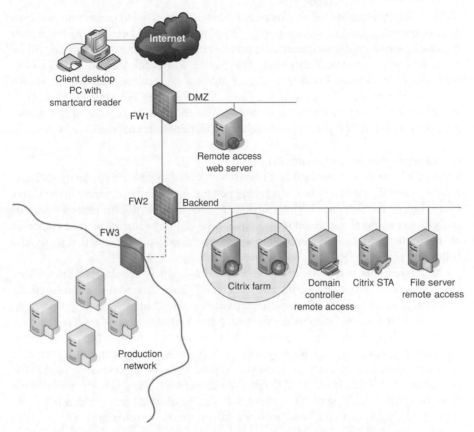

Figure 6.1   Network representation of the remote access facility.

The process of setting up a session

To ensure that only authorised users are allowed to set up a connection with the remote access facility, biometric identification is used, in combination with PKI certificates[2]. An authorised user is given a smartcard with a personal certificate. He will have to import the public part of that certificate onto his computer; the private part of that certificate is stored on an encrypted section that can only be decrypted by presenting the fingerprint that is also stored on the smartcard. This means that the user will need a smartcard reader that can read the users' fingerprint. This reader will be provided by Statistics Netherlands. Whenever the user wants to start a session, he will have to start an internet browser and type the https address of the remote access facility to try to initiate an SSL connection[3]. Since it is possible that multiple researchers will make use of the same physical computer at the institute to access the remote access facility, he will then be prompted to choose which certificate he wants to use. Obviously, he is only able to use his own certificate: the private part of his certificate is written

---

[2] PKI = Public key infrastructure.

[3] SSL = Secure sockets layer.

on the encrypted section of his smartcard. He will then have to present his finger to the fingerprint/smartcard reader in which the smartcard is inserted. If his fingerprint matches the one on the smartcard, the private part of his personal certificate will be released and sent to the Web server. This server will check the credentials of the user using the Domain Controller. If everything is correct, the user will be shown the login site of remote access, using Citrix MetaFrame (a Web Interface). Finally, he has to type in his user name and password to enter the main page of the remote access facility. On that page he will see the applications that he is allowed to use.

Checking output for confidentiality

Using the Remote Access facility a researcher is able to perform his analyses inter-actively: he will constantly see what is happening on the terminal server in the Citrix Farm. I.e., he is able to see his intermediate results and to adjust his analyses accord-ingly. At some point, he would like to have some of his results on his own computer at the institute. Since it is impossible to print or download the results directly, the following procedure is set up.

The researcher places the results in a specific directory within his own working environment. A program is constantly checking those directories and if anything is found in such a directory, that content will be placed on a secure ftp server. At the same time another program is constantly checking the secure ftp server for new content. If there is any, that program will move the content to a file server in the production environment. In this way, there will never be a direct connection between the working directory and the file server in the production environment. As soon as the output arrives on the file server of the production network, a signal is given to specific Statistics Netherlands' staff. The output will then be checked for confidentiality and sent to the corresponding researcher by email within reasonable time.

Glossary

In the field of Statistical Disclosure Control, much subject-related specific terminology is used. Unfortunately, there is no standard. Different researchers use different terms for the same subjects or even the same term for different topics. It would be very good to standardise this as much as possible.

In this view, Mark Elliot (University of Manchester), Anco Hundepool (Statistics Netherlands), Eric Schulte Nordholt (Statistics Netherlands), Jean-Louis Tambay (Statistics Canada) and Thomas Wende (Destatis, Germany) took the initiative in 2005 to compile a Glossary on Statistical Disclosure Control. A first version of the Glossary was presented at the UN-ECE worksession on SDC in Geneva in November 2005. Since then this glossary has been updated. You will find the current version of this glossary in this chapter.

This glossary can also be found at the CASC-website via http://neon.vb.cbs.nl/casc/glossary.htm. If you have comments to the glossary, you will find a link at the OnLine Glossary to submit your update.

The links in italics in this glossary refer to other term in this glossary. This handbook contains also an index, but to avoid misleading cross references we have not indexed this glossary.

A

Ambiguity rule: Synonym of *(p, q)-rule*.

Analysis server: A form of *remote data laboratory* designed to run analysis on data stored on a safe server. The user sees the results of their analysis but not the data.

Anonymised data: Data containing only *anonymised records*.

Anonymised record: A record from which *direct identifiers* have been removed.

Approximate disclosure: Approximate disclosure happens if a user is able to determine an estimate of a respondent value that is close to the real value. If the estimator is exactly the real value the disclosure is exact.

Statistical Disclosure Control, First Edition. Anco Hundepool, Josep Domingo-Ferrer, Luisa Franconi, Sarah Giessing, Eric Schulte Nordholt, Keith Spicer and Peter-Paul de Wolf.
© 2012 John Wiley & Sons, Ltd. Published 2012 by John Wiley & Sons, Ltd.

Argus: Two software packages for *Statistical Disclosure Control* are called Argus. μ-ARGUS is a specialised software tool for the protection of *microdata*. The two main techniques used for this are *global recoding* and *local suppression*. In the case of *global recoding*, several categories of a variable are collapsed into a single one. The effect of *local suppression* is that one or more values in an unsafe combination are suppressed, i.e. replaced by a missing value. Both *global recoding* and *local suppression* lead to a loss of information, because either less detailed information is provided or some information is not given at all.

τ-ARGUS is a specialised software tool for the protection of *tabular data*. τ-ARGUS is used to produce safe tables. τ-ARGUS uses the same two main techniques as μ-ARGUS: *global recoding* and *local suppression*. For τ-ARGUS, the latter consists of *suppression* of cells in a table. Various methods for cell suppression are available. Also rounding procedures can be applied.

Attribute disclosure: Attribute disclosure is *attribution* independent of *identification*. This form of disclosure is of primary concern to *NSIs* involved in *tabular data* release and arises from the presence of empty cells either in a released table or linkable set of tables after any *subtraction* has taken place. Minimally, the presence of an empty cell within a table means that an *intruder* may infer from mere knowledge that a population unit is represented in the table and that the *intruder* does not possess the combination of attributes within the empty cell.

Attribution: Attribution is the association or disassociation of a particular attribute with a particular population unit.

B

Barnardisation: A method of disclosure control for tables of counts that involves randomly adding or subtracting 1 from some cells in the table.

Blurring: Blurring replaces a reported value by an average. There are many possible ways to implement blurring. Groups of records for averaging may be formed by matching on other variables or by sorting on the variable of interest. The number of records in a group (whose data will be averaged) may be fixed or random. The average associated with a particular group may be assigned to all members of a group, or to the 'middle' member (as in a moving average). It may be performed on more than one variable with different groupings for each variable.

Bottom coding: See *top and bottom coding*.

Bounds: The range of possible values of a cell in a table of frequency counts where the cell value has been perturbed or suppressed. Where only margins of tables are released, it is possible to infer bounds for the unreleased joint distribution. One method for inferring the bounds across a table is known as the *Shuttle algorithm*.

C

Calculated interval: The interval containing possible values for a suppressed cell in a table, given the table structure and the values published.

Cell suppression: In *tabular data*, the cell *suppression* SDC method consists of *primary* and *complementary (secondary) suppression*. *Primary suppression* can be characterised as withholding the values of all *risky cells* from publication, which means that their value is not shown in the table but replaced by a symbol such as '×' to indicate the suppression. To reach the desired protection for *risky cells*, it is necessary to suppress additional non *risky cells*, which is called *secondary suppression*. The pattern of secondary/complementary suppressed cells has to be carefully chosen to provide the desired level of ambiguity for the *risky cells* with the least amount of suppressed information.

Complementary suppression: Synonym of *secondary suppression*.

Complete disclosure: Synonym of *exact disclosure*.

Concentration rule: A concentration rule is a rule to assess whether a cell is a *risky cell*, based on comparing the size of the individual contributions to the cell. Examples are the *dominance rule* and the *p%-rule*

Confidentiality edit: The confidentiality edit is a procedure developed by the US Census Bureau to provide protection in data tables prepared from the 1990 Census. There are two different approaches: one was used for the regular Census data; the other was used for the long-form data, which were filled by a sample of the population. Both techniques apply *statistical disclosure limitation* techniques to the *microdata* files before they are used to prepare tables. The adjusted files themselves are not released; they are used only to prepare tables. For the regular Census *microdata* file, the confidentiality edit involves '*data swapping*' or 'switching' of attributes between matched records from different geographical units. For small blocks, the Census Bureau increases the *sampling fraction*. After the *microdata* file has been treated in this way, it can be used directly to prepare tables and no further disclosure analysis is needed. For long-form data, *sampling* provides sufficient confidentiality protection, except in small geographic regions. To provide additional protection in small geographic regions, one household is randomly selected and a sample of its data fields are blanked and replaced by imputed values.

Controlled rounding: To solve the additivity problem, a procedure called controlled rounding was developed. It is a form of *random rounding*, but it is constrained to have the sum of the published entries in each row and column equal to the appropriate published marginal totals. Linear programming methods are used to identify a controlled rounding pattern for a table.

Controlled Tabular Adjustment (CTA): A method to protect *tabular data* based on the selective adjustment of cell values. *Sensitive cell* values are replaced by either of their closest safe values and small adjustments are made to other cells to restore the table additivity. Controlled tabular adjustment has been developed as an alternative to *cell suppression*.

Conventional rounding: A disclosure control method for tables of counts. When using conventional rounding, each count is rounded to the nearest multiple of a fixed base. For example, using a base of 5, counts ending in 1 or 2 are rounded down and replaced by counts ending in 0 and counts ending in 3 or 4 are rounded up and replaced by counts ending in 5. Counts ending between 6 and 9 are treated similarly. Counts with a last digit of 0 or 5 are kept unchanged. When rounding to base 10, a count ending in 5 may always be rounded up, or it may be rounded up or down based on a rounding convention. This is also known as unbiased rounding or banker's rounding.

D

Data divergence: The sum of all differences between two data sets (data–data divergence) or between a single data set and reality (data-world divergence). Sources of data divergence include: data ageing, response errors, coding or data entry errors, differences in coding and the effect of disclosure control.

Data intruder: A data user who attempts to disclose information about a population unit through *identification* or *attribution*.

Data intrusion detection: The detection of a *data intruder* through his behavior. This is most likely to occur through analysis of a pattern of requests submitted to a *remote data laboratory*. At present, this is only a theoretical possibility, but it is likely to become more relevant as *virtual safe settings* become more prevalent.

Data Intrusion Simulation (DIS): A method of estimating the probability that a *data intruder* who has matched an arbitrary population unit against a *sample unique* in a target *microdata* file has done so correctly.

Data protection: Data protection refers to the set of *privacy*-motivated laws, policies and procedures that aim to minimise intrusion into respondents' *privacy* caused by the collection, storage and *dissemination* of *personal data*.

Data swapping: A disclosure control method for *microdata* that involves swapping the values of variables for records that match on a representative *key*. In the literature, this technique is also sometimes referred to as 'multidimensional transformation'. It is a transformation technique that guarantees (under certain conditions) the maintenance of a set of statistics, such as means, variances and univariate distributions.

Data utility: A summary term describing the value of a given data release as an analytical resource. This comprises the data's analytical completeness and its analytical validity. *Disclosure control methods* usually have an adverse effect on data utility. Ideally, the goal of any disclosure control regime should be to maximise data utility while minimizing *disclosure risk*. In practice, disclosure control decisions are a trade-off between utility and *disclosure risk*.

Deterministic rounding: Synonym of *conventional rounding*.

Direct identifier: Synonym of *formal identifier*.

Direct identification: Identification of a statistical unit from its *formal identifiers*.

Disclosive cells: Synonym of *risky cells*.

Disclosure: Disclosure relates to the inappropriate *attribution* of information to a data subject, whether an individual or an organisation. Disclosure has two components: *identification* and *attribution*.

Disclosure by fishing: This is an attack method where an *intruder* identifies risky records within a target data set and then attempts to find population units corresponding to those records. It is the type of disclosure that can be assessed through a *special uniques analysis*.

Disclosure by matching: Disclosure by the linking of records within an *identification data set* with those in an *anonymised data set*.

Disclosure by response knowledge: This is disclosure resulting from the knowledge that a person was participating in a particular survey. If an *intruder* knows that a specific individual has participated in the survey, and that consequently his or her data are in the data set, *identification* and *disclosure* can be accomplished more easily.

Disclosure by spontaneous recognition: This means the recognition of an individual within the data set. This may occur by accident or because a *data intruder* is searching for a particular individual. This is more likely to be successful if the individual has a rare combination of characteristics which is known to the *intruder*.

Disclosure control methods: There are two main approaches to control the disclosure of confidential data. The first is to reduce the information content of the data provided to the external user. For the release of *tabular data* this type of technique is called *restriction-based disclosure control method*, and for the release of *microdata* the expression disclosure control by data reduction is used. The second is to change the data before the *dissemination* in such a way that the *disclosure risk* for the confidential data is decreased, but the information content is retained as much as possible. These are called *perturbation-based disclosure control methods*.

Disclosure from analytical outputs: The use of output to make *attributions* about individual population units. This situation might arise to users that can interrogate data but do not have direct access to them such as in a *remote data laboratory* or by using a *Research Data Centre*. One particular concern is the publication of residuals.

Disclosure limitation methods: Synonym of *disclosure control methods*.

Disclosure risk: A disclosure risk occurs if an unacceptably narrow estimation of a respondent's confidential information is possible or if *exact disclosure* is possible with a high level of confidence.

Disclosure scenarios: Depending on the intention of the *intruder*, his or her type of a priori knowledge and the *microdata* available, three different types of disclosure or disclosure scenarios are possible for *microdata*: *disclosure by matching*, *disclosure by response knowledge* and *disclosure by spontaneous recognition*.

Dissemination: Supply of data in any form whatever: publications, access to databases, microfiches, telephone communications, etc.

Disturbing the data: This process involves changing the data in some systematic fashion, with the result that the figures are insufficiently precise to disclose information about individual cases.

Dominance rule: Synonym of (n, k)-*rule*.

E

Exact disclosure: Exact disclosure occurs if a user is able to determine the exact attribute for an individual entity from released information.

F

Fishing: See *Disclosure by fishing*

Formal identifier: Any variable or set of variables which is structurally unique for every population unit, for example a population registration number. If the formal identifier is known to the *intruder*, *identification* of a target individual is directly possible for him or her, without the necessity to have additional knowledge before studying the *microdata*. Some combinations of variables such as name and address are pragmatic formal identifiers, where non-unique instances are empirically possible, but with negligible probability.

G

Global recoding: Problems of confidentiality can be tackled by changing the structure of data. Thus, rows or columns in tables can be combined into larger class intervals or new groupings of characteristics. This may be a simpler solution than the *suppression* of individual items, but it tends to reduce the descriptive and analytical value of the table. This protection technique may also be used to protect *microdata*.

Group Disclosure: When information about a (small) group can be disclosed collectively, this is called group disclosure. This can occur in frequency tables if all respondents of an identifiable group score on a single (sensitive) category.

H

HITAS: A heuristic approach to *cell suppression* in hierarchical tables.

I

Identification: Identification is the association of a particular record within a set of data with a particular population unit.

Identification data set: A data set that contains *formal identifiers*.

Identification data: Those *personal data* that allow *direct identification* of the data subject, and which are needed for the collection, checking and matching of the data, but are not subsequently used for drawing up statistical results.

Identification key: Synonym of *key*.

Identification risk: This risk is defined as the probability that an *intruder* identifies at least one respondent in the disseminated *microdata*. This identification may lead to the disclosure of (sensitive) information about the respondent. The risk of identification depends on the number and nature of *quasi-identifiers* in the *microdata* and in the a priori knowledge of the *intruder*.

Identifying variable: A variable that either is a *formal identifier* or forms part of a *formal identifier*.

Indirect identification: Inferring the identity of a population unit within a *microdata* release other than from *direct identification*.

Inferential disclosure: Inferential disclosure occurs when information can be inferred with high confidence from statistical properties of the released data. For example, the data may show a high correlation between income and purchase price of home. As the purchase price of a home is typically public information, a third party might use this information to infer the income of a data subject. In general, *NSIs* are not concerned with inferential disclosure for two reasons. First, a major purpose of statistical data is to enable users to infer and understand relationships between variables. If *NSIs* equated disclosure with inference, no data could be released. Second, inferences are designed to predict aggregate behaviour, not individual attributes, and thus often poor predictors of individual data values.

Informed consent: Basic ethical tenet of scientific research on human populations. Sociologists do not involve a human being as a subject in research without the informed consent of the subject or the subject's legally authorised representative, except as otherwise specified. Informed consent refers to a person's agreement to allow *personal data* to be provided for research and statistical purposes. Agreement is based on full exposure of the facts the person needs to make the decision intelligently, including awareness of any risks involved, of uses and users of the data, and of alternatives to providing the data.

Intruder: A data user who attempts to link a respondent to a *microdata* record or make *attributions* about particular population units from aggregate data. Intruders may be motivated by a wish to discredit or otherwise harm the *NSI*, the survey or the government in general, to gain notoriety or publicity, or to gain profitable knowledge about particular respondents.

K

Key: A set of *key variables*.

Key variable: A variable in common between two data sets, which may therefore be used for linking records between them. A key variable can either be a *formal identifier* or a *quasi-identifier*.

L

Licensing agreement: A permit, issued under certain conditions, for researchers to use confidential data for specific purposes and for specific periods of time. This agreement consists of contractual and ethical obligations, as well as penalties for improper disclosure or use of identifiable information. These penalties can vary from withdrawal of the license and denial of access to additional data sets to the forfeiting of a deposit paid prior to the release of a *microdata* file. A licensing agreement is almost always combined with the signing of a contract. This contract includes a number of requirements: specification of the intended use of the data; instruction not to release the *microdata* file to another recipient; prior review and approval by the releasing agency for all user outputs to be published or disseminated; terms and location of access and enforceable penalties.

Local recoding: A disclosure control technique for *microdata* where two (or more) different versions of a variable are used dependent on some other variable. The different versions will have different levels of coding. This will depend on the distribution of the first variable conditional on the second. A typical example occurs where the distribution of a variable is heavily skewed in some geographical areas. In the areas where the distribution is skewed minor categories may be combined to produce a courser variable.

Local suppression: Protection technique that diminishes the risk of recognition of information about individuals or enterprises by suppressing individual scores on *identifying variables*.

Lower bound: The lowest possible value of a cell in a table of frequency counts where the cell value has been perturbed or suppressed.

M

Macrodata: Synonym of *tabular data*.

Microaggregation: Records are grouped based on a proximity measure of variables of interest, and the same small groups of records are used in calculating aggregates for those variables. The aggregates are released instead of the individual record values.

Microdata: A microdata set consists of a set of records containing information on individual respondents or on economic entities.

Minimal unique: A combination of variable values that are unique in the *microdata* set at hand and contain no proper subset with this property (so it is a minimal set with the *uniqueness* property).

Modular approach: Synonym of *HITAS*.

N

NSI(s): Abbreviation for National Statistical Institute(s).

NSO(s): Abbreviation for National Statistical Organisation/Office(s). Synonym of *NSI*.

(*n*, *k*)-rule: A cell is regarded as confidential, if the *n* largest units contribute more than *k*% to the cell total, e.g. $n = 2$ and $k = 85$ means that a cell is defined as risky if the two largest units contribute more than 85% to the cell total. The *n* and *k* are given by the statistical authority. In some *NSIs*, the values of *n* and *k* are confidential. This rule is also known as *dominance rule*.

O

On-site facility: A facility that has been established on the premises of several *NSIs*. It is a place where external researchers can be permitted access to potentially disclosive data under contractual agreements which cover the maintenance of confidentiality, and which place strict controls on the uses to which the data can be put. The on-site facility can be seen as a '*safe setting*' in which confidential data can be analysed. The on-site facility itself would consist of a secure hermetic working and data-storage environment in which the confidentiality of the data for research can be ensured. Both the physical and the IT aspects of *security* would be considered here. The on-site facility also includes administrative and support facilities to external users, and ensures that the agreed conditions for access to the data were complied with.

Ordinary rounding: Synonym of *conventional rounding*.

Oversuppression: A situation that may occur during the application of the technique of *cell suppression*. This denotes the fact that more information has been suppressed than strictly necessary to maintain confidentiality.

P

Partial disclosure: Synonym of *approximate disclosure*.

Passive confidentiality: For foreign trade statistics, EU countries generally apply the principle of 'passive confidentiality', that is they take appropriate measures only at the request of importers or exporters who feel that their interests would be harmed by the *dissemination* of data.

Personal data: Any information relating to an identified or identifiable natural person (data subject). An identifiable person is one who can be identified, directly or indirectly. Where an individual is not identifiable, data are said to be anonymous.

Perturbation-based disclosure control methods: Techniques for the release of data that change the data before the *dissemination* in such a way that the *disclosure risk* for the confidential data is decreased but the information content is retained as far as possible. Perturbation-based methods falsify the data before publication by introducing an element of error purposely for confidentiality reasons. For example, an error can be inserted in the cell values after a table is created, which means that the error is introduced to the output of the data and will therefore be referred to as output perturbation. The error can also be inserted in the original data on the *microdata* level, which is the input of the tables one wants to create; the method will then be referred to as data perturbation - input perturbation being the better but uncommonly used expression. Possible perturbation methods are:

- *rounding*;

- perturbation, for example, by the addition of random noise or by the *Post Randomisation Method*;

- *disclosure control methods* for *microdata* applied to *tabular data*.

Population unique: A record within a data set which is unique within the population on a given *key*.

***p%*-rule:** A cell is regarded as confidential of an individual contribution to that cell can be estimated to within $p\%$. The $p\%$-rule is a special case of the (p, q)-*rule* where q is 100, meaning that from general knowledge any respondent can estimate the contribution of another respondent to within 100% (i.e. knows the value to be non-negative and less than a certain value which can be up to twice the actual value).

(p, q)-rule: It is assumed that out of publicly available information the contribution of one individual to the cell total can be estimated to within $q\%$ (q = error before publication); after the publication of the statistic the value can be estimated to within $p\%$ (p = error after publication). In the (p, q)-rule, the ratio p/q represents the information gain through publication. If the information gain is unacceptable, the cell is declared as confidential. The parameter values p and q are determined by the statistical authority and thus define the acceptable level of information gain. In some *NSIs*, the values of p and q are confidential. This rule is also known as the *prior-posterior rule*.

Post Randomisation Method (PRAM): Protection method for *microdata* in which the scores of a categorical variable are changed with certain probabilities into other scores. It is thus intentional misclassification with known misclassification probabilities.

Primary confidentiality: It concerns tabular cell data, whose *dissemination* would permit *attribute disclosure*. The two main reasons for declaring data to be primary confidential are:

- too few units in a cell;

- dominance of one or two units in a cell.

The limits of what constitutes 'too few' or 'dominance' vary between statistical domains.

Primary protection: Protection using *disclosure control methods* for all cells containing small counts or cases of dominance.

Primary suppression: This technique can be characterised as withholding all *disclosive cells* from publication, which means that their value is not shown in the table, but replaced by a symbol such as '×' to indicate the suppression. According to the definition of *disclosive cells*, in frequency count tables all cells containing small counts and in tables of magnitudes all cells containing small counts or representing cases of dominance have to be primary suppressed.

Prior-posterior rule: Synonym of the (p, q)-*rule*.

Privacy: Privacy is a concept that applies to data subjects while confidentiality applies to data. The concept is defined as follows: 'It is the status accorded to data which has been agreed upon between the person or organisation furnishing the data and the organisation receiving it and which describes the degree of protection which will be provided.' There is a definite relationship between confidentiality and privacy. Breach of confidentiality can result in disclosure of data which harms the individual. This is an attack on privacy because it is an intrusion into a person's self-determination on the way his or her *personal data* are used. Informational privacy encompasses an individual's freedom from excessive intrusion in the quest for information and an individual's ability to choose the extent and circumstances under which his or her beliefs, behaviours, opinions and attitudes will be shared with or withheld from others.

Probability based disclosures (approximate or exact): Sometimes although a fact is not disclosed with certainty, the published data can be used to make a statement that has a high probability of being correct.

Q

Quasi-identifier: Variable values or combinations of variable values within a data set that are not structural uniques but might be empirically unique, and therefore in principle uniquely identify a population unit.

R

Randomised response: Randomised response is a technique used to collect sensitive information from individuals in such a way that survey interviewers and those who process the data do not know which of two alternative questions the respondent has answered.

Random perturbation: This is a *disclosure control method* according to which a noise, in the form of a random value is added to the true value or, in the case of categorical variables, where another value is randomly substituted for the true value.

Random rounding: In order to reduce the amount of data loss that occurs with *suppression*, alternative methods have been investigated to protect *sensitive cells* in tables of frequencies. Perturbation methods such as random rounding and *controlled rounding* are examples of such alternatives. In random rounding cell values are rounded, but instead of using standard rounding conventions a random decision is made as to whether they will be rounded up or down. The rounding mechanism can be set up to produce unbiased rounded results.

Rank swapping: Rank swapping provides a way of using continuous variables to define pairs of records for swapping. Instead of insisting that variables match (agree exactly), they are defined to be close based on their proximity to each other on a list sorted on the continuous variable. Records which are close in rank on the sorted variable are designated as pairs for swapping. Frequently, in rank swapping, the variable used in the sort is the one that will be swapped.

Record linkage process: Process attempting to classify pairs of matches in a product space A×B from two files A and B into M, the set of true links, and U, the set of non-true links.

Record swapping: A special case of *data swapping*, where the geographical codes of records are swapped.

Remote access: On-line access to disclosive *microdata*. The connection is made such that no confidential data will leave the remote access system. The permitted users can use the data for analysis. But all results/output are checked before release.

Remote data laboratory: A virtual environment providing *remote execution* facilities.

Remote execution: Submitting scripts on-line for execution on disclosive *microdata* stored within an institute's protected network. If the results are regarded as *safe data*, they are sent to the submitter of the script. Otherwise, the submitter is informed that the request cannot be acquiesced. Remote execution may either work through submitting scripts for a particular statistical package such as SAS, SPSS or STATA which runs on the remote server or via a tailor made client system which sits on the user's desktop.

Research Data Centre: A controlled environment where researchers can analyse disclosive *microdata* without the possibility to export any confidential information. This can both be a physical environment at a NSI (on-site facility) or via a secure connection over the internet (remote access).

Research Data File: A microdata file that has been partially protected. It is meant to be made available for statistical analysis by trustworthy researchers at recognised research institutes. It is only protected against *disclosure by spontaneous recognition*. The researcher and the research institute have to sign a confidentiality agreement with the *NSI*.

Residual disclosure: Disclosure that occurs by combining released information with previously released or publicly available information. For example, tables for non-overlapping areas can be subtracted from a larger region, leaving confidential residual information for small areas.

Restricted access: Imposing conditions on access to the *microdata*. Users can either have access to the whole range of raw protected data and process individually the information they are interested in - which is the ideal situation for them - or their access to the protected data is restricted and they can only have a certain number of outputs (e.g. tables) or maybe only outputs of a certain structure. Restricted access is sometimes necessary to ensure that linkage between tables cannot happen.

Restricted data: Synonym of *safe data*.

Restriction based disclosure control method: Method for the release of *tabular data*, which consists in reducing access to the data provided to the external user. This method reduces the content of information provided to the user of the *tabular data*. This is implemented by not publishing all the figures derived from the collected data or by not publishing the information in as detailed a form as would be possible.

Risky cells: The cells of a table' which are non-publishable due to the risk of *statistical disclosure* are referred to as risky cells. By definition there are three types of risky cells: small counts, dominance and *complementary suppression* cells.

Risky data: Data are considered to be disclosive when they allow statistical units to be identified, either directly or indirectly, thereby disclosing individual information. To determine whether a statistical unit is identifiable, account shall be taken of all the means that might reasonably be used by a third party to identify the said statistical unit.

Rounding: Rounding belongs to the group of *disclosure control methods* based on output-perturbation. It is used to protect small counts in *tabular data* against disclosure. The basic idea behind this disclosure control method is to round each count up or down either deterministically or probabilistically to the nearest integer multiple of a rounding base. The additive nature of the table is generally destroyed by this process. Rounding can also serve as a recoding method for *microdata*.

R-U map: A graphical representation of the trade off between *disclosure risk* and *data utility*.

S

Safe data: *Microdata* or *macrodata* that have been protected by suitable *Statistical Disclosure Control* methods.

Safe setting: An environment such as a microdata lab whereby access to a disclosive data set can be controlled.

Safety interval: The minimal *calculated interval* that is required for the value of a cell that does not satisfy the *primary suppression* rule.

Sample unique: A record within a data set which is unique within that data set on a given *key*.

Sampling: In the context of disclosure control, this refers to releasing only a proportion of the original data records on a *microdata* file.

Sampling fraction: The proportion of the population contained within a data release. With simple random sampling, the sample fraction represents the proportion of population units that are selected in the sample. With more complex sampling methods, this is usually the ratio of the number of units in the sample to the number of units in the population from which the sample is selected. Sampling can be used as a data protection strategy: the lower the sampling fraction the smaller the probability that a sample unique is a population unique.

Scenario analysis: A set of pseudo-criminological methods for analysing and classifying the plausible risk channels for a data intrusion. The methods are based around first delineating the means, motives and opportunity that an *intruder* may have for conducting the attack. The output of such an analysis is a specification of a set of *keys* likely to be held by *data intruders*.

Scientific Use file: Synonym of *Researcher Data File*.

Secondary data intrusion: After an attempt to match between *identification* and *target data sets* an *intruder* may discriminate between non-unique matches by further direct investigations using additional variables.

Secondary disclosure risk: It concerns data which is not primary disclosive, but whose *dissemination*, when combined with other data permits the *identification* of a *microdata* unit or the disclosure of a unit's *attribute*.

Secondary suppression: To reach the desired protection for *risky cells*, it is necessary to suppress additional nonrisky cells, which is called secondary suppression or *complementary suppression*. The pattern of complementary suppressed cells has to be carefully chosen to provide the desired level of ambiguity for the *disclosive cells* at the highest level of information contained in the released statistics.

Security: An efficient *disclosure control method* provides protection against *exact disclosure* or unwanted narrow estimation of the attributes of an individual entity, in other words, a useful technique prevents *exact* or *partial disclosure*. The security level is accordingly high. In the case of *disclosure control methods* for the release of *microdata*, this protection is ensured if the *identification* of a respondent is not possible, because the *identification* is the pre-requisite for disclosure.

Sensitive category: A category of a variable is called sensitive, if that category contains sensitive information about an individual or enterprise.

Sensitive cell: Cell for which knowledge of the value would permit an unduly accurate estimate of the contribution of an individual respondent. Sensitive cells are identified by the application of a *dominance rule* such as the (n, k)-*rule* or the (p, q)-*rule* to their *microdata*.

Sensitive variables: Variables contained in a data record apart from the *key variables* that belong to the private domain of respondents who would not like them to be disclosed. There is no exact definition given for what a 'sensitive variable' is and therefore, the division into *key variables* and sensitive variables is somehow arbitrary. Some data are clearly sensitive such as the possession of a criminal record, one's medical condition or credit record, but there are other cases where the distinction depends on the circumstances, e.g. the income of a person might be regarded as a sensitive variable in some countries and as *quasi-identifier* in others, or in some societies the religion of an individual might count as a *key* and a sensitive variable at the same time. All variables that contain one or more *sensitive categories* are called sensitive variables.

Shuttle algorithm: A method for finding lower and upper cell *bounds* by iterating through dependencies between cell counts. There exist many dependencies between individual counts and aggregations of counts in contingency tables. Where not all individual counts are known, but some aggregated counts are known, the dependencies can be used to make inferences about the missing counts. The Shuttle algorithm constructs a specific subset of the many possible dependencies and recursively iterates through them in order to find *bounds* on missing counts. As many dependencies will involve unknown counts, the dependencies need to be expressed in terms of inequalities involving *lower* and *upper bounds*, rather than simple equalities. The algorithm ends when a complete iteration fails to tighten the *bounds* on any cell counts.

Special uniques analysis: A method of analysing the per-record risk of *microdata*.

Statistical confidentiality: The protection of data that relate to single statistical units and are obtained directly for statistical purposes or indirectly from administrative or other sources against any breach of the right to confidentiality. It implies the prevention of unlawful disclosure.

Statistical Data Protection (SDP): Statistical Data Protection is a more general concept which takes into account all steps of production. SDP is multidisciplinary and draws on computer science (data *security*), statistics and operations research.

Statistical disclosure: Statistical disclosure is said to take place if the *dissemination* of a statistic enables the external user of the data to obtain a better estimate for a confidential piece of information than would be possible without it.

Statistical Disclosure Control (SDC): Statistical Disclosure Control techniques can be defined as the set of methods to reduce the risk of disclosing information on individuals, businesses or other organisations. Such methods are only related to the *dissemination* step and are usually based on restricting the amount of or modifying the data released.

Statistical Disclosure Limitation (SDL): Synonym of *Statistical Disclosure Control*.

Sub-additivity: One of the properties of the (n, k)-*rule* or (p, q)-*rule* that assists in the search for complementary cells. The property means that the sensitivity of a union of disjoint cells cannot be greater than the sum of the cells' individual sensitivities (triangle inequality). Sub-additivity is an important property because it means that aggregates of cells that are not sensitive are not sensitive either and do not need to be tested.

Subtraction: The principle whereby an *intruder* may attack a table of population counts by removing known individuals from the table. If this leads to the presence of certain zeroes in the table then that table is vulnerable to *attribute disclosure*.

SUDA: A software system for conducting analyses on *population uniques* and special *sample uniques*. The *special uniques analysis* method implemented in SUDA for measuring and assessing *disclosure risk* is based on re-sampling methods and used by the ONS.

Suppression: One of the most commonly used ways of protecting *sensitive cells* in a table is via suppression. It is obvious that in a row or column with a suppressed *sensitive cell*, at least one additional cell must be suppressed, or the value in the *sensitive cell* could be calculated exactly by subtraction from the marginal total. For this reason, certain other cells must also be suppressed. These are referred to as *secondary suppressions*. While it is possible to select cells for *secondary suppression* manually, it is difficult to guarantee that the result provides adequate protection.

Swapping (or switching): Swapping (or switching) involves selecting a sample of the records, finding a match in the data base on a set of pre-determined variables and swapping all or some of the other variables between the matched records. Swapping (or switching) was illustrated as part of the *confidentiality edit* for tables of frequency data.

Synthetic data: An approach to confidentiality where instead of disseminating real data, synthetic data that have been generated from one or more population models are released.

Synthetic substitution: See *Controlled Tabular Adjustment*.

T

Table server: A form of *remote data laboratory* designed to release safe tables.

Tables of frequency (count) data: These tables present the number of units of analysis in a cell. When data are from a sample, the cells may contain weighted counts, where weights are used to bring sample results to the population levels. Frequencies may also be represented as percentages.

Tables of magnitude data: Tables of magnitude data present the aggregate of a 'quantity of interest' over all units of analysis in the cell. When data are from a sample, the cells may contain weighted aggregates, where quantities are multiplied by units' weights to bring sample results up to population levels. The data may be presented as averages by dividing the aggregates by the number of units in their cells.

Tabular data: Aggregate information on entities presented in tables.

Target dataset: An *anonymised data set* in which an *intruder* attempts to identify particular population units.

Threshold rule: Usually, with the threshold rule, a cell in a table of frequencies is defined to be sensitive if the number of respondents is less than some specified number. Some agencies require at least five respondents in a cell, others require three. When thresholds are not respected, an agency may restructure tables and combine categories or use *cell suppression*, *rounding* or the *confidentiality edit*, or provide other additional protection in order to satisfy the rule.

Top and bottom coding: It consists in setting top-codes or bottom-codes on quantitative variables. A top-code for a variable is an upper limit on all published values of that variable. Any value greater than this upper limit is replaced by the upper limit or is not published on the *microdata* file at all. Similarly, a bottom-code is a lower limit on all published values for a variable. Different limits may be used for different quantitative variables, or for different sub-populations.

U

Union unique: A *sample unique* that is also *population unique*. The proportion of *sample uniques* that are union uniques is one measure of file level *disclosure risk*.

Uniqueness: The term is used to characterise the situation where an individual can be distinguished from all other members in a population or sample in terms of information available on *microdata* records (or within a given *key*). The existence of uniqueness is determined by the size of the population or sample and the degree to which it is segmented by geographic information and the number and detail of characteristics provided for each unit in the data set (or within the *key*).

Upper bound: The highest possible value of a cell in a table of frequency counts where the cell value has been perturbed or suppressed.

V

Virtual safe setting: Synonym of *remote data laboratory*.

W

Waiver approach: Instead of suppressing *tabular data*, some agencies ask respondents for permission to publish cells even though doing so may cause these respondents' sensitive information to be estimated accurately. This is referred to as the waiver approach. Waivers are signed records of the respondents' granting permission to publish such cells. This method is most useful with small surveys or sets of tables involving only a few cases of dominance, where only a few waivers are needed. Of course, respondents must believe that their data are not particularly sensitive before they will sign waivers.

References

Abowd J.M. and Lanc J.I. 2004 New approaches to confidentiality protection: Synthetic data, remote access and research data centers. In *Privacy in Statistical Databases, PSD 2004* (eds. Domingo-Ferrer J. and Torra V.), vol. 3050 of *Lecture Notes in Computer Science*, pp. 282–289. Springer, Berlin/Heidelberg.

Abowd J.M. and Woodcock S.D. 2001 Disclosure limitation in longitudinal linked tables. In *Confidentiality, Disclosure and Data Access: Theory and Practical Applications for Statistical Agencies* (eds. Doyle P., Lane J.I., Theeuwes J.J. and Zayatz L.V.), pp. 215–278. North-Holland, Amsterdam.

Abowd J.M. and Woodcock S.D. 2004 Multiply-imputing confidential characteristics and file links in longitudinal linked data. In *Privacy in Statistical Databases, PSD 2004* (eds. Domingo Ferrer J. and Torra V.), vol. 3050 of *Lecture Notes in Computer Science*, pp. 290–297. Springer, Berlin/Heidelberg.

Abowd J.M., Stinson M. and Benedetto G. 2006 Final report to the social security administration on the sipp/ssa/irs public use file project. Technical report, U.S. Census Bureau Longitudinal Employer-Household Dynamics Program.

Aggarwal C.C. and Yu P.S. 2004 A condensation approach to privacy preserving data mining. In *Advances in Database Technology – EDBT 2004* (eds. Bertino E., Christodoulakis S., Plexousakis D., Christophides V., Koubarakis M., Böhm K. and Ferrari E.), vol. 2992 of *Lecture Notes in Computer Science*, pp. 183–199. Springer, Berlin/Heidelberg.

Aggarwal G., Feder T., Kenthapadi K., Motwani R., Panigrahy R., Thomas D. and Zhu A. 2005 Anonymizing tables. In *Proceedings of ICDT 2005* (eds. Eiter T. and Libkin L.), vol. 3363 of *Lecture Notes in Computer Science*, pp. 246–258. Springer, Berlin/Heidelberg.

Agrawal D. and Aggarwal C.C. 2001 On the design and quantification of privacy preserving data mining algorithms. *Proceedings of the 20th Symposium on Principles of Database Systems*. ACM, Santa Barbara CA.

An D. and Little R. 2007 Multiple imputation: an alternative to top coding for statistical disclosure control. *Journal of the Royal Statistical Society, Series A* **170**, 923–940.

Bacher J., Brand R. and Bender S. 2002 Re-identifying register data by survey data using cluster analysis: an empirical study. *International Journal of Uncertainty, Fuzziness and Knowledge Based Systems* **10**(5), 589–607.

Benedetti R. and Franconi L. 1998 Statistical and technological solutions for controlled data dissemination. *Pre-proceedings of New Techniques and Technologies for Statistics*, vol. 1, pp. 225–232. Eurostat, Luxemburg.

Box G.E.P. and Cox D.R. 1964 An analysis of transformations. *Journal of Royal Statistical Society, Series B 26*(22), 211–252.

Brand R. 2002a Microdata protection through noise addition. In *Inference Control in Statistical Databases* (ed. Domingo-Ferrer J.), vol. 2316 of *Lecture Notes in Computer Science*, pp. 97–116. Springer, Berlin/Heidelberg.

Brand R. 2002b Tests of the applicability of Sullivan's algorithm to synthetic data and real business data in official statistics. Deliverable of European Project IST-2000-25069 CASC, available at http://neon.vb.cbs.nl/casc.

Brand R., Domingo-Ferrer J. and Mateo-Sanz J.M. 2002 Reference data sets to test and compare SDC methods for protection of numerical microdata. Deliverable of European Project IST-2000-25069 CASC, available at http://neon.vb.cbs.nl/casc.

Brown D. 2004 Different approaches to disclosure control problems associated with geography *Work session on Statistical Data Confidentiality, Luxembourg, 7–9 April 2003*, pp. 250–262. Monographs in Official Statistics. Eurostat, Luxembourg. Available at http://www.unece.org/fileadmin/DAM/stats/documents/2003/04/confidentiality/wp.14.e.pdf.

Burridge J. 2003 Information preserving statistical obfuscation. *Statistics and Computing* **13**, 321–327.

Casciano C., Ichim D. and Corallo L. 2011 Sampling as a way to reduce risk and create a Public Use File maintaining weighted totals. Paper presented at the Joint UNECE/Eurostat work session on Statistical Data Confidentiality, Tarragona, 26–28 October 2011. Available at http://www.unece.org/fileadmin/DAM/stats/documents/ece/ces/ge.46/2011/04_Italy-Casciano.pdf.

Castro J. 2003 User's and programmer's manual of the network flows heuristics package for cell suppression in 1h2d tables. Technical Report DR 2003-07, Department of Statistics and Operations Research, Universitat Politècnica de Catalunya, Barcelona, Spain. Available at http://neon.vb.cbs.nl/casc/deliv/41D6_NF1H2D-Tau-Argus.pdf.

Castro J. 2006 Minimum-distance controlled perturbation methods for large-scale tabular data protection. *European Journal of Operational Research* **171**, 39–52.

Castro J. 2012 Recent advances in optimization techniques for statistical tabular data protection. *European Journal of Operational Research* **216**, 257–269.

Castro J. and Giessing S. 2006a Quality issues of minimum distance controlled tabular adjustment. Paper presented at the European Conference on Quality in Survey Statistics, Cardiff, 24–26 April 2006.

Castro J. and Giessing S. 2006b Testing variants of minimum distance controlled tabular adjustment. *Work session on Statistical Data Confidentiality, Geneva, 9–11 November 2005*, pp. 333–343 Monographs in Official Statistics. Eurostat, Luxembourg.

Castro J. and Gonzalez J.A. 2009 A package for l1 controlled tabular adjustment Paper presented at the Joint UNECE/Eurostat work session on Statistical Data Confidentiality, Bilbao, 2–4 December 2009.

Cator E., Hensbergen A. and Rozenholc Y. 2005 Statistical disclosure control using PRAM. *Proceedings of the 48th European Study Group Mathematics with Industry*, pp. 23–30. Delft University Press, Delft, The Netherlands.

Chatfield C. and Collins A.J. 1980 *Introduction to Multivariate Analysis*. Chapman and Hall, London.

Chaudhuri A. and Mukerjee R. 1988 *Randomized Response, Theory and Techniques*. Marcel Dekker Inc., New York.

Chen T.T. 1979 Analysis of randomized response as purposively misclassified data. *Proceedings of the section on survey research methods*, pp. 158–163 American Statistical Association, USA.

Cox L.H. 1981 Linear sensitivity measures in statistical disclosure control. *Journal of Planning and Inference* **5**, 153–164.

Cox L.H. 2001 Disclosure risk for tabular economic data. In *Confidentiality, Disclosure and Data Access: Theory and Practical Applications for Statistical Agencies* (eds. Doyle, Lane, Theeuwes and Zayatz), Chapter 8, North-Holland, Amsterdam, Netherlands.

Cox L.H. and Dandekar R.H. 2002 Synthetic tabular data – an alternative to complementary cell suppression. Manuscript available from ramesh.dandekar@eia.doe.gov or as http://mysite.verizon.net/vze7w8vk/syn_tab.pdf.

Cox L.H., Karr A.F. and Kinney S.K. 2011 Risk-utility paradigms for statistical disclosure limitation: how to think, but not how to act. *International Statistical Review* **79**(2), 160–183.

Cox L.H., Kelly J.P. and Patil R. 2004 Balancing quality and confidentiality for multivariate tabular data. *Privacy in Statistical Databases, PSD 2004*, vol. 3050 of *Lecture Notes in Computer Science*, pp. 87–98. Springer, Berlin/Heidelberg.

Cox L.H., Orelien J.G. and Shah B.V. 2006 A method for preserving statistical distributions subject to controlled tabular adjustment. In *Privacy in Statistical Databases, PSD 2006* (eds. Domingo-Ferrer J. and Franconi L.), vol. 4302 of *Lecture Notes in Computer Science*, pp. 1–11. Springer, Berlin/Heidelberg.

CRAN 2011 The Comprehensive R Archive Network. Available at http://cran.r-project.org.

Crystal.Ball 2004. Available at http://download.cnet.com/Crystal-Ball-2000-Pro/3000-18483_4-38546.html and http://www.oracle.com/us/products/applications/crystalball.

Daalmans J. and de Waal T. 2011 An improved formulation of the disclosure auditing problem for secondary cell suppression. *Transactions on Data Privacy* **3**, 217–251.

Dalenius T. 1977 Towards a methodology for statistical disclosure control. *Statistisk Tidskrift* **5**, 429–444.

Dalenius T. 1986 Finding a needle in a haystack – or identifying anonymous census records. *Journal of Official Statistics* **2**(3), 329–336.

Dalenius T. and Reiss S.P. 1978 Data-swapping: a technique for disclosure control (extended abstract). *Proceedings of the ASA Section on Survey Research Methods*, pp. 191–194. American Statistical Association, Washington DC.

Dandekar R., Cohen M. and Kirkendall N. 2002a Sensitive micro data protection using Latin hypercube sampling technique. In *Inference Control in Statistical Databases* (ed. Domingo-Ferrer J.), vol. 2316 of *Lecture Notes in Computer Science*, pp. 245–253. Springer, Berlin/Heidelberg.

Dandekar R., Domingo-Ferrer J. and Sebé F. 2002b Lhs-based hybrid microdata vs rank swapping and microaggregation for numeric microdata protection. In *Inference Control in Statistical Databases* (ed. Domingo-Ferrer J.), vol. 2316 of *Lecture Notes in Computer Science*, pp. 153–162. Springer, Berlin/Heidelberg.

de Waal A.G. and Willenborg L.C.R.J. 1995 Global recodings and local suppressions in microdata sets. *Proceedings of Statistics Canada Symposium'95*, pp. 121–132. Statistics Canada, Ottawa.

de Waal A.G. and Willenborg L.C.R.J. 1999 Information loss through global recoding and local suppression. *Netherlands Official Statistics* **14**, 17–20. Special issue on SDC.

de Wolf P.P. 2002 Hitas: a heustic approach to cell suppression in hierarchical tables. In *Inference Control in Statistical Databases 2002* (ed. Domingo-Ferrer J.), vol. 2316 of *Lecture Notes in Computer Science*. Springer, Berlin.

de Wolf P.P. 2007 Cell suppression in a special class of linked tables. *Work session on Statistical Data Confidentiality, Manchester, 17–19 December 2007*, pp. 220–226. Methodologies and Working papers. Eurostat, Luxembourg. Available at http://www.unece.org/fileadmin/DAM/stats/documents/2007/12/confidentiality/wp.21.e.pdf.

de Wolf P.P. and Giessing S. 2008 How to make the τ-ARGUS modular method applicable to linked tables. In *Privacy in Statistical Databases, PSD 2008* (eds. Domingo-Ferrer J. and Saygın Y.), vol. 5262 of *Lecture Notes in Computer Science*, pp. 227–238. Springer, Berlin/Heidelberg.

de Wolf P.P. and Hundepool A. 2010 Three ways to deal with a set of linked sbs tables using τ-ARGUS. In *Privacy in Statistical Databases, PSD2010* (eds. Domingo-Ferrer J. and Magkos E.), vol. 6344 of *Lecture Notes in Computer Science*, pp. 66–73. Springer, Heidelberg.

de Wolf P.P. and van Gelder I. 2004 An empirical evaluation of PRAM. Technical report, Statistics Netherlands. Discussion paper 04012, available at http://neon.vb.cbs.nl/casc.

de Wolf P.P., Gouweleeuw J.M., Kooiman P. and Willenborg L.C.R.J. 1998 Reflections on PRAM *Statistical Data Protection, Proceedings of the conference, Lisbon, 25–27 March 1998*, pp. 337–349. Eurostat, Luxembourg. Available at http://neon.vb.cbs.nl/casc/related/Sdp_98_2.pdf.

Defays D. and Nanopoulos P. 1993 Panels of enterprises and confidentiality: the small aggregates method. *Proceedings of 92 Symposium on Design and Analysis of Longitudinal Surveys*, pp. 195–204. Statistics Canada, Ottawa.

Di Consiglio L. and Polettini S. 2006 Improving individual risk estimators. In *Privacy in Statistical Databases, PSD 2006* (eds. Domingo-Ferrer J. and Franconi L.), vol. 4302 of *Lecture Notes in Computer Science*, pp. 241–256. Springer, Berlin/Heidelberg.

Di Consiglio L. and Polettini S. 2008 Use of auxiliary information in risk estimation. In *Privacy in Statistical Databases, PSD 2008* (eds. Domingo-Ferrer J. and Saygın Y.), vol. 5262 of *Lecture Notes in Computer Science*, pp. 213–225. Springer, Berlin/Heidelberg.

Domingo-Ferrer J. 2011 Risk-utility paradigms for statistical disclosure limitation: how to think, but not how to act – discussion: a science of statistical disclosure limitation? *International Statistical Review* **79**(2), 184–186.

Domingo-Ferrer J. and González-Nicolás U. 2010 Hybrid data using microaggregation. *Information Sciences* **180**(15), 2834–2844.

Domingo-Ferrer J. and Mateo-Sanz J.M. 1999 On resampling for statistical confidentiality in contingency tables. *Computers & Mathematics with Applications* **38**, 13–32.

Domingo-Ferrer J. and Mateo-Sanz J.M. 2002 Practical data-oriented microaggregation for statistical disclosure control. *IEEE Transactions on Knowledge and Data Engineering* **14**(1), 189–201.

Domingo-Ferrer J. and Solanas A. 2008 A measure of variance for hierarchical nominal attributes. *Information Sciences* **178**(24), 4644–4655. Erratum available at *Information Sciences*, **179**(20): 3732, 2009.

Domingo-Ferrer J. and Torra V. 2001a Disclosure protection methods and information loss for microdata. In *Confidentiality, Disclosure and Data Access: Theory and Practical Applications for Statistical Agencies* (eds. Doyle P., Lane J.I., Theeuwes J.J.M. and Zayatz L.), pp. 91–110. North-Holland, Amsterdam.

Domingo-Ferrer J. and Torra V. 2001b A quantitative comparison of disclosure control methods for microdata. In *Confidentiality, Disclosure and Data Access: Theory and Practical Applications for Statistical Agencies* (eds. Doyle P., Lane J.I., Theeuwes J.J.M. and Zayatz L.), pp. 111–134. North-Holland, Amsterdam.

Domingo-Ferrer J. and Torra V. 2003 Disclosure risk assessment in statistical microdata protection via advanced record linkage. *Statistics and Computing* 13(4), 343–354.

Domingo-Ferrer J. and Torra V. 2005 Ordinal, continuous and heterogenerous k-anonymity through microaggregation. *Data Mining and Knowledge Discovery* 11(2), 195–212.

Domingo-Ferrer J. and Torra V. 2008 A critique of k-anonymity and some of its enhancements. *Proceedings of ARES/PSAI 2008*, pp. 990–993. IEEE Computer Society, Los Alamitos, CA.

Domingo-Ferrer J., Drechsler J. and Polettini S. 2009 Report on synthetic data files. Technical report, Deliverable of Project ESSNET-SDC.

Domingo-Ferrer J., Mateo-Sanz J.M. and Torra V. 2001 Comparing SDC methods for microdata on the basis of information loss and disclosure risk. *Pre-proceedings of ETK-NTTS'2001 (vol. 2)*, pp. 807–826. Eurostat, Luxembourg.

Domingo-Ferrer J., Mateo-Sanz J.M., Oganian A. and Torres A. 2002 On the security of microaggregation with individual ranking: analytical attacks. *International Journal of Uncertainty, Fuzziness and Knowledge-Based Systems* 10(5), 477–492.

Domingo-Ferrer J., Sebé F. and Castellà J. 2004 On the security of noise addition for privacy in statistical databases. In *Privacy in Statistical Databases, PSD 2004* (eds. Domingo-Ferrer J. and Torra V.), vol. 3050 of *Lecture Notes in Computer Science*, pp. 149–161. Springer, Berlin/Heidelberg.

Domingo-Ferrer J., Sebé F. and Solanas A. 2008 A polynomial-time approximation to optimal multivariate microaggregation. *Computers & Mathematics with Applications* 55(4), 714–732.

Doyle P., Lane J.I., Theeuwes J.M.M. and Zayatz L. 2001 *Confidentiality, Disclosure and Data Access: Theory and Practical Application for Statistical Agencies*. Elsevier Science BV, North-Holland, Amsterdam.

Drechsler J. and Reiter J. 2008 Accounting for intruder uncertainty due to sampling when estimating identification disclosure risks in partially synthetic data. In *Privacy in Statistical Databases, PSD 2008* (eds. Domingo-Ferrer J. and Saygın Y.), vol. 5262 of *Lecture Notes in Computer Science*, pp. 227–238. Springer, Berlin/Heidelberg.

Drechsler J., Bender S. and Rässler S. 2008a Comparing fully and partially synthetic datasets for statistical disclosure control in the German IAB Establishment Panel. *Transactions on Data Privacy* 1(3), 105–130.

Drechsler J., Dundler A., Bender S., Rässler S. and Zwick T. 2008b A new approach for statistical disclosure control in the IAB establishment panel-multiple imputation for a better data access. *Advances in Statistical Analysis* 92(4), 439–458.

Duke-Williams O. and Rees P. 1998 Can census offices publish statistics for more than one small area geography? an analysis of the differencing problem in statistical disclosure. *International Journal of Geographical Information Science* 12, 579–605.

Duncan G. and Lambert D. 1986 Disclosure-limited data dissemination. *Journal of the American Statistical Association* 81(3), 10–27.

Duncan G., Keller-McNulty S. and Stokes S. 2001 Disclosure risk vs. data utility: the r-u confidentiality map. Technical Report LA-UR-01-6428, Los Alamos National Laboratory, Statistical Sciences Group, Los Alamos, New Mexico.

Duncan G.T. and Pearson R.W. 1991 Enhancing access to microdata while protecting confidentiality: prospects for the future. *Statistical Science* **6**, 219–239.

Duncan G.T., Elliot M. and Salazar-González J.J. 2011 *Statistical Confidentiality Principles and Practice* Statistics for Social and Behavioral Sciences. Springer, New York, Dordrecht, Heidelberg, London.

Dwork C. 2006 Differential privacy *Proceedings of 33rd International Colloquium on Automata, Languages and Programming (ICALP)*, vol. 4052 of *Lecture Notes in Computer Science*, pp. 1–12. Springer, Berlin.

Dwork C. 2011 A firm foundation for private data analysis. *Communications of the ACM* **54**(1), 86–95.

Elamir E.A.H. and Skinner C.J. 2006 Record level measures of disclosure risk for survey microdata. *Journal of Official Statistics* **22**(3), 525–539.

Elliot M.J. 2000 DIS: a new approach to the measurement of statistical disclosure risk. *International Journal of Risk Management* **2**(4), 39–48.

Elliot M.J. and Dale A. 1999 Scenarios of attack: a data intruder's perspective on statistical disclosure risk. *Netherlands Official Statistics* **14**, 6–10.

Elliot M.J., Lomax S., Mackey E. and Purdam K. 2010 Data environment analysis and the Key Variable Mapping System. In *Privacy in Statistical Databases, PSD 2010* (eds. Domingo-Ferrer J. and Magkos E.), vol. 6344 of *Lecture Notes in Computer Science*, pp. 138–145. Springer, Berlin/Heidelberg.

Elliot M.J., Manning A.M. and Ford R.W. 2002 A computational algorithm for handling the special uniques problem. *International Journal of Uncertainty, Fuzziness and Knowledge Based Systems* **5**(10), 493–509.

Elliot M.J., Manning A.M., Mayes K., Gurd J. and Bane M. 2006 SUDA: a program for detecting special uniques. *Work session on Statistical Data Confidentiality, Geneva, 9–11 November 2005*, pp. 353–362. Monographs in Official Statistics. Eurostat, Luxembourg. Available at http://www.unece.org/fileadmin/DAM/stats/documents/ece/ces/ge.46/2005/wp.44.e.pdf.

Elliot M.J., Skinner C.J. and Dale A. 1998 Special uniques, random uniques, and sticky populations: some counterintuitive effects of geographical detail on disclosure risk. *Research in Official Statistics* **1**(2), 53–67.

ESSnet on Common tools and harmonised methodology for SDC in the ESS 2012 Deliverables. Available at http://neon.vb.cbs.nl/casc/ESSNet2Deliverables.htm.

Eurostat 1990 Council regulation 1588/90. Available at http://eur-lex.europa.eu/LexUriServ/LexUriServ.do?uri=CELEX:31990R1588:EN:HTML.

Eurostat 1997 Council regulation 322/97. Available at http://eur-lex.europa.eu/LexUriServ/LexUriServ.do?uri=CELEX:31997R0322:EN:HTML.

Eurostat 2002 Commission regulation (ec) no 831/2002 concerning access to confidential data for scientific purposes. Available at http://eur-lex.europa.eu/LexUriServ/LexUriServ.do?uri=OJ:L:2002:133:0007:0009:EN:PDF.

Eurostat 2009 Commission regulation (EC) no 223/2009 on European statistics. Available at http://eur-lex.europa.eu/LexUriServ/LexUriServ.do?uri=OJ:L:2009:087:0164:0173:en:PDF.

Eurostat 2010 European Union Structure of Earning Survey access to microdata. Available at http://epp.eurostat.ec.europa.eu/portal/page/portal/microdata/ses.

Eurostat 2011a European statistics code of practice. Available at http://epp.eurostat.ec.europa. eu/portal/page/portal/quality/documents/KS-32-11-995-EN-C.pdf.

Eurostat 2011b European Union Labour Force Survey (EU LFS) access to microdata. Available at http://epp.eurostat.ec.europa.eu/portal/page/portal/microdata/lfs.

Eurostat 2011c Quality report of the European Union Labour Force Survey 2009. Available at http://epp.eurostat.ec.europa.eu/cache/ITY_OFFPUB/KS-RA-11-020/EN/KS-RA-11-020-EN.PDF.

Evans B.T., Zayatz L. and Slanta J. 1998 Using noise for disclosure limitation for establishment tabular data. *Journal of Official Statistics* **14**, 537–552.

Fellegi I.P. and Sunter A.B. 1969 A theory for record linkage. *Journal of the American Statistical Association* **64**(328), 1183–1210.

Fienberg S.E. 1994 A radical proposal for the provision of micro-data samples and the preservation of confidentiality. Technical Report 611, Carnegie Mellon University Department of Statistics.

Fienberg S.E. and Makov U.E. 1998 Confidentiality, uniqueness and disclosure limitation for categorical data. *Journal of Official Statistics* **14**(4), 385–397.

Fienberg S.E. and McIntyre J. 2004 Data swapping: variations on a theme by Dalenius and Reiss. In *Privacy in Statistical Databases, PSD 2004* (eds. Domingo-Ferrer J. and Torra V.), vol. 3050 of *Lecture Notes in Computer Science*, pp. 14–29. Springer, Berlin/Heidelberg.

Fienberg S.E., Makov U.E. and Steele R.J. 1998 Disclosure limitation using perturbation and related methods for categorical data. *Journal of Official Statistics* **14**(4), 485–502.

Fischetti M. and Salazar-González J.J. 2000 Models and algorithms for optimizing cell suppression problem in tabular data with linear constraints. *Journal of the American Statistical Association* **95**, 916–928.

Fischetti M. and Salazar-González J.J. 2003 Partial cell suppression: a new methodology for statistical disclosure control. *Statistics and Computing* **13**, 13–21.

Florian A. 1992 An efficient sampling scheme: updated Latin hypercube sampling. *Probabilistic Engineering Mechanics* **7**(2), 123–130.

Forbes A., Naylor J., Leaver V., Gare M., Hawkes T. and Camden M. 2009 Confidentiality plans for the 2011 censuses in the United Kingdom, Australia and New Zealand: a comparison. Paper presented at the Joint UNECE/Eurostat work session on Statistical Data Confidentiality, Bilbao, 2–4 December 2009. Available at http://www.unece.org/fileadmin/DAM/stats/documents/ece/ces/ge.46/2009/wp.28.e.pdf.

Foschi F. 2011 Risk for high dimensional business microdata. Paper presented at the Joint UNECE/Eurostat work session on Statistical Data Confidentiality, Tarragona, 26–28 October 2011. Available at http://www.unece.org/fileadmin/DAM/stats/documents/ece/ces/ge.46/2011/03_Italy-Foschi.pdf.

Franconi L. and Ichim D. 2007 Community Innovation Survey: comparable dissemination. *Work session on Statistical Data Confidentiality, Manchester, 17–19 December 2007*, pp. 11–23. Methodologies and Working papers. Eurostat, Luxembourg. Available at http://www.unece.org/fileadmin/DAM/stats/documents/2007/12/confidentiality/wp.2.e.pdf.

Franconi L. and Polettini S. 2004 Individual risk estimation in μ-ARGUS: a review. In *Privacy in Statistical Databases, PSD 2004* (eds. Domingo-Ferrer J. and Torra V.), vol. 3050 of *Lecture Notes in Computer Science*, pp. 262–272. Springer, Berlin/Heidelberg.

Franconi L. and Stander J. 2002 A model based method for disclosure limitation of business microdata. *Journal of the Royal Statistical Society D – Statistician* **51**, 1–11.

Franconi L., Ichim D. and Corallo L. 2011 Farm Structure Survey: consideration on the release of a European microdata file for research purposes. Joint UNECE/Eurostat work session on Statistical Data Confidentiality, Tarragona, 26–28 October 2011. Available at http://www.unece.org/fileadmin/DAM/stats/documents/ece/ces/ge.46/2011/36_Italy.pdf.

Fraser B. and Wooton J. 2006 A proposed method for confidentialising tabular output to protect against differencing *Work session on Statistical Data Confidentiality, Geneva, 9–11 November 2005*, pp. 299–302. Monographs in Official Statistics. Eurostat, Luxembourg. Available at http://www.unece.org/fileadmin/DAM/stats/documents/ece/ces/ge.46/2005/wp.35.e.pdf.

Frend J., Abrahams C., Groom P., Spicer K., Tudor C. and Forbes A. 2011 Statistical disclosure control for communal establishments in the UK 2011 census Paper presented at the Joint UNECE/Eurostat work session on Statistical Data Confidentiality, Tarragona, 26–28 October 2011. Available at http://www.unece.org/fileadmin/DAM/stats/documents/ece/ces/ge.46/2011/15_UK.pdf.

Frolova O., Fillion J.M. and Tambay J.L. 2009 CONFID2: Statistics Canada's new tabular data confidentiality software. *SSC Annual Meeting, Proceedings of the Survey Methods Section*. Statistics Canada, Ottawa, Canada.

Geurts J. 1992 Heuristics for cell suppression in tables. Technical report, Statistics Netherlands.

Giessing S. 2004 Survey on methods for tabular protection in Argus. In *Privacy in statistical databases, PSD 2004* (eds. Domingo-Ferrer J. and Torra V.), vol. 3050 of *Lecture Notes in Computer Science*, pp. 1–13. Springer, Berlin/Heidelberg.

Giessing S. 2008 Protection of tables with negative values. ESSnet report, Destatis. Available at http://neon.vb.cbs.nl/casc/ESSnet/PosNegReport.pdf.

Giessing S. 2009 Techniques for using τ-ARGUS modular on sets of linked tables. Paper presented at the Joint UNECE/Eurostat work session on Statistical Data Confidentiality, Bilbao, 2–4 December 2009. Available at http://www.unece.org/fileadmin/DAM/stats/documents/ece/ces/ge.46/2009/wp.35.e.pdf.

Giessing S. 2011 Post-tabular stochastic noise to protect skewed business data. Paper presented at the Joint UNECE/Eurostat work session on Statistical Data Confidentiality, Tarragona, 26–28 October 2011. Available at http://www.unece.org/fileadmin/DAM/stats/documents/ece/ces/ge.46/2011/47_Giessing.pdf.

Giessing S. and Repsilber D. 2002 Tools and strategies to protect multiple tables with the ghquar cell suppression engine. In *Inference Control in Statistical Databases* (ed. Domingo-Ferrer J.), vol. 2316 of *Lecture Notes in Computer Science*. Springer, Berlin/Heidelberg.

Giessing S., Dittrich S., Gehrling D., Krüger A., Merz F.J. and Wirtz H. 2006 Bericht der Arbeitsgruppe 'geheimhaltungskonzept des statistischen verbundes, pilotanwendung: Umsatzsteuerstatistik' (in German) Document for the meeting of the 'Ausschuss für Organisation und Umsetzung'.

Giessing S., Hundepool A. and Castro J. 2007 Rounding methods for protecting EU-aggregates. *Work session on Statistical Data Confidentiality, Manchester, 17–19 December 2007*, pp. 255–264. Methodologies and Working papers. Eurostat, Luxembourg. Available at http://www.unece.org/fileadmin/DAM/stats/documents/2007/12/confidentiality/wp.25.e.pdf.

Gomatam S. and Karr A. 2003 Distortion measures for categorical data swapping. Technical Report 131, National Institute of Statistical Sciences.

Gouweleeuw J.M., Kooiman P., Willenborg L.C.R.J. and de Wolf P.P. 1997 Post randomisation for statistical disclosure control: Theory and implementation. Technical report, Statistics Netherlands. Research paper no. 9731.

Gouweleeuw J.M., Kooiman P., Willenborg L.C.R.J. and de Wolf P.P. 1998a Post randomisation for statistical disclosure control: theory and implementation. *Journal of Official Statistics* **14**(4), 463–478.

Gouweleeuw J.M., Kooiman P., Willenborg L.C.R.J. and de Wolf P.P. 1998b The post randomisation method for protecting microdata. *Qüestiió-Quaderns d'Estadística i Investigació Operativa* **22**(1), 145–156.

Graham P. and Penny R. 2005 Multiply imputed synthetic data files. Technical report, University of Otago, Christchurch, New Zealand.

Greenberg B. 1987 Rank swapping for ordinal data. Washington, DC: U. S. Bureau of the Census (unpublished manuscript).

Gross B., Guiblin P. and Merrett K. 2004 Implementing the post randomisation method to the individual sample of anonymised records (SAR) from the 2001 Census. Office for National Statistics, available at http://www.ccsr.ac.uk/sars/events/2004-09-30/gross.pdf.

Hafner H. and Lenz R. 2006 Anonymisation of Linked Employer Employee datasets. On CD-ROM accompanying proceedings of Privacy in Statistical Databases, PSD 2006.

Hansen S.L. and Mukherjee S. 2003 A polynomial algorithm for optimal univariate microaggregation. *IEEE Transactions on Knowledge and Data Engineering* **15**(4), 1043–1044.

Härdle W. 1991 *Smoothing Techniques with Implementation in S*. Springer-Verlag, New York.

Heer G.R. 1993 A bootstrap procedure to preserve statistical confidentiality in contingency tables. In *Proceedings of the International Seminar on Statistical Confidentiality* (ed. Lievesley D.), pp. 261–271. Office for Official Publications of the European Communities, Luxembourg.

Heldal J. 2011 Anonymised integrated event history datasets for researchers. Joint UN-ECE/Eurostat work session on Statistical Data Confidentiality, Tarragona, 26–28 October 2011. Available at http://www.unece.org/fileadmin/DAM/stats/documents/ece/ces/ge.46/2011/29_Norway.pdf.

Höhne J. 2004 Varianten von Zufallsüberlagerung (in German). Working paper of the project 'Faktische Anonymisierung wirtschaftsstatistischer Einzeldaten'.

Hundepool A., van de Wetering A., Ramaswamy R., de Wolf P.P., Franconi L., Brand R. and Domingo-Ferrer J. 2008 μ-ARGUS *version 4.2 Software and User's Manual*. Statistics Netherlands, Voorburg NL. http://neon.vb.cbs.nl/casc/mu.htm.

Hundepool A., van de Wetering A., Ramaswamy R., de Wolf P.P., Giessing S., Fischetti M., Salazar-González J.J., Castro J. and Lowthian P. 2011 τ-ARGUS *3.5 user manual*. Statistics Netherlands, The Hague, The Netherlands.

Huntington D.E. and Lyrintzis C.S. 1998 Improvements to and limitations of Latin hypercube sampling. *Probabilistic Engineering Mechanics* **13**(4), 245–253.

Ichim D. 2008 Extensions of the re-identification risk measures based on log-linear models. In *Privacy in Statistical Databases, PSD 2008* (eds. Domingo-Ferrer J. and Saygın Y.), vol. 5262 of *Lecture Notes in Computer Science*, pp. 203–212. Springer, Berlin/Heidelberg.

Ichim D. 2009 Disclosure control of business microdata: a density based approach. *International Statistical Review* **77**, 196–211.

Ichim D. and Foschi F. 2011 Individual disclosure risk measure based on log-liner models *Proceedings of the 58th ISI Congress, Dublin, 21–26 August 2011*. Available at http://isi2011.congressplanner.eu/pdfs/650395.pdf.

Ichim D. and Franconi L. 2006 Calibration estimator for magnitude tabular data protection. On CD-ROM accompanying proceedings of Privacy in Statistical Databases, PSD 2006.

Ichim D. and Franconi L. 2007 Disclosure scenario and risk assessment: structure of earnings survey *Work session on Statistical Data Confidentiality, Manchester, 17–19 December 2007*, pp. 115–123. Methodologies and Working papers. Eurostat, Luxembourg. Available at http://www.unece.org/fileadmin/DAM/stats/documents/2007/12/confidentiality/wp.12.e.pdf.

Ichim D. and Franconi L. 2010 Strategies to achieve sdc harmonisation at European level: multiple countries, multiple files, multiple surveys. In *Privacy in Statistical Databases, PSD 2010* (eds. Domingo-Ferrer J. and Magkos E.), vol. 6344 of *Lecture Notes in Computer Science*, pp. 284–296. Springer, Berlin/Heidelberg.

ISI 1985 Declaration on professional ethics. http://www.isi-web.org/about-isi/professional-ethics/43-about/about/151-ethics1985.

ISI 2010 Declaration on professional ethics. http://www.isi-web.org/images/about/Declaration-EN2010.pdf.

IVEware 2011 IVEware: Imputation and Variance Estimation Software, version 0.2. Available at http://www.isr.umich.edu/src/smp/ive/.

Jaro M.A. 1989 Advances in record-linkage methodology as applied to matching the 1985 census of Tampa, Florida. *Journal of the American Statistical Association* **84**(406), 414–420.

Jewett R. 1993 Disclosure analysis for the 1992 economic census. Technical report. Economic Statistical Methods and Programming Division, Bureau of the Census, Washington, DC. Available at: http://www.census.gov/srd/sdc/Jewett.disc.econ.1992.pdf.

Kennickell A.B. 1999 Multiple imputation and disclosure protection: the case of the 1995 survey of consumer finances. In *Statistical Data Protection, SDP 98* (ed. Domingo-Ferrer J.), pp. 248–267. Office for Official Publications of the European Communities, Luxembourg.

Kim J.J. 1986 A method for limiting disclosure in microdata based on random noise and transformation. *Proceedings of the Section on Survey Research Methods*, pp. 303–308. American Statistical Association, Alexandria VA.

Kim J.J. 1990 Subpopulation estimation for the masked data. *Proceedings of the ASA Section on Survey Research Methods*, pp. 456–461. American Statistical Association, Alexandria VA.

Kooiman P., Willenborg L. and Gouweleeuw J. 1998 PRAM: a method for disclosure limitation of microdata. Technical report, Statistics Netherlands (Voorburg, NL).

Kuha J.T. and Skinner C.J. 1997 Categorical data analysis and misclassification In *Survey Measurement and Process Quality* (eds. Lyberg L., Biemer P., Collins M., de Leeuw E., Dippo C., Schwarz N. and Trewin D.), Chapter 28. John Wiley and Sons, New York.

Lambert D. 1993 Measures of disclosure risk and harm. *Journal of Official Statistics* **9**(3), 313–331.

Laszlo M. and Mukherjee S. 2005 Minimum spanning tree partitioning algorithm for microaggregation. *IEEE Transactions on Knowledge and Data Engineering* **17**(7), 902–911.

Laszlo M. and Mukherjee S. 2009 Approximation bounds for minimum information loss microaggregation. *IEEE Transactions on Knowledge and Data Engineering* **21**(11), 1643–1647.

Lenz R. 2010 *Methoden der Geheimhaltung wirtschaftsstatistischer Einzeldaten und ihre Schutzwirkung*. Statistisches Bundesamt, Wiesbaden.

Lenz R. and Vorgrimler D. 2005 Matching German turnover tax statistics. Technical Report FDZ-Arbeitspapier Nr. 4, Statistische Ämter des Bundes und der Länder-Forschungsdatenzentren.

Liew C.K., Choi U.J. and Liew C.J. 1985 A data distortion by probability distribution. *ACM Transactions on Database Systems* **10**, 395–411.

Little R.J.A. 1993 Statistical analysis of masked data. *Journal of Official Statistics* **9**, 407–426.

Little R.J.A., Liu F. and Raghunathan T.E. 2004 Statistical disclosure techniques based on multiple imputation. In *Applied Bayesian Modeling and Causal Inference from Incomplete-Data Perspectives* (eds. Gelman A. and Meng X.L.), pp. 141–152. Wiley, New York.

Lochner K., Bartee S., Wheatcroft G. and Cox C. 2008 A practical approach to balancing data confidentiality and research needs: the NHIS linked mortality files. In *Privacy in Statistical Databases, PSD 2008* (eds. Domingo-Ferrer J. and Saygın Y.), vol. 5262 of *Lecture Notes in Computer Science*, pp. 90–99. Springer, Berlin/Heidelberg.

Loeve A. 2001 Notes on sensitivity measures and protection levels. Technical report, Research paper, Statistics Netherlands. Available at http://neon.vb.cbs.nl/casc/related/marges.pdf.

Lucero J., Freiman M., Singh L., You J., DePersio M. and Zayatz L. 2011 The microdata analysis system at the U.S. Census Bureau. *SORT-Statistics and Operations Research Transactions* **35**, 77–98. Special issue on privacy in statistical databases.

Map O.T. 2007. Available at http://lehd.did.census.gov/.

Massell P. and Funk J. 2007 Recent developments in the use of noise for protecting magnitude data tables: balancing to improve data quality and rounding that preserves protection. *Proceedings of the Research Conference of the Federal Committee on Statistical Methodology*. Arlington, Virginia.

Mateo-Sanz J.M. and Domingo-Ferrer J. 1999 A method for data-oriented multivariate microaggregation. In *Statistical Data Protection, SDP 98* (ed. Domingo-Ferrer J.), pp. 89–99. Office for Official Publications of the European Communities, Luxembourg.

Mateo-Sanz J.M., Domingo-Ferrer J. and Sebé F. 2005 Probabilistic information loss measures in confidentiality protection of continuous microdata. *Data Mining and Knowledge Discovery* **11**(2), 181–193.

Mateo-Sanz J.M., Martínez-Ballesté A. and Domingo-Ferrer J. 2004a Fast generation of accurate synthetic microdata. In *Privacy in Statistical Databases, PSD 2004* (eds. Domingo-Ferrer J. and Torra V.), vol. 3050 of *Lecture Notes in Computer Science*, pp. 298–306. Springer, Berlin/Heidelberg.

Mateo-Sanz J.M., Sebé F. and Domingo-Ferrer J. 2004b Outlier protection in continuous microdata masking. In *Privacy in Statistical Databases, PSD 2004* (eds. Domingo-Ferrer J. and Torra V.), vol. 3050 of *Lecture Notes in Computer Science*, pp. 201–215. Springer, Berlin/Heidelberg.

Meindl B. 2009 Linking complementary cell suppression and the software R. Paper presented at the NTTS Conference, Brussels, 18–20 February 2009.

Meindl B. 2011 Package sdcTable. Available at http://cran.r-project.org/web/packages/sdcTable/sdcTable.pdf.

Mitra R. and Reiter J.P. 2006 Adjusting survey weights when altering identifying design variables via synthetic data. In *Privacy in Statistical Databases, PSD 2006* (eds. Domingo-Ferrer J. and Franconi L.), vol. 4302 of *Lecture Notes in Computer Science*, pp. 177–188. Springer, Berlin/Heidelberg.

Moore R. 1996 Controlled data swapping techniques for masking public use microdata sets. U. S. Bureau of the Census, Washington, DC. Available at: http://www.census.gov/srd/papers/pdf/rr96-4.pdf.

Muralidhar K. and Sarathy R. 2006 Data shuffling: a new masking approach for numerical data. *Management Science* **52**(5), 658–670.

Muralidhar K. and Sarathy R. 2008 Generating sufficiency-based non-synthetic perturbed data. *Transactions on Data Privacy* **1**(1), 17–33. Available at http://www.tdp.cat/issues/tdp.a005a08.pdf.

National Center for Health Statistics 2011 National health interview survey linked mortality public use file. Available at http://www.cdc.gov/nchs/data_access/data_linkage/mortality/nhis_linkage_public_use.htm.

Nayak T.K., Sinha B. and Zayatz L. 2010 Statistical properties of multiplicative noise masking for confidentiality protection. Technical report, Statistical Research Division Research Report Series (Statistics 2010-05). U.S. Census Bureau. Available at http://www.census.gov/srd/papers/pdf/rrs2010-05.pdf.

Oganian A. 2010 Multiplicative noise protocols. *Privacy in Statistical Databases, PSD 2010*, vol. 6344 of *Lecture Notes in Computer Science*, pp. 107–117. Springer, Berlin.

Oganian A. 2011 Multiplicative noise for masking numerical microdata with constraints. *SORT-Statistics and Operations Research Transactions* **35**, 99–112. Special issue on privacy in statistical databases.

Oganian A. and Domingo-Ferrer J. 2001 On the complexity of optimal microaggregation for statistical disclosure control. *Statistical Journal of the United Nations Economic Comission for Europe* **18**(4), 345–354.

ONS 2011 Registrars general's agreement. Available at http://www.ons.gov.uk/ons/guide-method/census/2011/the-2011-census/the-2011-census-project/national-statistican-and-registrars-agreement—september-2011.pdf.

Paas G. 1988 Disclosure risk and disclosure avoidance for microdata. *Journal of Business and Economic Statistics* **6**, 487–500.

Pagliuca D. and Seri G. 1999 Some results of individual ranking method on the system of enterprise accounts annual survey. Esprit SDC Project, Deliverable MI-3/D2.

Parzen E. 1962 On estimation of a probability density and mode. *Annals of Mathematical Statistics* **35**, 1065–1076.

Pink B. 2009 Principles and guidelines on confidentiality aspects of data integration UNECE United Nations Economic commission for Europe. Available at http://www.unece.org/stats/publications/Confidentiality_aspects_data_integration.pdf.

Polettini S. 2004 Some remarks on the individual risk methodology. *Work session on Statistical Data Confidentiality, Luxembourg, 7–9 April 2003*, pp. 299–311. Monographs in Official Statistics. Eurostat, Luxembourg. Available at http://www.unece.org/fileadmin/DAM/stats/documents/2003/04/confidentiality/wp.18.e.pdf.

Polettini S., Franconi L. and Stander J. 2002 Model based disclosure protection. In *Inference Control in Statistical Databases* (ed. Domingo-Ferrer J.), vol. 2316 of *Lecture Notes in Computer Science*, pp. 83–96. Springer, Berlin/Heidelberg.

Raghunathan T.E., Lepkowski J.M., van Hoewyk J. and Solenberger P. 2001 A multivariate technique for multiply imputing missing values using a series of regression models. *Survey Methodology* **27**, 85–96.

Raghunathan T.E., Reiter J.P. and Rubin D.B. 2003 Multiple imputation for statistical disclosure limitation. *Journal of Official Statistics* **19**(1), 1–16.

Reiss S.P. 1984 Practical data-swapping: the first steps. *ACM Transactions on Database Systems* **9**, 20–37.

Reiss S.P., Post M.J. and Dalenius T. 1982 Non-reversible privacy transformations. *Proceedings of the ACM Symposium on Principles of Database Systems*, pp. 139–146. ACM, Los Angeles, CA.

Reiter J.P. 2002 Satisfying disclosure restrictions with synthetic data sets. *Journal of Official Statistics* **18**(4), 531–544.

Reiter J.P. 2003 Inference for partially synthetic, public use microdata sets. *Survey Methodology* **29**, 181–188.

Reiter J.P. 2004 Simultaneous use of multiple imputation for missing data and disclosure limitation. *Survey Methodology* **30**, 235–242.

Reiter J.P. 2005a Releasing multiply-imputed, synthetic public use microdata: an illustration and empirical study. *Journal of the Royal Statistical Society, Series A* **168**, 185–205.

Reiter J.P. 2005b Significance tests for multi-component estimands from multiply-imputed, synthetic microdata. *Journal of Statistical Planning and Inference* **131**(2), 365–377.

Reiter J.P. and Drechsler J. 2010 Releasing multiply-imputed synthetic data generated in two stages to protect confidentiality. *Statistica Sinica* **20**(1), 405–422.

Reiter J.P. and Mitra R. 2009 Estimating risks of identification disclosure in partially synthetic data. *Journal of Privacy and Confidentiality* **1**(1), 99–110.

Repsilber R.D. 1994 Preservation of confidentiality in aggregated data. Paper presented at the Second International Seminar on Statistical Confidentiality, Luxembourg, 1994.

Repsilber R.D. 2002 Sicherung persönlicher Angaben in tabellendaten *Landesamt für Datenverarbeitung und Statistik NRW, Ausgabe 1/2002 (in German)* Statistische Analysen und Studien Nordrhein-Westfalen.

Rinott Y. and Shlomo N. 2006 A generalized negative binomial smoothing model for sample disclosure risk estimation. In *Privacy in Statistical Databases, PSD 2006* (eds. Domingo-Ferrer J. and Franconi L.), vol. 4302 of *Lecture Notes in Computer Science*, pp. 82–93. Springer, Berlin/Heidelberg.

Rinott Y. and Shlomo N. 2007 A smoothing model for sample disclosure risk estimation. *Tomography, Networks and Beyond*, vol. 54 of *IMS Lecture Notes – Monograph series Complex Datasets and Inverse Problems*, pp. 161–171.

Ritchie F. 2007a Disclosure detection in research environments in practice. *Work session on Statistical Data Confidentiality, Manchester, 17–19 December 2007*, pp. 399–406. Methodologies and Working papers. Eurostat, Luxembourg. Available at http://www.unece.org/fileadmin/DAM/stats/documents/2007/12/confidentiality/wp.37.e.pdf.

Ritchie F. 2007b Statistical disclosure control in a research environment. Technical Report Mimeo, Office for National Statistics.

Robertson D. 1993 Cell suppression at Statistics Canada. Proceedings Annual Research Conference, U.S. Bureau of the Census.

Robertson D. 1994 Automated disclosure control at Statistics Canada. Paper presented at the Second International Seminar on Statistical Confidentiality, Luxembourg, 1994.

Robertson D. 2000 Improving Statistics Canada's cell suppression software (confid). In *Proceedings in Computational Statistics 2000* (eds. Bethlehem J.G. and van der Hejden P.G.M.), vol. COMPSTAT Proceedings in Computational Statistics 2000. Physica-Verlag, Heidelberg/ New York.

Roque G.M. 2000 *Masking Microdata Files with Mixtures of Multivariate Normal Distributions* PhD thesis University of California at Riverside.

Rosemann M. 2003 Erste ergebnisse von vergleichenden untersuchungen mit anonymisierten und nicht anonymisierten einzeldaten am beispiel der kostenstrukturerhebung und der umsatzsteuerstatistik. In *Anonymisierung wirtschaftsstatistischer Einzeldaten* (eds. Ronning G. and Gnoss R.), pp. 154–183. Statistisches Bundesamt, Wiesbaden, Germany.

Rosenblatt M. 1956 Remarks on some non-parametric estimates of a density function. *Annals of Mathematical Statistics* **27**, 642–669.

Rubin D.B. 1987 *Multiple Imputation for Nonresponse in Surveys*. Wiley, New York.

Rubin D.B. 1993 Discussion of statistical disclosure limitation. *Journal of Official Statistics* **9**(2), 461–468.

Salazar-González J.J. 2008 Statistical confidentiality: optimization techniques to protect tables. *Computers and Operations Research* **35**, 1638–1651.

Salazar-González J.J., Staggermeier A. and Bycroft C. 2006 Controlled rounding implementation. *Work session on Statistical Data Confidentiality, Geneva, 9–11 November 2005*, pp. 303–308. Monographs in Official Statistics. Eurostat, Luxembourg. Available at http://www.unece.org/fileadmin/DAM/stats/documents/ece/ces/ge.46/2005/wp.36.e .pdf.

Samarati P. 2001 Protecting respondents' identities in microdata release. *IEEE Transactions on Knowledge and Data Engineering* **13**(6), 1010–1027.

Samarati P. and Sweeney L. 1998 Protecting privacy when disclosing information: k-anonymity and its enforcement through generalization and suppression. Technical report, SRI International.

Sande G. 1984 Automated cell suppression to preserve confidentiality of business statistics. *Statistical Journal of the United Nations* **2**, 33–41.

Sande G. 1999 Structure of the ACS automated cell suppression system. *Proceedings of the Joint Eurostat/UNECE work session on Statistical Confidentiality, Thessaloniki, 8–10 March 1999*, pp. 105–121. Available at http://www.unece.org/fileadmin/DAM/stats/documents/1999/03/confidentiality/9.e.pdf.

Sande G. 2002 Exact and approximate methods for data directed microaggregation in one or more dimensions. *International Journal of Uncertainty, Fuzziness and Knowledge-Based Systems* **10**(5), 459–476.

Sarathy R. and Muralidhar K. 2011 Evaluating Laplace noise addition to satisfy differential privacy for numeric data. *Transactions on Data Privacy* **4**(1), 1–17.

Schmid M. 2006 Estimation of a linear model under microaggregation by individual ranking. *Allgemeines Statistisches Archiv* **90**(3), 419–438.

Schmid M. and Schneeweiss H. 2009 The effect of microaggregation by individual ranking on the estimation of moments. *Journal of Econometrics* **153**(2), 174–182.

Schmid M., Schneeweiss H. and Küchenhoff H. 2007 Estimation of a linear regression under microaggregation with the response variable as a sorting variable. *Statistica Neerlandica* **61**(4), 407–431.

Schmidt K. and Giessing S. 2011 A SAS-tool for managing secondary cell suppression on sets of linked tables by τ-ARGUS modular. Poster presented at the NTTS conference, Brussels, 22–24 February 2011. Available at http://www.cros-portal.eu/sites/default/files/PS1 Poster 7.pdf.

Schneeweiss H., Rost D. and Schmid M. 2011 Probability and quantile estimation from individually micro-aggregated data. *Metrika* 1–22. Doi: 10.1007/s00184-011-0349-5.

Sebé F., Domingo-Ferrer J., Mateo-Sanz J.M. and Torra V. 2002 Post-masking optimization of the tradeoff between information loss and disclosure risk in masked microdata sets. In *Inference Control in Statistical Databases* (ed. Domingo-Ferrer J.), vol. 2316 of *Lecture Notes in Computer Science*, pp. 163–171. Springer, Berlin/ Heidelberg.

Shlomo N. and Young C. 2006a Information loss measures for frequency tables. *Work session on Statistical Data Confidentiality, Geneva, 9–11 November 2005*, pp. 277–289. Monographs in Official Statistics. Eurostat, Luxembourg. Available at http://www.unece.org/fileadmin/DAM/stats/documents/ece/ces/ge.46/2005/wp.33.e.pdf.

Shlomo N. and Young C. 2006b Statistical disclosure control methods through a risk-utility framework. In *Privacy in Statistical Databases, PSD 2006* (eds. Domingo-Ferrer J. and Franconi L.), vol. 4302 of *Lecture Notes in Computer Science*, pp. 68–81. Springer, Berlin/Heidelberg.

Silverman B.W. 1982 Kernel density estimation using the fast Fourier transformation. *Applied Statistics* **31**, 93–97.

Singh A.C., Yu F. and Dunteman G.H. 2004 MASSC: a new data mask for limiting statistical information loss and disclosure. *Work session on Statistical Data Confidentiality, Luxembourg, 7–9 April 2003*, pp. 373–394 Monographs in Official Statistics. Eurostat, Luxembourg.

SIPP 2007 U.S. Survey of Income and Program Participation. Available at http://www.census.gov/sipp/synth_data.html.

Skinner C. 2008 Assessing disclosure risk for record linkage. In *Privacy in Statistical Databases, PSD 2008* (eds. Domingo-Ferrer J. and Saygın Y.), vol. 5262 of *Lecture Notes in Computer Science*, pp. 166–176. Springer, Berlin/Heidelberg.

Skinner C.J. and Holmes D.J. 1998 Estimating the re-identification risk per record in microdata. *Journal of Official Statistics* **14**, 361–372.

Skinner C.J. and Shlomo N. 2008 Assessing disclosure risk in survey microdata using log-linear models. *Journal of the American Statistical Association* **103**, 989–1001.

Solanas A. and DiPietro R. 2008 A linear-time multivariate microaggregation for privacy protection in uniform very large data sets. In *5th International Conference on Modeling Decisions in Artificial Intelligence-MDAI 2008* (eds. Torra V. and Narukawa Y.), vol. 5285 of *Lecture Notes in Computer Science*, pp. 203–214. Springer, Berlin.

Solanas A. and Martínez-Ballesté A. 2006 V-MDAV: a multivariate microaggregation with variable group size. In *COMPSTAT 2006-Proceedings of the 17th Symposium in Computational Statistics* (eds. Rizzi A. and Vichi M.), pp. 917–926. Physica Verlag, Heidelberg.

Solanas A., Martínez-Ballesté A., Domingo-Ferrer J. and Mateo-Sanz J.M. 2006 A 2^d-tree-based blocking method for microaggregating very large data sets. *Proceedings of the 1st International Conference on Availability, Reliability and Security-ARES 2006*, pp. 922–928. IEEE Computer Society, Los Alamitos, CA.

Soria-Comas J. and Domingo-Ferrer J. 2011 Optimal random noise distributions for differential privacy. manuscript.

Stuart A. and Ord J.K. 1994 *Kendall's Advanced Theory of Statistics, Volume 1: Distribution Theory (6th Edition)*. Arnold, London.

Sullivan G.R. 1989 *The Use of Added Error to Avoid Disclosure in Microdata Releases*. PhD thesis Iowa State University.

Sweeney L. 2002a Achieving k-anonymity privacy protection using generalization and suppression. *International Journal of Uncertainty, Fuzziness and Knowledge Based Systems* **10**(5), 571–588.

Sweeney L. 2002b k-anonimity: a model for protecting privacy. *International Journal of Uncertainty, Fuzziness and Knowledge Based Systems* **10**(5), 557–570.

Templ M. 2008 Statistical disclosure control for microdata using the R-package sdcMicro. *Transactions on Data Privacy* **1**(2), 67–85.

Templ M. and Meindl B. 2008 Robust statistics meets SDC: new disclosure risk measures for continuous microdata masking. In *Privacy in Statistical Databases, PSD 2008* (eds. Domingo-Ferrer J. and Saygın Y.), vol. 5262 of *Lecture Notes in Computer Science*, pp. 177–189. Springer, Berlin/Heidelberg.

Templ M. and Petelin T. 2009 A graphical user interface for microdata protection which provides reproducibility and interactions: the sdcMicro GUI. *Transactions on Data Privacy* **2**(3), 207–224.

Tendick P. 1991 Optimal noise addition for preserving confidentiality in multivariate data. *Journal of Statistical Planning and Inference* **27**, 341–353.

Tendick P. and Matloff N. 1994 A modified random perturbation method for database security. *ACM Transactions on Database Systems* **19**, 47–63.

Thomas W., Gregory A. and Hamilton A. 2011 Metadata standards to support controlled access to microdata. Joint UNECE/Eurostat work session on Statistical Data Confidentiality, Tarragona, 26–28 October 2011. Available at http://www.unece.org/fileadmin/DAM/stats/documents/ece/ces/ge.46/2011/41_Thomas.pdf.

Torra V. 2004 Microaggregation for categorical variables: a median based approach. In *Privacy in Statistical Databases, PSD 2004* (eds. Domingo-Ferrer J. and Torra V.), vol. 3050 of *Lecture Notes in Computer Science*, pp. 162–174. Springer, Berlin/Heidelberg.

Torra V. and Domingo-Ferrer J. 2003 Record linkage methods for multidatabase data mining. In *Information Fusion in Data Mining* (ed. Torra V.), pp. 101–132. Springer, Germany.

Trewin D. 2007 Principles and guidelines of good practice for managing statistical confidentiality and microdata access. Available at http://www.unece.org/fileadmin/DAM/stats/documents/tfcm/1.e.pdf.

Trottini M. 2003 *Decision models for data disclosure limitation*. PhD thesis Carnegie Mellon University.

Trottini M., Franconi L. and Polettini S. 2006 Italian Household Expenditure Survey: a proposal for data dissemination. In *Privacy in Statistical Databases, PSD 2006* (eds. Domingo-Ferrer J. and Franconi L.), vol. 4302 of *Lecture Notes in Computer Science*, pp. 318–333. Springer, Berlin/Heidelberg.

Truta T., Fotouhi F. and Barth-Jones D. 2006 Global disclosure risk for microdata with continuous attributes. In *Privacy and Technologies of Identity* (eds. Strandburg K. and Raicu D.S.), pp. 350–363. Springer, US.

UNECE 2001 Statistical data confidentiality in the transition countries: 2000/2001 winter survey. Paper presented at joint UNECE/Eurostat work session on Statistical Data Confidentiality, Skopje, 14–16 March 2001. Available at http://www.unece.org/fileadmin/DAM/stats/documents/2001/03/confidentiality/43.e.pdf.

U.S. Bureau of Census 2007 Statistical disclosure control (SDC) checklist. Available at hhttp://www.census.gov/srd/sdc/.

Vale S. 2011 Update on UNECE principles and guidelines Paper presented at the Joint UNECE/Eurostat work session on Statistical Data Confidentiality, Tarragona, 26–28 October 2011. Available at http://www.unece.org/fileadmin/DAM/stats/documents/ece/ces/ge.46/2011/33_UNECE.pdf.

van den Hout A. 2000 *The Analysis of Data Perturbed by PRAM*. Delft University Press, Delft, The Netherlands.

van den Hout A. 2004 *Analyzing misclassified data: randomized response and post random-ization*. PhD thesis Utrecht University.

van den Hout A. and Elamir E.A.H. 2006 Statistical disclosure control using post randomisation: variants and measures for disclosure risk. *Journal of Offcial Statistics* **22**, 711–731.

van den Hout A. and van der Heijden P.G.M. 2002 Randomized response, statistical disclosure control and misclassification, a review. *International Statistical Review* **70**, 269–288.

Warner S.L. 1965 Randomized response: a survey technique for eliminating evasive answer bias. *Journal of the American Statistical Association* **60**, 63–69.

Willenborg L.C.R.J. and de Waal A.G. 1996 *Statistical Disclosure Control in Practice*. Springer-Verlag, New York.

Willenborg L.C.R.J. and de Waal A.G. 2001 *Elements of Statistical Disclosure Control*. Springer-Verlag, New York.

Winkler W.E. 1999 Re-identification methods for evaluating the confidentiality of analytically valid microdata. *Statistical Data Protection, Proceedings of the conference, Lisbon, 25–27 March 1998*, pp. 319–335. Eurostat, Luxembourg. (Journal version in *Research in Official Statistics*, vol. 1, no. 2, pp. 50–69, 1998. Also available at http://neon.vb.cbs.nl/casc/related/Sdp_98_2.pdf).

Winkler W.E. 2004a Masking and re-identification methods for public-use microdata: overview and research problems. In *Privacy in Statistical Databases, PSD 2004* (eds. Domingo-Ferrer J. and Torra V.), vol. 3050 of *Lecture Notes in Computer Science*, pp. 231–246. Springer, Berlin/Heidelberg.

Winkler W.E. 2004b Re-identification methods for masked microdata. In *Privacy in Statistical Databases, PSD 2004* (eds. Domingo-Ferrer J. and Torra V.), vol. 3050 of *Lecture Notes in Computer Science*, pp. 216–230. Springer, Berlin/Heidelberg.

Woodcock S.D. and Benedetto G. 2007 Distribution-preserving statistical disclosure limitation. Available at SSRN: http://ssrn.com/abstract=931535.

Yancey W.E., Winkler W.E. and Creecy R.H. 2002 Disclosure risk assessment in perturbative microdata protection. In *Inference Control in Statistical Databases 2002* (ed. Domingo-Ferrer J.), vol. 2316 of *Lecture Notes in Computer Science*, pp. 135–152. Springer, Berlin/Heidelberg.

Zayatz L. 2007 New implementations of noise for tabular magnitude data, synthetic tabular frequency and microdata, and a remote microdata analysis system. *Work session on Statistical Confidentiality, Manchester, 17–19 December 2007*, pp. 147–157 Methodologies and Working papers. Eurostat, Luxembourg.

Zwick T. 2005 Continuing vocational training forms and establishment productivity in Germany. *German Economic Review* **6**(2), 155–184.

Author index

Abowd, J.M., 85–86, 111
Aggarwal, C.C., 106
Aggarwal, G., 67–68, 92
Agrawal, D., 106
An, D., 86

Bacher, J., 42, 50
Benedetti, R., 41, 44
Benedetto, G., 82
Box, G.E.P., 167
Brand, R., 54–55, 57, 94, 114
Brown, D., 190
Burridge, J., 84, 90

Casciano, C., 38, 111
Castro, J., 158, 160, 165, 167–169
Cator, E., 77
Chatfield, C., 101
Chaudhuri, A., 74
Chen, T.T., 74
Collins, A.J., 101
Cox, D.R., 167
Cox, L.H., 5, 148, 155, 161, 165

Daalmans, J., 155, 158
Dale, A., 38
Dalenius, T., 4, 37, 72
Dandekar, R., 83, 86–87, 101, 165
de Waal, A.G., 11, 33, 36, 51, 53, 73, 109, 137, 155, 158, 191

de Wolf, P.P., 74, 77, 131–132, 162–163, 181
Defays, D., 60–61, 63–64
DiConsiglio, L., 44, 49
DiPietro, R., 65
Domingo-Ferrer, J., 5–7, 43, 50, 57, 60–61, 63–65, 67, 69–72, 74, 79, 90–91, 93–94, 102–103, 105, 108
Doyle, P., 191
Drechsler, J., 82–83, 85–86
Duke-Williams, O., 190
Duncan, G.T., 2–3, 5, 11, 39, 54, 158
Dwork, C., 6

Elamir, E.A.H., 41, 47, 76
Elliot, M.J., 38, 41, 49–50
Evans, B.T., 165

Fellegi, I.P., 50
Fienberg, S.E., 39, 72, 83
Fischetti, M., 153, 158–159, 164
Florian, A., 83
Forbes, A., 207
Foschi, F., 42, 49
Franconi, L., 41, 44–45, 86, 111, 123–126, 130, 165
Fraser, B., 206
Frend, J., 205
Frolova, O., 159, 168, 171
Funk, J., 165

Geurts, J., 153
Giessing, S., 132, 148, 160, 162–165, 167–169, 181
Gomatam, S., 200
Gonzalez, J.A., 169
González-Nicolás, Ú., 90, 93–94
Gouweleeuw, J.M., 74, 76
Graham, P., 86
Greenberg, B., 72
Gross, B., 74

Hafner, H., 124, 126
Hansen, S.L., 61, 64–66
Härdle, W., 108
Heer, G.R., 74
Heldal, J., 130
Höhne, J., 58
Holmes, D.J., 39, 41
Hundepool, A., 45, 47, 52–53, 66, 70, 72, 93, 109, 111, 117, 138, 149, 151, 160, 163, 167–168, 181, 201
Huntington, D.E., 83

Ichim, D., 43, 49, 111, 123–126, 130, 165

Jaro, M.A., 50
Jewett, R., 160, 168

Karr, A., 200
Kennickell, A.B., 85
Kim, J.J., 56
Kooiman, P., 109
Kuha, J.T., 76

Lambert, D., 38–39
Lane, J.I., 86, 111
Laszlo, M., 60–61
Lenz, R., 11, 60, 124, 126
Liew, C.K., 79–80
Little, R.J.A., 84, 86, 116
Lochner, K., 129–130
Loeve, A., 148
Lucero, J., 237
Lyrintzis, C.S., 83

Makov, U.E., 39
Martínez-Ballesté, A., 60, 65
Massell, P., 165
Mateo-Sanz, J.M., 42, 60–61, 64–65, 69–70, 74, 84, 104, 107–108
Matloff, N., 56
McIntyre, J., 72
Meindl, B., 42, 168–169
Mitra, R., 85–86
Moore, R., 72
Mukerjee, R., 74
Mukherjee, S., 60–61, 64–66
Muralidhar, K., 7, 73, 87–90, 94

Nanopoulos, P., 60–61, 63–64
Nayak, T.K., 165

Oganian, A., 59–61, 65, 70
Ord, J.K., 104

Paass, G., 36
Pagliuca, D., 60
Parzen, E., 108
Pearson, R.W., 54
Penny, R., 86
Petelin, T., 93, 113
Pink, B., 18
Polettini, S., 44–45, 49, 86

Raghunathan, T.E., 82, 115–116
Rees, P., 190
Reiss, S.P., 72
Reiter, J.P., 33, 81–82, 85–86, 116
Repsilber, R.D., 160–161, 163, 166
Rinott, Y., 48, 49
Ritchie, F., 213
Robertson, D., 159
Roque, G.M., 56
Rosemann, M., 60, 64
Rosenblatt, M., 108
Rubin, D.B., 80–81

Salazar-González, J.J., 153, 158–159, 164, 201
Samarati, P., 4–6, 61

Sande, G., 60–61, 64, 159, 168
Sarathy, R., 7, 73, 87–90, 94
Schmid, M., 63
Schmidt, K., 181
Schneeweiss, H., 63
Sebé, F., 86, 103
Seri, G., 60
Shlomo, N., 48–49, 199–201
Silverman, B.W., 108
Singh, A.C., 78
Skinner, C.J., 39, 41, 43, 47–48, 50, 76
Solanas, A., 60, 65–66, 91
Soria-Comas, J., 7
Stander, J., 86
Stuart, A., 104
Sullivan, G.R., 54, 57
Sunter, A.B., 50
Sweeney, L., 5–6, 61

Templ, M., 42, 53, 93, 113
Tendick, P., 56
Thomas, W., 36
Torra, V., 5–6, 43, 50, 60–61, 63, 65, 67, 70, 72, 91, 93–94, 102–103, 105, 108

Trewin, D., 2, 16
Trottini, M., 104, 111
Truta, T., 42

Vale, S., 19
van Gelder, I., 77
van den Hout, A., 74, 76
van der Heijden, P.G.M., 74
Vorgrimler, D., 60

Warner, S.L., 74
Willenborg, L.C.R.J., 11, 33, 36, 51, 53, 73, 109, 137, 191
Winkler, W.E., 33, 43, 50, 65, 100
Woodcock, S.D., 82, 86
Wooton, J., 206

Yancey, W.E., 103
Young, C., 48, 199–201
Yu, P.S., 92

Zayatz, L., 165
Zwick, T., 83

Subject index

μ-ARGUS, 45, 52, 53, 66, 67, 72, 93, 109, 111, 113, 115
μ-Approx, 67, 69, 91
τ-ARGUS, 138, 168, 183, 201

additive noise, 54
analytical
    interest, 100
    validity, 100
approximation heuristic, 65, 67
attribute, 3
attribute disclosure, 2, 37, 72, 73, 185
audit, 153, 172
authorisations, 148
average
    mean, 91
    median, 91
    modal value, 91

backtracking procedure, 161
barnardisation, 199
best-fit density function, 79
blocking, 66
bottom coding, 53

c-means, 72
calibration, 78
calibration probability, 76
cardinality, 52
CASC project, 57, 94

CASC reference microdata sets, 94
cell suppression, 157, 193
'Census' dataset, 94
Cholesky decomposition, 84
CIF, 169
Citrix, 210
coalition of respondents, 143
column relation, 153
communality, 101
concentration rules, 142
condensation, 92
conditional distribution, 82
CONFID2, 168
confidential outcome variables, 4
contingency table, 109
controlled rounding, 183, 197, 201
controlled tabular adjustment, 165
correlation coefficient, 105, 224
correlation matrix, 101
covariance, 105
covariance matrix, 59, 101
cover table, 162
CTA, 165

data
    categorical, 34, 39, 51
        nominal, 34
        ordinal, 34
    continuous, 33, 41, 51

Statistical Disclosure Control, First Edition. Anco Hundepool, Josep Domingo-Ferrer, Luisa Franconi, Sarah Giessing, Eric Schulte Nordholt, Keith Spicer and Peter-Paul de Wolf.
© 2012 John Wiley & Sons, Ltd. Published 2012 by John Wiley & Sons, Ltd.

data distortion by probability
   distribution, 79
data shuffling, 73
data swapping, 72
DataFERRETT, 237
density function, 107
differential privacy, 6
disclosure by differencing, 187
disclosure risk, 23, 30, 31, 38, 184
disclosure scenario, 30, 31, 37, 40
distribution fitting software, 80
dominance rule, 140, 142, 143

'EIA' dataset, 94
entropy, 109
Euclidean distance, 91
exact disclosure, 155

feasibility interval, 152
foreign trade rule, 149
frequency tables, 3, 183, 217
fully synthetic data, 78, 79
   bootstrap, 83
   empirical examples, 82
   IPSO, 84
   Latin Hypercube Sampling (LHS),
      83
   multiple imputation, 80
   sufficiency-based, 90

generalisation, 52
generic synthetic data generator, 91
Gibbs sampling, 82
global method, 161
global recoding, 52, 157
global risk, 31, 38, 41
goodness of fit, 79
graph-based algorithm, 67
graphs, 222
group centroid, 61
group disclosure, 186

hierarchical nominal variance, 91
hierarchical tables, 132
histogram smoothing, 108

holdings, 149
hybrid data, 79, 86
   microaggregation-based, 90–93
   recordwise combination, 86
   sufficiency-based, 87–88
   vs synthetic data, 99
hypercube method, 160

identification, 184
identifier, 3
identity disclosure, 2, 36
imputation model
   linear, 82
   logit, 82
individual attribute disclosure, 185, 190
individual ranking, 63
individual risk, 31, 38, 44, 45, 47
inequality constraints, 60
inferential disclosure, 37, 155
information loss, 34, 35, 100, 137, 166,
      183, 199
   bounded measures, 103
   categorical data, 108
   contingency tables, 109
   continuous data, 101
   direct comparison, 108
   entropy-based, 109
   global, 161
   probabilistic measures, 104
   unbounded measures, 101
Information Preserving Statistical
   Obfuscation (IPSO), 84
insider knowledge, 155
integer linear programming (ILP)
   approach, 158
intruder, 134, 138, 139
intruder scenarios, 138
ISI Declaration on Professional Ethics,
   11
item non-response, 80
IVEware, 115
   SYNTHESISE, 116

k-anonymity, 5, 61
k-modes, 72

k-partition, 61, 67
kernel density estimation, 82, 108
key variable, 4, 53
Kolmogorov–Smirnov test, 79

linear programming approach, 159
linear regression, 222
linear relation, 152
linear relationship, 152
linked tables, 132
LISSY, 210
local suppression, 53
lognormal noise, 59

magnitude tables, 3, 131, 218
majority rule, 72
marginal distribution, 73
Markov matrix, 75, 110
masking, 33
   methods, 33
   non-perturbative, 33, 51
   perturbative, 33, 53
MASSC, 78
matrix masking, 54
Maximum Distance to Average Vector
   (MDAV), 67
maximum likelihood estimate, 84
MDAV-generic, 91
micro agglomeration, 78
microaggregate, 60
microaggregation, 60
   μ-Approx, 67
   aggregation step, 91
   approximation heuristic to optimal,
      65, 67
   categorical, 72
   data-oriented, 64
   exact optimal, 65, 66
   fixed-size, 64
   heuristic, 65
   individual ranking, 63, 65
      parameter inference, 63
   linear complexity, 65
   MDAV, 67
   multivariate, 63–65

one-dimensional projection, 63
   partition step, 91
   quadratic complexity, 65, 71
   sum of z-scores, 63
   univariate, 63, 65, 66
   variable-size, 64
Microdata Analysis System (MAS),
   237
   cutpoints, 238
   regression, 238
   synthetic residual plots, 238
   universe, 238
Microdata File for Researchers (MFR),
   29, 31, 111, 119
microdata set, 3
Microdata Under Contract (MUC), 29
microhybrid procedure, 91, 93
   $\mathbb{R}$-*microhybrid*, 93
minimum frequency rule, 140, 141
misclassification, 74
missing data, 80
model-based disclosure protection, 86
mu-argus, *see* μ-ARGUS
multi cell disclosure, 155, 158
multiple imputation (MI), 80, 115
   complexity, 83
   full, 81, 82
   partial, 84, 85
   software, 83
multiple regression model, 84
multiplicative noise, 57
   preserving 1st and 2nd moments,
      58
   preserving various constraints, 59
multivariate normal distribution, 84

(n, k)-dominance rule, *see* dominance
   rule
negative contributions, 148
network flow heuristic, 159
noise addition, 54
   correlated, 55
   linear transformation, 56
   non-linear transformation, 57
   uncorrelated, 55

noise masking
    additive, 54
    multiplicative, 57
nominal average, 91
nominal distance, 91
non-confidential outcome variable, 4
non-linear regression, 223
non-perturbative masking, 33, 51
normality assumption, 106
NP-hardness, 65
numeric average, 91

OnTheMap, 85
ordinal average, 91
ordinal distance, 91

(*p*, *q*) rule, *see* prior-posterior
    rule
p% rule, 140, 142, 143
partially synthetic data, 78, 82,
    84
perturbative masking, 33, 53
positivity constraints, 60
post-masking optimisation, 103
post-randomisation method, 74
post-tabular, 193
PRAM, 74, 109
    correction methods, 76
    invariant PRAM, 77
    when to use, 77
PRAM-matrix, 75
pre-tabular, 191
principal component, 101
principles-based model, 214
prior-posterior rule, 146
probabilistic information loss, 104
protection level, 154
Public Use File (PUF), 28, 29, 31

quantile, 105
quasi-identifier, 4, 28, 37, 39, 41, 47,
    53, 84

R-U map, 5
RADL, 210

random rounding, 196
randomised response, 74
rank swapping, 72
re-identification, 6, 36, 53, 61, 84
re-identification risk, 39, 43
record linkage, 6, 50
    distance-based, 94
    match, 94
remote access, 208
remote execution, 208
request rule, 149
re-sampling, 74
Research Data Centre, 209
residuals, 223
respondent, 3
risk
    global, 31
    individual, 31
risk strata, 78
risk-utility map, 6
risk-utility models, 6
rounding, 73, 183, 194
    controlled, 183, 197, 201
    multivariate, 73
    random, 196
    univariate, 73
rounding point, 73
rounding set, 73
row relation, 153
rule of thumb, 216

safety bounds, 154
safety rules, 138, 140
Sample of Anonymised Records
    (SAR), 74
sample statistic, 104
sampling, 51
sampling weights, 44, 150
score formula, 103
sdcMicro, 45, 48, 93, 113
sdcTable, 168
secondary cell suppression problem,
    158
secondary risk assessment, 152
sensitive cells, 138

sensitivity measures, 140
sensitivity of variables, 139
sensitivity rules, 140
sequential regression multiple
        imputation, 82
shortest-path, 65, 66
simple random sampling, 107
simulated data, 78
singleton, 155
singleton disclosure, 155, 158
SME dataset, 70
SME example, 61
SRMI, 82, 115
Steiner vertex, 68
sub-table, 133
sub-sampling, 78
substitution, 78
sum of squares
    total, 61
    within-groups, 61
suppression pattern, 157
synthetic data, 33, 78
    bootstrap, 83
    full, 79
    IPSO, 84
    Latin Hypercube Sampling (LHS),
        83
    limited data utility, 99
    partial, 82, 84
    re-identification, 98
    sufficiency-based, 90
    transparency, 99
    vs hybrid data, 99

table redesign, 192
table relation, 153
tau-argus, *see* τ-ARGUS
top coding, 53
transition probability, 75, 76
transparency, 6, 99

unique combinations, 33, 39
unique match, 51
uniqueness, 78
unsafe combination, 53
upper and lower protection levels, 154

variable, 3
    categorical, 4, 39
    confidential outcome, 4
    continuous, 4, 41
    key, 4, 78
    nominal, 4
    non-confidential outcome, 4
    ordinal, 4
variance
    sample correlation, 107
    sample covariance, 107
    sample mean, 107
    sample quantile, 107
    sample statistic, 107
    sample variance, 107

waivers, 148
within-groups homogeneity, 61

z-score, 59

WILEY SERIES IN SURVEY METHODOLOGY

Established in Part by WALTER A. SHEWHART and SAMUEL S. WILKS

Editors: Mick P. Couper, Graham Kalton, Lars Lyberg, J.N.K. Rao, Norbert Schwarz, Christopher Skinner

The *Wiley Series in Survey Methodology* covers topics of current research and practical interests in survey methodology and sampling. While the emphasis is on application, theoretical discussion is encouraged when it supports a broader understanding of the subject matter.

The authors are leading academics and researchers in survey methodology and sampling. The readership includes professionals in, and students of, the fields of applied statistics, biostatistics, public policy, and government and corporate enterprises.

ALWIN • Margins of Error: A Study of Reliability in Survey Measurement

BETHLEHEM • Applied Survey Methods: A Statistical Perspective

* BIEMER, GROVES, LYBERG, MATHIOWETZ, and SUDMAN • Measurement Errors in Surveys

BIEMER and LYBERG • Introduction to Survey Quality

BRADBURN, SUDMAN, and WANSINK • Asking Questions: The Definitive Guide to Questionnaire Design—For Market Research, Political Polls, and Social Health Questionnaires, *Revised Edition*

BRAVERMAN and SLATER • Advances in Survey Research: New Directions for Evaluation, No. 70

CHAMBERS and SKINNER (editors) • Analysis of Survey Data

COCHRAN • Sampling Techniques, *Third Edition*

CONRAD and SCHOBER • Envisioning the Survey Interview of the Future

COUPER, BAKER, BETHLEHEM, CLARK, MARTIN, NICHOLLS, and O'REILLY (editors) • Computer Assisted Survey Information Collection

COX, BINDER, CHINNAPPA, CHRISTIANSON, COLLEDGE, and KOTT (editors) • Business Survey Methods

* DEMING • Sample Design in Business Research

DILLMAN • Mail and Internet Surveys: The Tailored Design Method

FULLER • Sampling Statistics

GROVES and COUPER • Nonresponse in Household Interview Surveys

GROVES • Survey Errors and Survey Costs

GROVES, DILLMAN, ELTINGE, and LITTLE • Survey Nonresponse

GROVES, BIEMER, LYBERG, MASSEY, NICHOLLS, and WAKSBERG • Telephone Survey Methodology

GROVES, FOWLER, COUPER, LEPKOWSKI, SINGER, and TOURANGEAU • Survey Methodology

* HANSEN, HURWITZ, and MADOW • Sample Survey Methods and Theory, Volume 1: Methods and Applications

* HANSEN, HURWITZ, and MADOW • Sample Survey Methods and Theory, Volume II: Theory

HARKNESS, van de VIJVER, and MOHLER • Cross-Cultural Survey Methods

HUNDEPOOL, DOMINGO-FERRER, FRANCONI, GIESSING, NORDHOLT, SPICER and de WOLF • Statistical Disclosure Control

KALTON and HEERINGA • Leslie Kish Selected Papers

KISH • Statistical Design for Research

* KISH • Survey Sampling

KORN and GRAUBARD • Analysis of Health Surveys

*Now available in a lower priced paperback edition in the Wiley Classics Library.

LEPKOWSKI, TUCKER, BRICK, DE LEEUW, JAPEC, LAVRAKAS, LINK, and SANGSTER (editors) • Advances in Telephone Survey Methodology

LESSLER and KALSBEEK • Nonsampling Error in Surveys

LEVY and LEMESHOW • Sampling of Populations: Methods and Applications, *Fourth Edition*

LYBERG, BIEMER, COLLINS, de LEEUW, DIPPO, SCHWARZ, TREWIN (editors) • Survey Measurement and Process Quality

MAYNARD, HOUTKOOP-STEENSTRA, SCHAEFFER, VAN DER ZOUWEN • Standardization and Tacit Knowledge: Interaction and Practice in the Survey Interview

PORTER (editor) • Overcoming Survey Research Problems: New Directions for Institutional Research, No. 121

PRESSER, ROTHGEB, COUPER, LESSLER, MARTIN, MARTIN, and SINGER (editors) • Methods for Testing and Evaluating Survey Questionnaires

RAO • Small Area Estimation

REA and PARKER • Designing and Conducting Survey Research: A Comprehensive Guide, *Third Edition*

SARIS and GALLHOFER • Design, Evaluation, and Analysis of Questionnaires for Survey Research

SÄRNDAL and LUNDSTRÖM • Estimation in Surveys with Nonresponse

SCHWARZ and SUDMAN (editors) • Answering Questions: Methodology for Determining Cognitive and Communicative Processes in Survey Research

SIRKEN, HERRMANN, SCHECHTER, SCHWARZ, TANUR, and TOURANGEAU (editors) • Cognition and Survey Research

SUDMAN, BRADBURN, and SCHWARZ • Thinking about Answers: The Application of Cognitive Processes to Survey Methodology

UMBACH (editor) • Survey Research Emerging Issues: New Directions for Institutional Research No. 127

VALLIANT, DORFMAN, and ROYALL • Finite Population Sampling and Inference: A Prediction Approach